W9-CAI-561

# RESEARCH IN ORGANIZATIONAL BEHAVIOR

*Volume 2* • 1980

# RESEARCH IN ORGANIZATIONAL BEHAVIOR

*An Annual Series of Analytical Essays and Critical Reviews*


*Editors:* BARRY M. STAW
*Graduate School of Management*
*Northwestern University*

LARRY L. CUMMINGS
*Graduate School of Business*
*University of Wisconsin*


---


VOLUME 2 • 1980


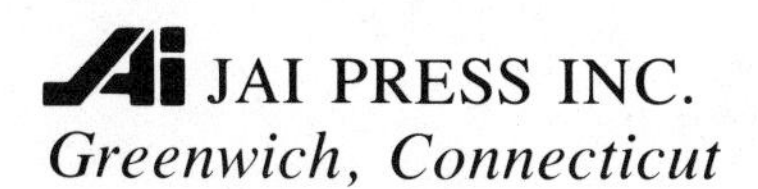


JAI PRESS INC.
*Greenwich, Connecticut*

*165 West Putnam Avenue*
*Greenwich, Connecticut 06830*


*ISBN NUMBER: 0-89232-099-0*

*Manufactured in the United States of America*

# CONTENTS

# EDITORIAL STATEMENT

This volume, like the first in our series, brings together chapters across a large variety of topics in organizational behavior. The unit of analysis for the chapters ranges from microscopic to macroscopic and their orientation ranges from critical reviews of established research topics to more speculative conceptual pieces. As in the case of Volume I, invited contributors were provided a large measure of flexibility in presenting their material. Our role as editors has been one of seeking clarification, coherence, and strength of argumentation rather than a traditional filtering or gate-keeping function. In this way, we have attempted to construct a series of essays which may be lively and have the potential to stimulate future research.


Barry M. Staw
*Evanston, Illinois*
Larry L. Cummings
*Madison, Wisconsin*
*Series Editors*

# RESEARCH IN ORGANIZATIONAL BEHAVIOR

*Volume 2* • 1980

# CONSTRUCT VALIDITY IN ORGANIZATIONAL BEHAVIOR

Donald P. Schwab,[1]

UNIVERSITY OF WISCONSIN-MADISON

ABSTRACT

The paper begins with a review of construct validity and its meaning within scientific and applied settings. Methods for assessing construct validity are discussed. Illustrative measures and experimental manipulations that have been employed in the organizational behavior literature are used to illustrate consequences of failure to pay sufficient attention to construct validation issues. Finally, recommendations are made for the proper sequencing of construct validation and substantive investigation in the development of knowledge.

**Research in Organizational Behavior, Volume 2, pages 03–43**

**ISBN: 0-89232-099-0**

# I. INTRODUCTION

Oversimplify for a moment and assume only two kinds of research investigations are performed. One type focuses on covariation between measures of *different* constructs. Studies with titles containing, "effects of," "impacts on," "consequences of," and "correlates of," are illustrative of this type. Relationships between independent and dependent variables are the focus of such investigations. Hereafter this sort of research will be called *substantive*. However, substantive research constitutes only one part of the research process. An equally important set of research issues involves the relationship between the results obtained from measures and the concepts or constructs the measures are purported to assess. Studies with the latter focus will hereafter be called *construct validation* research.

The present essay was motivated by a conclusion reached in my reading of the organizational behavior literature. Specifically, substantive relative to construct validation research has been overemphasized in organizational behavior. As a consequence, our knowledge of substantive relationships is not as great as is often believed, and (more speculatively) not as great as would be true if the idea of construct validity received greater attention. The purpose of the present essay then is to encourage investigators to emphasize construct validity more strongly in their research activities. Especially I hope to motivate investigators to sequence their research activities so that construct validity is considered before substantive research is performed.

To accomplish these objectives the essay has been divided into three major sections. The next (Section II) is largely tutorial; in it definitions are provided and the relationship between substantive and construct validation research is explored through a simple model. Emphasis is placed on the implications of the latter for the former. To a considerable extent this section is a summarization of the thinking of others. A novel portion of the section is the distinction made between construct validity as it pertains to scientific research concerns versus applied research concerns. This distinction is pursued in the following section (III) which considers methods to obtain evidence about the validity of one's measures or manipulations. Emphasis is placed on the importance of conceptual as well as empirical procedures for obtaining construct validity. A recommendation is also made that differentiation between constructs be given greater priority in construct validation efforts in organizational behavior. The final major section (IV) illustrates issues raised previously with two well-known organizational behavior measurement procedures. One of these is a traditional survey instrument and the other is an experimental manipulation. The aim is to show some dysfunctional consequences of the typical emphasis on substantive validity without adequate concern for construct validity.

## II. CONSTRUCT VALIDITY

As a recently evolved area of study, organizational behavior draws much of its intellectual sustenance from economics, management science, psychology, and sociology. Of the four, psychology has undoubtedly been the major source of knowledge about measurement. The specific set of issues to be considered here is most frequently discussed using the rubric "construct validity." It is on this body of knowledge that my discussion draws most heavily. In the following subsets of this section *construct validity* and related terms are defined. A qualified distinction is made between construct validity for scientific versus applied purposes.

As currently used, the term "construct validity" first appeared in "Technical recommendations for psychological tests and diagnostic techniques," published in *Psychological Bulletin Supplement* (1954). The generally acknowledged seminal work on construct validity was presented by Cronbach and Meehl (1955). It should be noted, however, that the essential elements of construct validity were articulated long before 1955 and by nonpsychologists [see, in particular, Northrop (1959), *The logic of modern physics,* first published in 1928]. Moreover, the central tenets of construct validity have been advocated without reference to the psychological nomenclature (e.g., Blalock, 1968).

As Cronbach and Meehl developed the issues, reference was explicitly made only to psychological tests. "A construct is some postulated attribute of people, assumed to be reflected in test performance (p. 283)." In the present paper the term *construct* is broadened to include *any* variable (i.e., entity capable of assuming two or more values) of a mental or conceptual nature.[2] Thus overt individual behaviors (such as performance activities), group characteristics (such as cohesion), organizational characteristics (such as centralization), managerial activities (such as delegation), and leadership behaviors (such as consideration), can all appropriately be thought of as constructs.

The term *construct validity* has, until recently, been exclusively considered with reference to measuring scales or instruments of various sorts, as indicated, usually in psychological tests. Such a perspective is unnecessarily restrictive since the issues involved are as appropriately considered when evaluating the adequacy of experimental or quasi-experimental manipulations as they are in evaluating measurement scales (see also Cook and Campbell, 1976).

### *A. Definitions*

Throughout this paper a construct will be thought of as referring to a conceptual variable. Relationships between variables will be defined in several ways depending on whether the variables involved in the relationships are conceptual, operational, or a mixture of the two. *Construct validity is defined as representing the correspondence between a con-*

*struct (conceptual definition of a variable) and the operational procedure to measure or manipulate that construct.* From this definition it is acceptable to think of construct validity as representing the correlation coefficient between the construct and the measure. This conceptualization is similar to Northrop's (1959) *epistemic correlation,* defined as the ". . . relation joining an unobserved component of anything designated by a concept . . . to its directly inspected component . . ." (p. 119). It is also similar to Nagle (1953, p. 274) who defined *relevancy* (construct validity as used here) as ". . . the hypothetical correlation coefficient between the criterion used (measure) and the ultimate criterion (construct)." Departures from construct validity coefficients of 1.00 are due to *contamination* (variance in the measure not present in the construct) and/or *deficiency* (variance in the construct not captured by the measure).[3]

The main danger of defining construct validity in correlational terms is that it may be interpreted as suggesting that the construct is real in some operational way and that some real measure of it is obtainable. Such an interpretation would be incorrect. As Nunnally (1967) stated, "[t]he problem (of regarding the construct as real) is not that of searching for a needle in the haystack, but that of searching for a needle that is not in the haystack" (p. 97). It is imperative that the conceptual nature of the construct be kept in mind: the construct is nothing more or less than our mental definition of a variable.

A corollary danger of the correlational analogy is if one thinks that the correlation between the measure and the construct is related to the processes used to assess construct validity empirically. The latter are necessarily *indirect* since they can only involve observations on measures and relationships between measures. Indeed, it is probable that commentators such as Marx (1963) and Messick (1975) defined construct in terms of relationships between observables because of the need to examine such relationships when investigating construct validity. In this paper the empirical steps used to investigate the construct validity of a measure (and the theoretical reasoning underlying those steps) are viewed as a process and hence will be referred to as *construct validation.*

### *B. Scientific and Applied Orientations*

The importance of construct validity and the methods used to assess it depend to some extent on which of two types of substantive research (i.e., relationships between measures of different constructs) is being undertaken. One pertains to the role of construct validity for scientific advancement. Within this framework constructs are of interest only if they are connected to other constructs. Measures of such constructs then are interesting only if they actually covary with measures of other constructs.

Failure of the measure to relate to measures of other constructs (assuming the latter are accepted as valid) leads to the modification of the measure (and hence the construct), modification of the theory connecting the construct to the other constructs, or to the abandonment of both.

A representation of the issues regarding construct validity for substantive research with a scientific orientation is shown in Figure 1. It indicates various linkages between the simplest sort of scientific research question; namely between one independent and one dependent variable. Examples from organizational behavior include relationships between: (1) role conflict and job satisfaction; (2) leadership style and employee tenure; (3) task design and organizational climate; (4) punishment and employee absenteeism; and (5) organizational shape and commitment. These variables are shown at two levels of abstraction, conceptual (I and D) and operational (I′ and D′).[4] The vertical linkages (II′ and DD′) represent construct validity, while the horizontal linkages (ID and I′D′) constitute substantive research linkages as discussed earlier.

In a scientific context the construct validity of I′ and D′ are approximately equal. This equality can be seen by considering several implications of Figure 1. In scientific research we theorize and understand entirely at the conceptual level. Thus, for example, we hypothesize that variation in D is a function of variation in I. When we test such a hypothesis, however, we necessarily use specific empirical indicators or measures. In Figure 1 the empirical validity of the D = f(I) hypothesis would be assessed by observing the covariation between I′ and D′. If the observation of I′ preceded D′ in time we could call our procedure *predictive* validation (or if I′ was a manipulation, an experiment or quasi-experiment); if the observations were obtained at approximately the same time the procedure would be called *concurrent* validation. A critical point, however, is that the *only* relationship in Figure 1 which can be observed empirically is between I′ and D′.

Yet the inferences we make are not conducted at the operational level. Whatever we observe when we look at I′D′ covariation, we again cast

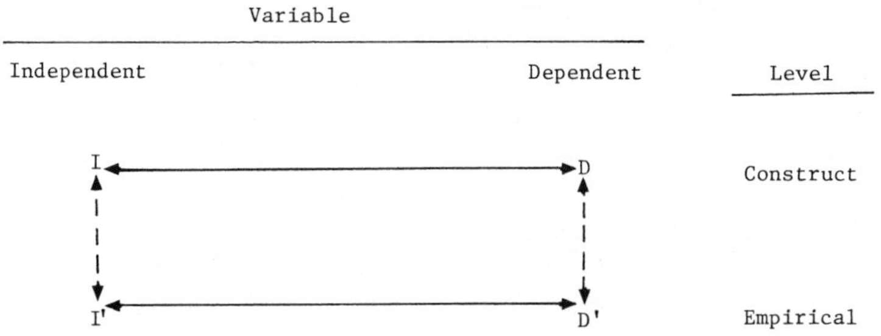

*Figure 1.* Representation of Construct and Empirical Validity

into a conceptual framework. If I′D′ are positively correlated in some sample we think (and write in our scientific reports) about the positive relationship between I and D. In testing scientific hypotheses, therefore it is essential that the results obtained from the I′ operation correspond to our way of thinking about the construct I and that the results obtained from the D′ operation correspond to our way of thinking about the construct D (i.e., that I′ and D′ be construct valid). Unless these validities are high, the substantive statements about ID will be correct only by chance. *Thus construct validity of both I′ and D′ are as necessary to scientific knowledge as is empirical validity.*

A second type of substantive research problem occurs where the investigator is primarily concerned with accounting for variance in the dependent variable. Here the emphasis (and it is important to recognize that the distinction drawn is one of emphasis, not of kind) is on predicting a dependent construct. In organizational behavior, dependent variable prediction, per se, occurs most frequently in applied situations. Examples might include predicting future work performance at the time of employment application, or the prediction of salary variation among a group of organizations.

The difference between scientific and applied orientations, however, is not determined by the choice of independent or dependent variables. Rather, the distinction is determined by the relative importance accorded the independent and dependent constructs by the investigator. A scientific orientation places approximately equal importance on both constructs and hence substantive validity depends similarly on the construct validity of both. An applied orientation (as defined here) places greatest emphasis on the construct validity of the dependent variable. The investigator (and the organizational sponsor of the applied research) views the problem as one of predicting the dependent variable and not necessarily of what it is that is providing the prediction.

Figure 2 illustrates the importance of construct validity when prediction is at issue. Again, the only observed relation is between I′ and D′. Now, however, the issue is what inferences I′D′ has for I′D; not what inferences I′D′ has for ID (the scientific question). For example, how well does the observed relation between our selection predictor and a measure of performance represent the relation between the predictor and the construct assessed by the performance measure?

The answer to this question depends only on the construct validity of the dependent variable. If the correlation between I′D′ is known, and a correlation between DD′ (construct validity) is *assumed* ($r_{DD'}$ can never be known) it is possible to specify the limits of the hypothetical correlation I′D from McNemar's (1969, p. 185) formula:[5]

$$r_{I'D} = r_{I'D'} r_{DD'} \pm \sqrt{1 - r^2_{I'D'} - r^2_{DD'} + (r^2_{I'D'} r^2_{DD'})} \qquad (1).$$

*Figure 2.* Applied Prediction Design

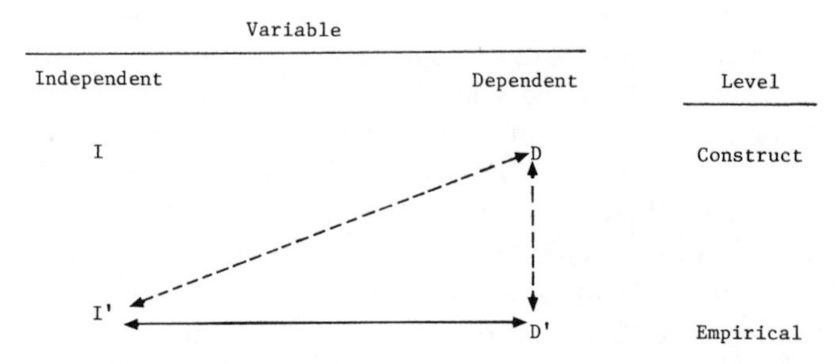


Formula 1 was used to generate the correlational limits for $r_{I'D}$ for the illustrative construct ($r_{DD'}$) and empirical ($r_{I'D'}$) validities shown in Table 1. Note several characteristics of the table. First, the construct validities reported are high. Second, the empirical validities are typical of findings obtained in much organizational behavior research. Given these values, the range of the I′D correlation coefficients is in all instances substantial, especially when construct validity ($r_{DD'}$) departs even moderately from 1.0. The implication is clear: unless we can be confident that a dependent measure has extremely high construct validity, we cannot be confident that the observed relationship ($r_{I'D'}$) approximates the relationship we are trying to estimate in the applied problem. Even the direction of the relationship between I′D may be erroneously estimated by I′D′ if construct validity does not approximate 1.0.[6]

There is another difference between scientific and applied orientations (again of degree). Specifically, in the scientific context, construct validation is often a sequential process. The scientist typically begins with a construct, probably ill defined. She/he suspects (hypothesizes) that this construct is related to other constructs in some sort of theoretical model which is probably also ill defined. At this point a measure of the construct is typically developed.

It is recommended that construct validity become the major priority of

*Table 1.* Illustrative I′D Correlation Limits Given Observed I′D′ and Hypothetical DD′ Correlation Coefficients

| | $r_{I'D'}$ | | |
|---|---|---|---|
| $r_{DD'}$ | .20 | .40 | .60 |
| .99 | .00 – .39 | .27 – .53 | .48 – .71 |
| .95 | −.12 – .50 | .10 – .66 | .33 – .81 |
| .90 | −.25 – .61 | −.04 – .76 | .19 – .89 |
| .80 | −.43 – .75 | −.23 – .87 | −.06 – 1.00 |

the scientist at this time. Initial construct validation will likely lead to modification of the instrument *and* perhaps in the investigator's definition of the construct. The sequential and interactive nature of construct validation as recommended for the scientific context is shown in Figure 3. We start with a construct (C) that leads us to its measurement (C′). C′, after whatever adjustments are made on it as a result of empirical research, ultimately defines the construct (Ĉ) that becomes a part of our body of knowledge. It is desirable that when the construct reaches the Ĉ state it be thought of, and communicated, in terms which closely approximate the outcomes of its operational procedures.[7]

An alternative strategy is to use the measure in substantive research without construct validation. In such cases the measure will usually not be modified. Nevertheless if interesting covariation between the measure and other measures is found it is likely to be retained and the findings are likely to become a part of our "body of knowledge." This strategy (substantive without construct validity) as we will see in Section IV, can be highly dysfunctional.

The modification of constructs which is characteristic of science during construct validation is not as applicable in applied settings. In the latter, the dependent construct is often established *normatively;* often by someone other than the investigator, such as the organizational sponsor of the research. As a result, little flexibility in the definition of the construct is warranted in applied settings. If, for example, the organization seeks to change the organizational contribution of its members through training, it makes no sense to measure an alternative construct such as "employee perceptions of amount learned" (unless it can be assumed that the two are highly related) in the program simply because a measure of organizational contribution is difficult or impossible to obtain. If the organization is interested in contribution, a measure of an alternative construct will not suffice no matter how elegant it might be. Moreover, it is inappropriate to modify or change a measure of contribution presumed to be construct

*Figure 3.* Construct Validation as a Sequential Interactive Process

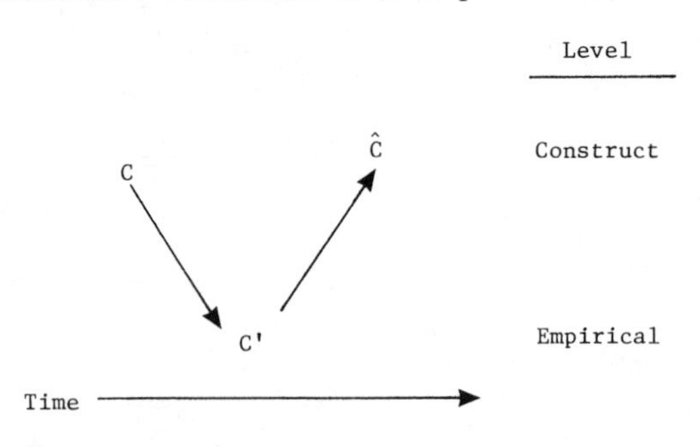

valid simply because the training program was ineffective in achieving change on that measure.

The position taken here admittedly overstates the rigor of the measurement process as it is typically implemented in applied settings. For example, in measuring organizational contribution we are not likely to specify the construct so precisely that the construct validity of a single measure could be shown to be demonstrably superior to one or several alternative measures (i.e., each measure is likely to be somewhat deficient and/or contaminated). Indeed, it has been recommended that multiple measures of change in contribution be observed as a mechanism for capturing important consequences that might not be picked up by a single measure (Campbell, Dunnette, Lawler, & Weick, 1970, 280–285).

Nevertheless, it is important to distinguish conceptually between scientific concerns, where covariation is primary, and applied concerns where the dependent construct is critical. A special responsibility is placed on the investigator when developing and evaluating dependent measures in the latter context. In terms of Figure 3, dependent measures for applied purposes place an even greater premium on making C′ conform to C than in scientific paradigms.

The distinction I have tried to draw between a scientific and applied orientation is also implied in the work by earlier commentators representing these alternative perspectives. Psychologists writing about construct validity (e.g., Campbell, 1960; Cronbach & Meehl, 1955; Nunnally, 1967) have ordinarily had a scientific orientation and have, therefore, viewed construct validation as an interactive process involving changes through time in both construct and measure. There is also an extensive literature relevant to organizational behavior that has developed on a more applied plane. It has been aimed at developing measures of organizational effectiveness. In this literature the construct has often been referred to as the *ultimate criterion* (Thorndike, 1949). Individuals such as Nagle (1953) who have written about construct validity with this applied perspective have emphasized the importance of making the measure conform to the construct. They have been appropriately not sympathetic to modifying the ultimate criterion as a mechanism for resolving measurement problems.

## III. ASSESSING CONSTRUCT VALIDITY

In the preceding section a simple model of construct and substantive validity was presented. It was additionally argued that *scientific* (i.e., the search for relationships) and *applied* (i.e., where the dependent construct assumes priority) orientations differ somewhat in terms of the requirements they impose on the construct validation process.

The present section contains a brief exposition of methods to be considered for assessing the construct validity of a measuring instrument or of an experimental manipulation. The issues considered here are complex largely because the variable of concern—the construct—cannot be directly assessed empirically. As a consequence, any empirical investigation can make only indirect inferences about construct validity. Moreover, it follows from the previous section that any single construct validation procedure can only lead to rejection or recommendations for modification of a measure or manipulation, not acceptance. Here construct validation is analogous to theory validation in the sense that an empirical test can be viewed as necessary (the measure or manipulation is not construct valid if it fails the test), but not sufficient. At best empirical tests can serve ". . . to state . . . the *degree* of validity the [measure] is presumed to have" (Cronbach & Meehl, 1955, p. 290, emphasis added). Implied in that statement is the need to identify ways in which the validity of the measure is unknown.

A related point which makes construct validation a difficult process results because the appropriate validational procedures depend on the nature of the construct specified and on the hypothetical linkages it is presumed to have with other constructs. Thus, beyond certain rudimentary psychometric techniques that have fairly wide applicability, a "cookbook" guide to construct validation procedures is of limited value. This caveat should be borne in mind when reading the remainder of the present section.

### *A. Conceptual Issues in Construct Validation*

*Role of Definition.* Occasionally a set of operations may be found that yield interesting relationships with measures of other constructs. In such situations, *meaning* (i.e., construct definition) would be attributed only post hoc. Typically, however, one begins with a theoretical perspective so that constructs and hence definitions are assumed. The purpose of this section is to briefly explicate how these definitions can be made more useful from a construct validation perspective.

The definition phase involves several steps. First, of course, is the need to specify the nature of the construct. What meaning is to be attributed to it? What would a measure of it reflect? What variance is *not* included in the construct? It appears especially important in the study of organizational behavior to specify the appropriate level of analysis when defining constructs. For example, does the construct represent a structural property of organizations, perceptions of the property (and if so, by whom), or employee affect toward the property? Much confusion has been created because the construct referent has not been made clear in the definition and/or in moving from definition to measurement.

An illustration of this confusion involves the definition and measurement of *organizational climate*. Early conceptualizations (e.g., Forehand & Gilmer, 1964) defined "climate" as a structural property of organizations that was hypothesized to influence employee attitudes and behaviors. Measures of the construct, however, have often assessed employee perceptions of "climate" (e.g., Litwin & Stringer, 1968).[8] Perhaps not surprisingly, some more recent conceptualizations have defined climate in perceptual terms (e.g., Hellriegel & Slocum, 1974; Schneider & Hall, 1972) so that a recent review now differentiates between structural and perceptual climate constructs (Payne & Pugh, 1976).

To a considerable extent such problems occur because measures of organizational characteristics are often difficult to obtain, while measurement through the ubiquitous employee survey is relatively easy (whatever violence such surveys may do to the validity of the measurement outcomes). James and Jones (1974) noted the dysfunctional aspects of allowing ease of measurement to dictate construct definition. While their comments are made regarding organizational climate, their point is broadly applicable to measurement in organizational behavior.

> Only after the conceptual boundaries of organizational climate are spelled out should the measurement and operationalization become matters of concern. In other words, the definition should guide measurement rather than available tools and psychometric limitations serving to delimit the definition (p. 1108).

Related to the need to specify the definitional parameters of the construct is the need to identify the probable psychometric properties of measures designed to assess the constructs. Is the construct unidimensional? Is it stable? Specification of probable psychometric characteristics is helpful in formulating tests to investigate the adequacy of measures and ultimately the constructs themselves. Such specification cannot be undertaken frivolously, however, if my earlier statement that measures can only be shown invalid, and not valid, is taken seriously. If the probable stability of some construct is unknown, for example, it should not be specified as a part of the definition. Nor should stability be used as a test of the construct.[9]

*Role of Theory.* If the construct is defined from a normative perspective the definitional obligation ends with a specification of what is included in the construct domain and the probable psychometric properties of its measurement. Illustrative of this orientation has been the measurement of job performance to be used as criteria for personnel decisions. Much energy has been devoted to the question of how performance is to be viewed vis-à-vis organizational goals. Specifications of performance ". . . range from the description of actual behavior, through evaluation of results, to estimates of the effects upon the organization and society" (Smith, 1976, p. 750). This is clearly a definitional perspective. Additional effort has been devoted to psychometric issues such as dimensionality, reliability, and errors of measurement (Barrett, 1966; Cummings & Schwab, 1973; Dunnette, 1966; Guion, 1965; Smith, 1976; Thorndike, 1949). Relatively little concern, however, has been shown to the relationships that performance may have to other constructs as a basis for providing evidence on the construct validity of the performance measures, per se. Concern for correlates of performance in such personnel activities as selection takes the construct validity of the performance measure as given.

If, however, the construct is defined for scientific purposes an additional step is required. Specifically, it is important to specify probable (hypothetical) linkages between the construct of interest and measures of other constructs.[10] Now this is admittedly theory building pure and simple. It is, in the terminology of Cronbach & Meehl (1955), specifying the *nomological net*. Such theorizing serves two important purposes for construct validation. First, specifying interconstruct linkages can serve to provide clarification of the construct under consideration as suggested by Messick (1975). Second, specification of interconstruct linkages can serve as a valuable input in establishing construct validation procedures (see the next section).

In short, specification of definition and hypothetical linkages (within a theoretical framework) can serve to specify the parameters of the construct. The definition should spell out true variance, and (by implication at the very least) variance to be regarded as deficient and contaminated. Specifying theoretical linkages serves the additional purpose of providing validational tests spanning constructs.

*B. Empirical Approaches to Construct Validation*

Definition of the construct and its location within a theoretical framework precede specification of empirical assessments of construct validity. There are, nonetheless, certain psychometric procedures that are so commonly appropriate that they deserve some attention in this essay. The following sections briefly discuss several of these. The discussion differentiates between two classes of construct validation procedures, those dealing primarily with (a) measures or manipulations of the construct under consideration, and (b) measures of the focal construct and of alternative constructs. In the discussion which follows it will be seen that this differentiation is roughly analogous to the distinction made between the role of definition (focal construct) and the role of theory (relationships among constructs) in specifying the construct initially.

*Measures of the Focal Construct.* *Reliability/generalizability.* Much effort has been devoted to the advancement of reliability theory as originally formulated by Spearman. As a construct, reliability refers to the ratio of "true" to total variance in a set of parallel measurements obtained on an individual.

$$\text{Reliability, } r_{11} = \frac{\sigma_t^2}{\sigma^2} \tag{2}$$

True variance, $\sigma_t^2$ is defined as systematic variance so that by definition error (nonreliable) variance cannot be construct valid. However, $\sigma_t^2$ may reflect systematic contamination and/or fail to capture construct variance (and hence be deficient). Thus, as has often been stated, reliability is necessary for validity (construct validity in this case), but it is not sufficient. (Actually this statement applies for empirical as well as construct validity since unreliability attenuates the observed relationship between measures of alternative constructs vis-à-vis the relation to be expected if the measurements were reliable.)

Reliability theory and the development of estimating procedures have been largely the purview of investigators concerned with measuring individual characteristics, especially human abilities. As a consequence, a set of items comprising a particular instrument is appropriately viewed as a sample from a much larger domain of items that could be included. Not surprisingly then, investigators concerned with reliability have tended to focus on developing indices of internal consistency or estimates of the generalizability of scores across parallel instrument forms. Thus, Cronbach's alpha, Hoyt's coefficient, Kuder-Richardson formulas, and the Spearman-Brown prophecy formula, all assess the correspondence of a measurement with a parallel form of measurement (where time is assumed not to operate as an error source).[11] Failure to achieve adequate internal consistency indicates that the scores are idiosyncratic to the particular items comprising the instrument. Thus internal consistency estimates are important for measures of unidimensional constructs (or factors of multidimensional constructs) whenever it is reasonable to assume that the items contained in the scale represent a larger domain of potential items.

But there are many potential error sources besides item sampling that may generate unreliability of measurement. Two in particular have been considered as a part of orthodox reliability theory and deserve mention in connection with a discussion of construct validity in the study of organizational behavior. One arises when it is desired to obtain an assessment of some organizational or personal property from an *external* observer. Examples include measurements of employee performance, leader behavior, task structure, worth of jobs, and organizational climate (if climate is defined as a structural rather than a perceptual property). In such situations the observer becomes a potentially important source of error variance. Estimates of interobserver reliability are then necessary to assess the degree to which total variance is a function of the individual(s) providing the measurements.

A substantial amount of evidence has been generated on the reliability to be expected in certain situations where external observations are used as measurements. This evidence suggests that interobserver unreliability serves as an important limitation on the measurements typically obtained. The literature on the measurement of employee performance through performance appraisal procedures, for example, indicates that observers are an important and pervasive source of error (e.g., Borman, 1978; Schwab, Heneman and Decotiis, 1975). Moreover, there is a very extensive literature that has developed on the sources of unreliability in appraisals which include characteristics of the appraisal situation, the observer, the performer, and various interactions between these sources (for a brief, but excellent review of this literature, see Cascio, 1978, pp. 336–339).

A corresponding body of knowledge has not been generated about other organizational behavior constructs assessed through external observers. Measures of leader behavior serve as an interesting example because the analogy to employee performance is so apparently close. The Ohio State Leadership Behavior Description Questionnaire (LBDQ), which is certainly the most extensively used measure in substantive research (see reviews by Kerr & Schriesheim, 1974; Kerr, Schriesheim, Murphy & Stogdill, 1974; Korman, 1966; Stogdill, 1974), was originally designed to have observers (typically subordinates) provide estimates of the leader's "true" behavior. As such, variance in observer estimates of a single leader should be viewed as interobserver unreliability. Yet Kerr & Schriesheim's (1974) review of the LBDQ's psychometric properties found that internal consistency estimates of the scales and stability (across time) have been about the only types of reliability analysis reported.[12]

A second type of reliability that is important for a variety of organizational behavior constructs is consistency across time, typically called *stability*. Stability becomes an issue if the construct is to be predictive of, or predicted by, some other construct. Thus, if satisfaction is hypothesized to be predictive of subsequent turnover, one must assume that satisfaction is reasonably stable, at least over the time interval from satisfaction measurement to turnover measurement. Lack of stability would serve to attenuate the expected relationship.

In a purely predictive sense any form of instability, including true construct change, would be viewed as a source of error (i.e., leading to prediction attenuation) variance. To be construct valid, however, measures must be sensitive to true change in the construct. This in turn, poses a problem for measurement when the investigator seeks to use the construct to make substantive predictions. If the construct is dynamic, measures that capture only stable variance are deficient. In such situations the construct must be redefined to include only its stable components (and hence probably lose much of its inherent value), or it must not be viewed as a predictor beyond the time it ceases to be stable.

Traditionally, reliability analyses have focused on the sampling of items, time, and observers as sources of error. And, typically only one source of variance has been considered in any given study. A much less restrictive view (in terms of error sources) has been proposed by Cronbach, Rajaratnam, & Gleser (1963) and Cronbach, Gleser, Nanda, & Rajaratnam (1972) who have recast reliability theory into a theory of *generalizability*. Their model is developed in analysis of variance terms where subjects' mean score across all conditions serves as the *universe score* and intraclass correlation coefficient becomes the estimate of generalizability. The procedure allows for the simultaneous (through n-way ANOVA) analysis of multiple sources of measurement variance including, but not restricted to, those examined in traditional reliability analysis. [An elaborated discussion of the procedure can be found in Campbell, 1976. See, Katerberg, Smith & Hoy (1977) for an illustration of the technique investigating three sources of variance in two measures of job satisfaction.]

The problem of reliability (or generalizability) for construct validity then is to specify, a priori, the domain over which the measurement results are to be generalized. Appropriate reliability estimates depend on the nature of the construct definition. Usually, however, multiple estimates will be required. An advantage of the ANOVA model proposed in generalizability theory is that these multiple sources can be examined simultaneously.

*Convergence.* Convergent validity, the extent to which responses from alternative measurements of the same construct share variance, has become a popular criterion in the organizational behavior literature for assessing construct validity since Campbell & Fiske (1959) proposed the multitrait-multimethod matrix methodology. While evidence of convergence is desirable (for a reason to be articulated shortly), its presumed implications for construct validity have been much overemphasized. At the outset it may be useful to think of convergence as a subset of generalizability (Cronbach, *et al.*, 1963). That is, measurement methods can be viewed as one source of variance and their impact on the universe score can be estimated. The advantage of such a perspective is that it puts convergence of results from alternative methods in the same domain as other sources of measurement variance. In most circumstances convergence will be desired (the construct is defined to span measurement procedures), but in others it may not (the construct is procedure specific).

There are several reasons why convergence as specified by Campbell and Fiske provides only limited evidence about construct validity. In the first place, of the two criteria for showing evidence of convergence (". . . the entries in the validity diagonal should be significantly different from zero and sufficiently large . . ." p. 125), the first is largely a function of sample size and the second is easily misinterpretable by the investigator who seeks to conclude that convergence has been established. Thus, for example, Lawler (1967) concluded correlation coefficients as small as .52 (representing less than 30 percent common variance) showed "good convergent validity (p. 374)," but elsewhere (Lawler, 1966, p. 158) concluded a correlation coefficient of .56 showed adequate discrimination between two different constructs.

At a more fundamental level, for convergence to represent construct validity one must assume that the methods are uncorrelated. That is, the convergent variance must not reflect method variance. Now there are many reasons why alternative methods will yield correlated results when responses are obtained from individuals (e.g., Cronbach, 1950; Jackson & Messick, 1958; Smith, 1975, pp. 136–140; Webb, Campbell, Schwartz & Sechrest, 1966, pp. 19–21), especially if these measurements are obtained at a moment in time. If method variance is probable, method and "construct" variance must be decomposed through some sort of factor (Jackson, 1969) or path (Althauser & Heberlein, 1970; Alwin, 1974) analytic technique (for an example of this methodology using an organizational behavior illustration see, Kalleberg & Kluegel, 1975).

Even if the problem of method variance has been accounted for, however, convergence cannot be assumed to represent only common construct variance. The shared variance may still represent contamination vis-à-vis the construct definition. Moreover, the shared variance may still exclude construct variance and hence be deficient. In short, convergence is no more sufficient as a test of construct validity than is reliability.

Nevertheless, convergence is worth examining because it is often useful to know the extent to which alternative procedures, whatever their common method variance, yield similar results. If two or more measures do not converge and if it is assumed that method variance, if any, adds rather than detracts from the resulting correlation coefficient, it can be concluded that at least one of the procedures is not providing construct valid results. Thus, convergence can still be viewed as necessary to construct validity.

Evidence of convergence is important to enhancing knowledge about organizational behavior because there has been an incredible proliferation of measures presumably assessing the same constructs. Schriesheim & Kerr (n.d.), for example, report having found over 120 "scales" of leadership reported in the literature between 1960 and 1975. Assessments of convergence can be of some value in estimating the likelihood that substantive results are measure specific. Note, however, that even here the value of convergence is limited because alternative measures seldom yield high correlation coefficients. Thus, as we have already noted in Formula 1, little may be known about the relationship to be expected between measures A and B (where A and B represent different constructs) when the relationships between AA′ and A′B are known.[13]

*Factor analysis.* Reliability/generalizability analysis and convergent validation are procedures for answering specific questions about measures of a construct. As indicated, their value to construct validation depends on the definition of, and the theoretical model surrounding, the construct. Factor analysis, alternatively, is a mathematical technique that can be used for a variety of purposes. It is often associated with construct validation because of its frequent use in instrument development.

Factor analysis can be helpful to construct validation at several points. It can be helpful initially as a way of identifying tentative dimensions and in suggesting items for deletion and areas where items should be added to the instrument. Use of factor analysis in this way, however, must be done carefully and several cautions are worth noting. First, factor analysis does not identify dimensions of a construct. It only identifies dimensions within the items included in the analysis. When items are chosen haphazardly, often lifted from other instruments, factor analysis becomes a poor substitute for the definitional stage recommended above.

Second, as with other multivariate techniques, the results of factor analysis are likely to be sample specific. Some of the problem can be ameliorated by having a large sample relative to the number of items factored (a minimum ratio of 10:1 is often recommended). A related problem, however, is not solved by sample size alone. Specifically, within sample heterogeneity (such as age, sex, and job level) can influence the resulting factor structure and hence the results can only be generalized to a population with those unique combinations of characteristics. A recent study by Dunham, Aldag & Brief (1977) illustrates this possibility. They obtained substantially different factor solutions to Hackman & Oldham's (1975) Job Diagnostic Survey across 20 subsamples of employees in five organizations.

Thus, investigators must be careful to develop items within a definitional framework that has been established prior to the investigation. Developmental samples must be chosen to represent the population of subjects on which subsequent substantive research is to be performed. When items are deleted, or especially when they are added, as a result of a factor analysis, the "new" scale should be administered to another sample and factored again before the scale is used as a substantive research instrument.

The record of using factor analysis in organizational behavior as a preliminary means of assessing items and dimensions has not been inspiring. As an example, Sims, Szilagyi & Keller (1976) factored 23 items (taken primarily from other questionnaires) of affective responses about job characteristics obtained on a heterogeneous sample of medical personnel. Presumably they used a stepwise procedure although this, or the decision rule for determining the appropriate number of factors, is not reported. Sims, et al. (1976) next performed a congruency analysis between various occupational subgroups (where in one case the ratio of responses to items was less than 2.5:1) and the factor loading obtained on the entire sample. Besides the erroneous inflation of congruency coefficients resulting from the lack of independence between samples, Sims, et al., erroneously concluded that the congruency coefficients (which refer to the correspondence of factor *loadings*) provided information about the stability of the factor *structure* (Nunnally, 1967, p. 367). [Despite the inflation, less than 60 percent of the congruency coefficients exceeded .90.] Although Sims, et al. (1976), subsequently modified the instrument substantially (dropping two items and adding 14 new ones), Sims & Szilagyi (1976) used job characteristic responses obtained from the original sample to perform a substantive study.

A second situation where factor analysis can have construct validation implications is to test hypotheses about an already developed scale. Two levels of hypotheses can be tested. First, the investigator can ask if the number of dimensions generated by the factor analysis replicates the number of dimensions specified, a priori. Usually investigators have both forced a factor solution to the number of a priori dimensions and compared that with a stepwise solution. Whether the a priori items load on the appropriate factors is then examined by an "eyeball" inspection. Examples in the organizational behavior literature include Dunham (1976), Dunham, et al. (1977), and Herman & Hulin (1973). A second and more rigorous method of testing a developed scale involves specification of both the number of dimensions *and* the specific items that are hypothesized to load on the dimensions. Although infrequently used in organizational behavior (see Dunham, 1975 for an exception), procedures are described in Nunnally (1967, 333–347) and Schönemann (1966).

*Measures of Alternative Constructs*

The procedures discussed above are often helpful for obtaining information about measures of a single construct. From a construct validity perspective, such procedures are more useful in invalidating measures during an early stage than they are in providing information about the nature of the construct. For example, the domain over which the construct is to be reliable/generalizable should be specified, a priori, and the measure should be tested against that criterion.

The orientation of this section is somewhat broader because a fully developed theory linking a construct within a nomological net is not likely to be specified before the measure or manipulation is developed and tested, at least in a preliminary fashion (Cronbach & Meehl, 1955; Nunnally, 1967). Some theoretical linkages will likely be specified, a priori, however, and can be used as construct validation criteria much as the psychometric methods discussed earlier. Other linkages will undoubtedly be observed (induction) or hypothesized (deduction) after the measure or manipulation has demonstrated initial evidence of some construct validity (or more strictly, given the arguments above, not been invalidated by previous tests). Theory specified prior to measure development and theory evolving from tests of measures frequently blend together in practice. Nevertheless, a priori theory requires emphasis. Indeed the most important implication of construct validity ". . . is the increased emphasis which . . . (it) . . . places upon the role of theory in . . . validation" (Jessor & Hammond, 1957, p. 161). In this context theory is viewed as a framework against which to test a given measure or manipulation.

*Empirical designs.* Viewed in this context nearly any empirical investigation can potentially be viewed as a study of construct validity depending on the nomological network specified. More importantly, *whether an empirical study is interpreted as having construct or substantive validation implications depends entirely on the assumptions the investigator makes about the measures employed and about the veracity of the hypotheses linking the constructs studied.*

Consider again a linkage in which an independent construct I is hypothesized to covary with a dependent construct D, and that measures exist for both (I′ and D′) as in Figure 1. In Section II it was argued that an empirically observed relationship between I′D′ has implications for the hypothesized relationship ID if, and only if, I′ and D′ were assumed to validly represent I and D. If, alternatively, the hypothesis connecting I and D is assumed true, and the construct validity of either measure is accepted, the *same* design can be used to make inferences about the construct validity of the other measure. The same reasoning can be extended to experimental research designs. Suppose that $I^m$ (an experimental manipulation) is substituted for I′ above. If ID is assumed true then differences in mean scores on D′ as a function of $I^m$ variation can be viewed as a test of the construct validity of (1) D′, if $I^m$ is assumed construct valid, or (2) $I^m$, if D′ is assumed construct valid.

The circularity of making construct or substantive validation inferences is both obvious and distressing, especially when one adds uncertainty to the linkages accepted as given. Consider first the case of negative results (i.e., no significant I′D′ covariation). Was the failure to observe the expected I′D′ or $I^mD'$ relationship a failure of (1) the ID hypothesis, (2) the I′ measure or $I^m$ manipulation, (3) the D′ measure, (4) some combination of the three, or (5) a sampling artifact (lack of statistical power)? A single empirical assessment which obtained an insignificant empirical relationship would obviously tell us little one way or the other. The situation is not markedly improved when positive results are obtained. While significant I′D′ or $I^mD'$ covariation is not inconsistent with the validation expectations (construct or substantive), the number of conditions which must hold (alpha replacing the issue of power in the latter case) before the study has validation implications is substantial. Moreover, positive results run the additional risk of being due to the fact that the measures reflect the same construct, be it I or D. (More will be said about this possibility below.)

Unfortunately, there are no short-cut solutions to the circularity problem presented. A single assessment of the construct validity of any measure must necessarily be examined within the context of uncertain knowledge about the validity of measures of other constructs and imperfect confidence in the hypotheses as they apply to the sample investigated. The problem can be mitigated, however, by performing a series of assessments examining the measure or manipulation against criteria established from multiple hypotheses, measures of alternative constructs, and samples. Such an orientation requires that the investigator view construct validation as a continuing process rather than a one-time or limited term project. It also requires the highest level of ingenuity in formulating research designs that will efficiently speak to the validation questions of the measure or manipulation under consideration. Obviously, the development of these designs will be substantially aided if the construct is embedded in a well-articulated nomological network.

Great care must be taken in sequencing the research program when investigating the construct validity of a measure or manipulation by comparing it with measures of alternative constructs. Positive results in one study and negative results in another, given a hypothesis, become difficult to interpret if more than one factor in the design (sample, procedure, or measures) is allowed to vary between studies. In my judgment, the organizational behavior literature has been hindered in both substantive and construct validation research by holding too few variance sources constant across studies. As a consequence, it is frequently impossible to explain the disparate findings obtained in different investigations. In terms of enhancing our knowledge it would be desirable if some of the entrepreneurial spirit that characterizes much of the research in the field would be subordinated to the mundane task of more, nearly literal, replications.

*Differentiation among constructs.* Issues regarding convergence of alternative measures of a single construct have been considered in substantial detail by those concerned with construct validity. A related issue has not been attended to with as much care, though it is at least as important. Specifically, what evidence is there that instruments which purportedly measure *different* constructs in fact do so?[14]

I was reminded of this latter issue recently when reading a review of job involvement research by Rabinowitz & Hall (1977). Their definition of job involvement (emphasis on identification with work and work success) appears somewhat differentiated from standard definitions of satisfaction (focus on affect toward job and work environment).[15] The definitional differences, however, are not reflected in empirical findings linking measures of job involvement to measures of other variables. Their summary of these relationships is reported in Table 2. The variables labeled personal and situational covary with job involvement much as they covary with measures of job satisfaction (Quinn, Staines & McCullough, 1974; Vroom, 1964). Moreover, job involvement appears to relate to the outcome variables in Table 2 of absenteeism, performance, and turnover in almost the same way as job satisfaction covaries with these variables (Brayfield & Crockett, 1955; Porter & Steers, 1973; Schwab & Cummings, 1970; Vroom, 1964).[16]

Now it might be argued that the modest relationships reported in Table 2 between measures of job involvement and job satisfaction (median r = .35) demonstrate that they reflect identifiably different constructs. Such an argument is not compelling, however, when it is recognized that both constructs are purportedly multidimensional and that alternative measures of the same construct often do not covary highly. For example, in comparing four commonly labeled satisfaction scales of Smith, Kendall & Hulin's (1969) Job Descriptive Index (JDI) and Weiss, Dawis, England & Lofquist's (1967) Minnesota Satisfaction Questionnaire (MSQ), Gillet & Schwab (1975) found a median heterotrait-heteromethod correlation coefficient (i.e., between different facets of satisfaction measured with different instruments) of only .21. Thus, job involvement measures appear

*Table 2.* Summary of Correlates of Job Involvement

| *Correlates* | *Number of Studies* | *Approximate Magnitude of Relationship* |
|---|---|---|
| Personal characteristics: | | |
| Age | 8 | .25 |
| Education | 6 | 0 and + |
| Sex | 1 | 0 |
| Internal locus of control | 2 | + |
| Tenure | 5 | 0 and + |
| Community size | 2 | + |
| Protestant Ethic | 2 | .30 |
| Higher order needs | 2 | .20 |
| Marital status | 3 | 0 |
| Situational characteristics: | | |
| Participation in decision making | 4 | .50 |
| Job characteristics | 7 | .30 |
| Job level | 4 | 0 |
| Leader behavior | 2 | 0 |
| Social factors | 4 | + |
| Outcomes: | | |
| Job satisfaction | | |
| Work | 3 | .40 |
| Promotion | 3 | .30 |
| Supervision | 3 | .35 |
| People | 3 | .35 |
| Pay | 3 | .15 |
| Company | 1 | .40 |
| Performance | 7 | 0 |
| Turnover | 4 | .25 |
| Absenteeism | 2 | 0 and .45 |
| Success | 2 | + |

Note: Where correlations were not reported in the literature, +, −, or 0 are used to indicate the relationships found.

Reprinted from Rabinowitz & Hall (1977), with the permission of the American Psychological Association.

to correlate more highly with job satisfaction facets than alternative satisfaction facets (measured with different instruments) correlate with each other. It might also be noted that the off-diagonal heteromethod correlations between the JDI and MSQ scales observed by Gillet & Schwab were lower than the correlations between all five JDI scales and: (1) Rizzo, House & Lirtzman's (1970) measure of role ambiguity and conflict (median r = .28, signs reversed, Schuler, Aldag & Brief, 1977); and (2) eleven scales of the perceived work environment scale (median r = .32, Newman, 1977). Findings such as these raise questions about the appropriateness of identifying as measures of new constructs instruments whose outcomes correlate more highly with satisfaction measures than alternative measures and facets of satisfaction.[17]

Construct redundancy poses a problem if we take parsimony in scientific explanation seriously. Tessor & Krauss (1976) have suggested a number of empirical procedures that may be used to obtain information on whether two measurements (A′ and B′) reflect different constructs. These include assessments of: (1) asymmetric relationships (A′ leads to B′ or B′ leads to A′, but not both); (2) nonmonotonicity between A′ and B′; and (3) interactions between some third measurement, C′, and either A′ or B′ in producing B′ or A′. The main burden again is at the theoretical level to identify the appropriate procedures for a given circumstance.

In hypothesizing linkages between "new" constructs and other constructs investigators would be well advised to specify how the construct will differ from other potentially similar constructs. For example, in thinking about organizational climate as a perceptual variable, questions such as the following. (1) Is it an antecedent or consequence of job satisfaction? (i.e., what is the causal ordering, if any?). (2) Are satisfaction and perceived climate consequences of different organizational characteristics, or different organizational personal characteristic interactions. (3) Do perceptions of climate and satisfaction lead to different behavioral consequences. (4) Will perceived climate increase explained variance in outcome variables beyond that accounted for by satisfaction, and if so, why? Specification of such differential linkages would be most helpful in choosing appropriate empirical procedures to assess construct validity.

In commenting on the state of the sociological literature, Blalock (1968) argued that an outright moratorium on new constructs could inhibit much needed advancement in the field. His observation undoubtedly also applies to organizational behavior. Nevertheless, in my view, organizational behaviorists have been overly eager to abandon familiar constructs in their zeal to increase the variance explained in outcome variables such as employee performance and turnover. The modest relationships obtained have too quickly been interpreted as resulting because the constructs fail to relate (i.e., a theoretical failure), when in fact, the observed relationships probably reflect more a failure of measurement (e.g., low reliability in both independent and dependent variables) than of theory. More care at the outset in specifying likely nomological net differences when new constructs are specified might appropriately reduce the proliferation in hypothesized explanatory variables.

## IV. ILLUSTRATIONS FROM ORGANIZATIONAL BEHAVIOR

Since measurement is pervasive in the research process, an attempted assessment of the "state of the art" in any subject area as broad as organizational behavior would transcend the scope of an essay, and perhaps even a monograph. At the same time, however, a selected evaluation may be helpful as a way of judging the implications of the typical emphasis on substantive validity to the virtual exclusion of construct validity. To illustrate some of the problems that such an orientation can produce I have chosen two examples, one representing a survey instrument and one illustrating an experimental manipulation. Beyond representing two quite different research contexts where construct validity is an issue, the choice of Porter's need satisfaction questionnaire (hereafter referred to as the NSQ) and Adam's overreward manipulation (hereafter referred to as OM) deserve a bit of explanation.

Neither was chosen because it represents atypically poor research methodology. To the contrary, both reflect greater conceptual creativity than is usual in the organizational behavior literature. More important in my choice is the fact that both the NSQ and OM have been subject, post hoc, to a fair amount of research that can legitimately be thought of as construct validation. While the validation research would better have preceded the substantive research, a partial evaluation of each, in terms of the concern of this essay, is possible. To an extent, the empirically founded criticisms that can be made of the NSQ and OM are due to the fact that Porter and Adams have had a sufficient impact on the field to warrant the interest of other investigators. An evaluation of the NSQ and OM is also appropriate since both have had a significant impact on the substantive conclusions made in the organizational behavior literature.

*A. Need Satisfaction Questionnaire*

The construct job satisfaction has been subjected to an unusually heavy amount of substantive research (Crites, 1969; Herzberg, Mausner, Peterson & Capwell, 1957; Locke, 1969; 1976; Vroom, 1964). Measures of satisfaction have also been abundant, ranging from Herzberg, Mausner & Snyderman's (1959) story-telling methodology to standardized questionnaire formats (e.g., Smith, et al., 1969; Weiss, et al., 1967). These have been subjected to varying amounts of construct validation evidence.[18] The NSQ evolved out of an interest in measuring job satisfaction within the context of Maslow's (1943, 1954) need hierarchy.

*The Questionnaire.* The first research regarding the NSQ was described by Porter (1961). He developed 13 items to measure a modified form of Maslow's hierarchy. Porter excluded Maslow's physiological category and added autonomy needs which were arbitrarily placed between esteem and actualization. The resulting "scales" in the hierarchal order specified were security (one item), social (two items), esteem (three items), autonomy (four items), and actualization (three items). Three types of information were requested for each item: (1) "How much is there now?", (2) "How much should there be?", (3) "How important is this to me?".

Responses for each type of information were obtained on a 1 (minimum) to 7 (maximum) scale format. In the first paper Porter measured need deficiencies by recording the percentages of respondents who reported (2) "How much should there be?" to be greater than (1) "How much is there now?". In subsequent papers (Porter, 1962; 1963a; 1963b) he subtracted responses (1) from (2).[19]

*Evaluation of the Questionnaire.* Butler (1954) has criticized psychological research because measures tend not to be developed within a theoretical framework. In this respect, the NSQ is unusual because it is carefully linked to theory at two points. First, the dimensions follow a theory that was widely accepted at the time but essentially untested. Most of the research which casts doubt on the validity of Maslow's hierarchy has been generated since the development of the NSQ (see reviews by Miner and Dachler, 1973; Wahba and Bridwell, 1976). Second, the difference score format corresponds conceptually to most current theorizing about the nature of job satisfaction. That is, satisfaction is seen as the result of an evaluation of what the individual perceives s/he is obtaining against some standard or expectation of what s/he wants (Locke, 1969; 1976; Smith, et al., 1969). Thus, the NSQ was developed within a theoretical framework as recommended above.

Nevertheless, the relatively unique features of the NSQ might alert one to the need to engage in some construct validation efforts prior to its use as a substantive research instrument. For example, to what extent do the items within hierarchy categories define homogeneous dimensions? How reliable are deficiency scores? To what extent do deficiency scores yield information not contained in the separate components? These are important questions given what is known to be problematical about difference scores (e.g., Cronbach & Furby, 1970; Gulliksen, 1950). To what extent do the results of the NSQ converge with the results obtained from more conventional measurement procedures? Early investigations using the NSQ focused exclusively on substantive issues.[20] Questions of reliability and other construct validation issues were not addressed until a substantial body of substantive conclusions relating satisfaction to structural characteristics (such as job level, line versus staff, span of control and organizational size and shape) were amassed and, insofar as the validity of the NSQ was concerned, uncritically accepted (e.g., Cummings & El Salmi, 1968; Porter & Lawler, 1965).

Subsequently, researchers began to investigate the NSQ with a construct validation perspective. Wall & Payne (1973), using the reasoning of Werts & Linn (1970) and Cronbach & Furby (1970), showed that the deficiency scores are not independent of the component parts. Thus, for example, the relationships observed between deficiency scores and job level (e.g., Cummings & El Salmi, 1970; Porter, 1961; 1962) likely reflect primarily differences in "what is there now" across job levels. Corroborating evidence for this was obtained by Imparato (1972) who found the *location* of the difference score to account for more satisfaction variance (using JDI scores as the criterion) than the *magnitude* of the difference score. Wallace & Berger (1973) obtained only marginal internal consistency estimates (median r = .60) on the four multiple item NSQ scales. Wanous and Lawler (1972) found responses from the difference format to have the lowest absolute correlations with (1) the mean of responses to the question "how satisfied are you with these aspects of your job" (for 23 facets), and (2) a single item assessing overall satisfaction, from among eight scaling formats investigated. Low convergence has been obtained between the NSQ and the JDI (Herman & Hulin, 1973), and an alternative measure of satisfaction based on Maslow's need hierarchy (Schneider & Alderfer, 1973).[21]

Other construct validation evidence has been obtained from studies that have factor analyzed the NSQ. Payne (1970) factored eight modified NSQ items spanning all five hierarchical levels using two scaling formats, deficiency scores, and importance scores. All but one of the deficiency items had their highest loadings on one factor. Lawler and Hall (1970) factored six autonomy and actualization items from the NSQ (scored in deficiency format) with items designed to tap job involvement and intrinsic motivation (scored with alternative formats). All deficiency items loaded on one factor. These two studies suggest the possibility of an overall methods factor for deficiency scoring. Waters and Roach (1973) factored all NSQ items with a measure of overall satisfaction and respondents' job level. Ten of the 13 items had highest loadings on a single factor. Schneider and Alderfer (1973) obtained a two-factor solution to the 13 hierarchical items of the NSQ. Simple structure was only marginally approximated and no clear differentiation on the basis of a priori scales was obtained. Roberts, Walter & Miles (1971) factored 12 (some modified) of the 13 NSQ items plus seven additional items. Thus this study is not a direct assessment of the NSQ. Nevertheless, the 12 items did not define factors as specified, a priori. Finally, Herman and Hulin (1973) factored the 13 NSQ items. They concluded that a one-factor solution was most appropriate suggesting a single methods or trait factor. When Herman and Hulin forced a five-factor solution, no correspondence was obtained between the a priori dimensions and the factor analytic solution.

*Discussion*

At least 18 substantive studies were published using the NSQ to measure satisfaction. These studies examined relationships between the NSQ hierarchical scales and various organizational and personal characteristics. Inferences were typically drawn from the results about the veracity of the hierarchy, or the hierarchy was taken as valid. Given that subsequent construct validation evidence suggests that the NSQ fails to measure the need hierarchy as intended, it is difficult to see how this stream of research made a contribution to our knowledge about these relationships.

*B. Equity Overreward Manipulation*

A number of theorists have posited forms of equity theory (e.g., Adams, 1963a, 1965; Homans, 1961; Jacques, 1961; Patchen, 1961). Adams' is the most rigorously formulated, the best known, and has received by far the most empirical investigation in an organizational contest (see reviews by Campbell & Pritchard, 1976; Carrell & Dittrich, 1978; Goodman & Friedman, 1971; Lawler, 1968; Pritchard, 1969).[22] The theory develops a balance model of motivation that hypothesizes individuals seek equity (a sense of justice or fairness) in exchange relationships such as employment. In such situations it is hypothesized that an individual attempts to equate, through cognitive or behavioral responses, the ratio of one's own *inputs* (what one brings to the exchange) to *outcomes* (what one obtains from the exchange) with the inputs/outcomes ratio of a reference person(s). All variables in the exchange, including reference person(s), are assumed to be perceptually determined by the individual. When the ratios are in balance, the individual is hypothesized to experience equity and a maintenance of cognitions and behaviors is predicted. When the ratios are not in balance, perceived inequity is hypothesized and it is then predicted that the individual will seek to modify behavior or cognitions to establish equity.

Equity theory differs most dramatically from alternative cognitive motivation theories such as goal-setting (Locke, 1968) or expectancy (Vroom, 1964) in its hypotheses regarding overreward inequity. If the individual experiences overreward inequity (i.e., his/her inputs/outcomes ratio is more favorable than referant's ratio), s/he is hypothesized to make cognitive or behavioral adjustments to establish equity just as in the case of underreward inequity. This hypothesis leads to some interesting (given an outcome maximizing perspective) predictions. If, for example, subjects are financially overrewarded and if their only behavioral method for changing inputs is to influence quantity or quality of production, the theory predicts, relative to equitably rewarded subjects:

(1) Given payment by time, higher quantity or quality of production will result.

(2) Given payment by quantity of production, higher quality and lower quantity will result (Adams, 1965).

*The Manipulation.* To induce feelings of overreward, Adams & Rosenbaum (1962) told male student applicants for part-time employment that they were not qualified for the job to be performed. This manipulation was performed under the guise of an employment interview where an experimenter (acting as the interviewer) reviewed the applicant's qualifications and then told the applicant he was not qualified for the job. The experimenter also feigned a call to the placement service in the applicant's presence to complain about the quality of the applicant. The experimenter then said he would have to hire the applicant anyway, that the applicant should pay close attention to the job instructions, and that he would have to pay the applicant at the advertised rate (presumably equitable *if* the applicant was qualified).

Overreward inequity was thus supposed to have been established because the subject was expected to feel that the pay received was more than his qualifications justified relative to more qualified subjects. Other subjects were told they were qualified for the task, also were offered the advertised rate, and hence were presumably equitably rewarded. No manipulation check was reported, but the results supported the predictions of equity theory. A number of other studies have been performed using essentially the same overreward manipulation. Adams (1963b), Adams & Jacobsen (1964), Arrowood (1961), Goodman & Friedman (1968), and Wood & Lawler (1970) all obtained results supportive of equity theory's overreward hypotheses. Lawler, Koplin, Young & Fadem (1968) also found support for the theory, but only in the first day of an experiment that extended over three days. Friedman & Goodman (1967) did not obtain supportive results.

*Evaluation of the Manipulation.* The construct of inequity is reasonably well defined in Adams' version of the theory. Moreover, the nomological network is specified in greater detail than is typical of most organizational behavior theories. Thus, there is a basis for evaluating whether the manipulation (as operationalized by the denigration of subjects' qualification for the task) validly represents the construct of overreward inequity. Obviously, some sort of a manipulation check would have been potentially enlightening. However, a manipulation check as ordinarily operationalized only protects against deficiency (did the manipulation tap the construct?); it typically does not protect against contamination (did the manipulation tap other constructs?). Manipulation checks can, of course, be developed to assess both contamination and deficiency. See the discussion of Evan & Simmons' (1969) study below.

While the terminology of construct validity has not been used, questions were quickly raised about both sources of invalidity in the overpayment manipulation (OM). Lawler (1968b) suggested that OM might induce feelings of job insecurity and that increased performance might be due to a desire for continued employment. Both Lawler (1968b) and Pritchard

(1969) suggested that OM might be interpreted by subjects as an attack on their self-esteem. Higher performance in the OM condition might thus be due to the subjects confirming their self-esteem by showing the experimenter they could perform successfully. Opsahl & Dunnette (1966) suggested that the higher quality obtained from the piece-rate rewarded subjects exposed to OM might have resulted because that manipulation emphasized the need to do a good job. All of these then are explanations suggesting OM was either varying constructs in addition to inequity (contamination) or only other constructs (contamination and deficiency).

A number of other experiments have been conducted that implicitly address the construct validity of OM. Generally speaking, studies that have used alternative manipulations for inducing overreward inequity have obtained less supportive results than those obtained from OM. Neither Andrews (1967) who manipulated incentive pay levels directly, nor Pritchard, Dunnette & Jorgenson (1972) who attempted to induce inequity by changing pay policies during the course of an experiment, obtained significant performance differences between overrewarded and equitably rewarded subjects. However, it is not entirely clear whether the manipulations in these studies were successful either. Subjects in Andrews' study recommended that the job should pay different amounts as a function of the treatment conditions they were in. However, the significance of these differences was not reported. Moreover, the relationship of the recommendation to feelings of equity is not determinant. Pritchard, et al., found that one of their two overpaid groups actually reported feeling insignificantly less paid than one equity group. In addition, all subjects reported feeling underpaid on an "absolute" scale, although with the exception noted, the order of perceived payment was as specified by the manipulation.

Evan & Simmons (1969) performed two experiments. In the first, overpayment was induced as a fortuitous circumstance and a manipulation check on perceptions of pay was supportive of the manipulation. In this study a manipulation check for the contaminating effects of variation in perceived competence was also conducted. No differences in perceived competence as a function of the payment manipulation were obtained. In the second experiment overpayment was induced in a similar fashion. However, a manipulation check suggested that subjects did not experience feelings of pay equity and inequity as manipulated. In neither experiment did overrewarded subjects perform differently than equitably rewarded subjects.

Other studies have varied OM somewhat in an attempt to elicit further information about its meaning. Adams & Jacobsen (1964), Goodman & Friedman (1968), and Gordon & Lowin (1965, as reported in Lawler, 1968b) used OM on two groups, one conventionally and one where subjects were told their pay was to be lower than the standard rate because their qualifications were lower (reduced dissonance group). Theoretically, the reduced dissonance group was hypothesized to perform as an equitably paid group. To further cloud the validational issue, reduced dissonance subjects in the Adams and Jacobsen, and in the Goodman and Friedman studies performed as hypothesized by equity theory. In the Gordon and Lowin study, however, there were no differences between the performance of the OM and reduced dissonance groups.

Lawler (1968a) created two overreward conditions, one using OM and the other using an overpaid by circumstance manipulation. Only the OM group performed as hypothesized by the theory in comparison to an equitably paid group. There were no performance differences between the overpaid by circumstance and equitably paid groups. Weiner (1970) also used two overreward manipulations, OM and overpaid by circumstance. He found that the OM group only conformed to equity theory performance hypotheses in a task involving valued abilities; when the task did not involve such abilities, there were no performance differences between the OM and equitably paid groups. However, the performance hypothesis was supported for the overrewarded group by circumstance.

An attempt to directly assess the meaning of OM has been performed by Andrews & Valenzi (1970). They had OM acted out in front of 80 subjects who were first asked in an open-ended question format how they would react had they been the applicant. None of the subjects provided a pay inequity response to this question. However, nearly 75 percent responded in terms of the self-image of the applicant. In a subsequently administered fixed alternative questionnaire, self-image was chosen as the most likely reaction and wage inequity the least likely reaction from among the 13 alternatives provided. It is, of course, not clear whether observer perceptions of how subjects would respond reflects how subjects actually respond to the manipulation.[23]

*Discussion*

Most early investigations of equity theory were viewed by investigators as substantive theoretical tests. It is clear in retrospect, however, that except for the earliest studies, investigators have been wrestling with the question of construct validity. Do OM and/or alternative manipulations reflect overreward inequity as defined by Adams? To the extent that they do not, we know no more about the validity of overrewards in employment situations than when Adams formulated his original hypothesis.

It is clear that OM results in larger performance effects than the alternative manipulations tested. Whether this is true because OM is contaminated, as has essentially been argued by Campbell & Pritchard (1976), Lawler (1968b) and Pritchard (1969), or because alternative manipulations are deficient, as has been argued by Goodman & Friedman (1971), is not entirely settled. Nor is it certain that a greater concern with construct validity at the outset would necessarily have truncated the extensive research process that has produced so much ambiguity. It seems probable, however, that had investigators been thinking explicitly in construct validity terms, greater concern would have been attended to the quality of the manipulations from the outset. Substantive research on, and conclusions about, equity theory might now look quite different as a consequence.

### C. *Conclusions*

The NSQ and OM were obviously not chosen randomly as illustrations. To the contrary, they were deliberately selected to emphasize my initial premise, namely that an emphasis on substantive validity without proper sensitivity to construct validity is dysfunctional for substantive knowledge. Thus, no inference should be drawn from this essay that all research efforts in organizational behavior have been characterized by an inordinate emphasis on substantive research issues. Neither should one conclude necessarily that greater a priori concern with construct validity would inevitably lead to greater substantive knowledge. There are many constraints operating besides suboptimal measurements (suboptimal design, suboptimal theory, etc.). It is safe to conclude, however, that some potentially erroneous substantive knowledge could be avoided if measures that are subsequently shown to be invalid were so identified before extensive substantive research is undertaken.

## V. SUMMARY

Given the nature of the measurement issues in organizational behavior, it was argued that a construct validation orientation is a necessary and major element in the research process. Construct validity refers to the correspondence between the results obtained from a measuring instrument and the meaning attributed to those results. Substantive validity (i.e., the relationship between constructs) depends on inferences made about the construct validity of the measures or manipulations employed.

Construct validity is essential to organizational behavior as a scientific and also an applied field. The validation issues and appropriate procedures differ somewhat between the two. In a scientific context, the construct validity of both independent and dependent variables is fundamental since one does not, in the abstract, take precedence over the other. An observed substantive relationship will lead to erroneous inferences regardless of whether the independent or dependent variable is measured invalidly. Also, constructs are amenable to modification within a scientific context if, in so doing, more useful relationships between variables are observed. Finally, research in a scientific context provides the investigator with greater validation alternatives to the extent that a nomological net has been specified which can be used to provide criteria for validation assessments.

In applied settings the dependent variable is clearly the most important. This to some extent frees the researcher of the need to be as concerned with the construct validity of the independent variables. At the same time, however, the investigator is not free in applied contexts to change the definition of the dependent variable to enhance relationships with independent variables. Moreover, the nomological network that can be valuable in scientific settings plays only a limited role in aiding construct validation efforts in applied contexts.

Since the criterion in construct validity is conceptual, direct tests are not possible. As a consequence, construct validity must be inferred on the basis of indirect assessments. In addition, research results that have construct validation implications (e.g., convergence) can be viewed as necessary, but not sufficient. Psychometric evaluations can be valuable construct validation aids, but also are not of themselves sufficient. Depending on the nature of the nomological net, nearly any research can be interpreted within a construct validation framework. "Construct validity is not to be identified solely by particular investigative procedures, but by the orientation of the investigation" (Cronbach & Meehl, 1955, p. 282).

Organizational behavior has suffered because investigators have not accorded construct validity the same deference as substantive validity. Illustrations include Porter's measure of satisfaction and Adams' overreward manipulation. While both constructs are embedded in an extensive nomological net, instrumentation was not initially assessed in construct validation terms. As a consequence, substantive conclusions have been generated that may not be warranted. Future organizational behavior research will benefit by giving construct validity at least as much attention as substantive validity.

## FOOTNOTES

1. Special thanks are extended to Herbert G. Heneman, III and the editors for making critical comments on an earlier draft of the entire manuscript. I am also indebted to J. Stacy Adams and Lyman W. Porter for their comments on Section IV. I accept full responsibility for any remaining errors of fact or judgment.

2. This conceptualization is congruent with Northrop's (1959) earlier discussion of the idea.

3. Cook & Campbell (1976) refer to the former as "surplus construct irrelevancies" and the latter as "construct underrepresentation."

4. The former corresponds to Margenau's (1950) C plane; the latter to his P (perceptual plane).

5. This formula is typically discussed in the context of estimating the relationships to be expected between two test scores (A with B) when the relationship of each test to another (C) is known (e.g., McCornack, 1956). The model is equally applicable to the problem at hand. Recognize the hypothetical character of the model in relation to this problem, however, since the correlations $r_{I'D}$ $r_{DD'}$ are never empirically known.

6. The problem is exacerbated in the scientific case (i.e., where construct to construct relationships are of interest) because departures from construct validity coefficients of 1.0 on either the independent or dependent variable side contribute to uncertainty in the $r_{I'D'}$ estimate of $r_{ID}$.

7. Note that it is the results (i.e., the values obtained on the operational indicator) of the operational procedure, not the procedure itself which needs to conform to the construct (Tenopyr, 1977). That the *procedure* appears valid (face validity) is not necessary to construct validity. Indeed, to require that the procedure conform to the construct may be dysfunctional to obtaining construct valid results.

8. The latter orientation has led some investigators to suggest that climate is nothing more than a synonym for job satisfaction (Guion, 1973, Johanesson, 1973).

9. In this vein, Smith, Kendall & Hulin wisely did not emphasize the stability of the Job Descriptive Index since they argued that job satisfaction is a construct that may change through time (1969 p. 75).

10. My distinction between applied and scientific orientations vis-à-vis the role of definition and theory, is analogous to Campbell's (1960) distinction between *trait validity* (theorizing ". . . goes no further than indicating a hypothetical syndrome, trait or personality dimension," p. 547) and *nomological validity* (when a measure is evaluated ". . . in a formal theoretical network," p. 547).

11. They need not all provide identical estimates, however, if the true score is multidimensional (Campbell, 1976).

12. It should be noted that the theory underlying the LBDQ, namely that leaders exhibit a consistent (across subordinates) behavior has been challenged by research that has concluded subordinate behavior influences leader behavior (Farris & Lim, 1969; Greene, 1975; Kavanaugh, 1971; Lowin & Craig, 1968) and by Graen's dyadic leadership theory (e.g., Dansereau, Graen & Haga, 1975). These alternative conceptualizations of leadership obviously would view within leader variance differently than is implied by the theory underlying the development of the LBDQ. In the alternative formulations, within leader variance would be a confounded combination of error and true (to the extent that leaders behave differently with different subordinates) variance.

13. On the other hand, substantive validities do not necessarily differ when nonconvergent measures are employed because of the generally low relationships observed *between* constructs. As an illustration, Schwab & Wallace (1974) observed a convergent correlation coefficient of only .56 between the pay satisfaction scales of the Job Descriptive Index and the Minnesota Satisfaction Questionnaire; yet found similar relationships when correlating these two scales with a set of personal and organizational characteristics.

14. Unlike convergence, which is applicable to both normative and scientific concerns, differentiation is usually important only when the latter is at issue. For example, it will ordinarily not be important to know whether an ability test chosen to predict a normatively specified performance criterion is tapping something conceptually different from the domain of the performance measure, as long as it is predictive.

15. See also Locke (1976) who elaborated on the conceptual differences between satisfaction and involvement.

16. Actually, it would not be appropriate to argue that the two measures reflect different constructs even if they relate differentially to other constructs unless it is additionally assumed that the two measures are equally reliable and that they assess unidimensional constructs (Tesser & Krauss, 1976).

17. For another example see a recent study by Pierce & Dunham (1978). They found a high correspondence between commonly employed measures of macro-organizational constructs (routinization, centralization, formalization and predictability) and measures of a micro-organizational construct (task characteristics).

18. For sheer volume, the methodology of Herzberg, et al. (1959) has been subjected to perhaps more research than any alternative measurement technique. Much of this research has indirectly focused on the question of construct validity. The major issues have been so exhaustively discussed elsewhere (Bockman, 1971; Dunnette, Campbell & Hakel, 1967, Grigaliunas & Wiener, 1974; Herzberg, 1966; House & Wigdor, 1967; King, 1970; Locke, 1976; Whitsett & Winslow, 1967), however, that I have chosen to focus on another measurement procedure.

19. Importance responses have not been directly incorporated in the measure of satisfaction. This was probably a wise decision since attempts to weight satisfaction scores by importance have generally not proved to be useful (Blood, 1971; Ewen, 1967; Mikes & Hulin, 1968; Mobley & Locke, 1970; Quinn & Mangione, 1973).

20. Cummings & El Salmi (1970), Edel (1966), El Salmi & Cummings (1968), Eran (1966), Hall & Lawler (1970), Heller & Porter (1966), Lawler & Porter (1967), Miller (1966), Mitchell (1970), Paine, Carroll & Leete (1966), Porter (1961, 1962, 1963a, 1963b), Porter & Lawler (1964), Porter & Mitchell (1967), Porter & Siegel (1965), Rhinehart, Barrell, DeWolfe, Griffin & Spaner (1969).

21. Evan (1969) is sometimes cited as a study which assessed the congruence of the NSQ with the JDI (e.g., Herman & Hulin, 1973). However since Evan modified both items and scaling format as specified by Porter, it seems inappropriate to regard the study as an assessment of the NSQ.

22. More recently Adams' version of the theory has been elaborated on by Walster, Berscheid and Walster (1973) and the equity paradigm has been considered in a variety of nonorganizational situations (see a review by Adams & Freedman, 1976).

23. I am indebted to J. Stacy Adams for this observation.

## REFERENCES


Adams, J. S. (1963a) "Toward an understanding of inequity." *Journal of Abnormal and Social Psychology*, 47, 422–436.

——— (1963b) "Wage inequities, productivity and work quality." *Industrial Relations*, 3(1), 9–16.

——— (1965) "Injustice in social exchange." In L. Berkowitz (ed.) *Advances in experimental social psychology*. Vol. 2, New York: Academic Press, 267–299.

——— & Freedman, S. (1976) "Equity theory revisited: Comments and annotated bibliography." In L. Berkowitz and E. Walster (eds.) *Advances in experimental social psychology*. Vol. 9, New York: Academic Press, 43–90.

——— & Jacobsen, P. R. (1964) "Effects of wage inequalities on work quality." *Journal of Abnormal and Social Psychology*, 69, 19–25.

——— & Rosenbaum, W. B. (1962) "The relationship of worker productivity to cognitive dissonance about wage inequities." *Journal of Applied Psychology*, 46, 161–164.

Althauser, R. P. & Heberlein, T. A. (1970) "Validity and the multitrait-multimethod matrix." In E. F. Borgotta & F. W. Bornstedt (eds.) *Sociological methodology 1970*. San Francisco: Jossey-Bass, 151–169.

Alwin, D. (1974) "An analytic comparison of four approaches to the interpretation of relationships in the multitrait-multimethod matrix." In H. Costner (ed.) *Sociological methodology 1973-74*. San Francisco: Jossey-Bass.

Andrews, I. R. (1967) "Wage inequity and job performance: An experimental study." *Journal of Applied Psychology,* 51, 39–45.

——— & Valenzi, E. R. (1970) "Overpay inequity or self-image as a worker: A critical examination of an experimental induction procedure." *Organizational Behavior and Human Performance,* 5, 266–276.

Arrowood, A. J. (1961) "Some effects on productivity of justified and unjustified levels of rewards under public and private conditions." Unpublished doctoral dissertation, University of Minnesota.

Barrett, R. S. (1966) *Performance rating.* Chicago: Science Research Associates.

Blalock, H. M., Jr. (1968) "The measurement problem: A gap between the languages of theory and research." In H. M. Blalock, Jr. and A. B. Blalock (eds.) *Methodology in social research.* New York: McGraw-Hill, 5–27.

Blood, M. R. (1971). "The validity of importance." *Journal of Applied Psychology,* 55, 487–488.

Bockman, V. M. (1971) "The Herzberg controversy." *Personnel Psychology,* 24, 155–189.

Borman, W. C. (1978) "Exploring upper limits of reliability and validity in job performance ratings." *Journal of Applied Psychology,* 63, 135–144.

Brayfield, A. H. & Crockett, W. H. (1955) "Employee attitudes and employee performance." *Psychological Bulletin,* 52, 396–424.

Butler, J. M. (1954) "The use of a psychological model in personality testing." *Educational and Psychological Measurement,* 14, 77–89.

Campbell, D. T. (1960) "Recommendations for APA test standards regarding construct, trait, or discriminant validity." *American Psychologist,* 15, 546–553.

——— & Fiske, D. W. (1959) "Convergent and discriminant validation by the multitrait-multimethod matrix." *Psychological Bulletin,* 56, 81–105.

Campbell, J. P. (1976) "Psychometric theory." In M. D. Dunnette (ed.) *Handbook of industrial and organizational psychology.* Chicago: Rand McNally, 185–222.

———, Dunnette, M. D., Lawler, E. E., III & Weick, K. E., Jr. (1970) *Managerial behavior, performance, and effectiveness.* New York: McGraw-Hill.

——— & Pritchard, R. D. (1976) "Motivation theory in industrial and organization psychology." In M. D. Dunnette (ed.) *Handbook of industrial and organizational psychology.* Chicago: Rand McNally, 63–130.

Carrell, M. R. & Dittrich, J. E. (1978) "Equity theory: The recent literature, methodological considerations, and new directions." *Academy of Management Review,* 3, 202–210.

Cascio, W. F. (1978). *Applied psychology in personnel management.* Reston, Va.: Reston.

Cook, T. D. & Campbell, D. T. (1976) "The design and conduct of quasi-experiments and true experiments in field settings." In M. D. Dunnette (ed.) *Handbook of industrial and organizational psychology.* Chicago: Rand McNally, 223–326.

Crites, J. O. (1969) *Vocational psychology: The study of vocational behavior and development.* New York: McGraw-Hill.

Cronbach, L. J. (1950) "Further evidence on response sets and test design." *Educational and Psychological Measurement,* 10, 3–31.

——— & Furby, L. (1970) "How we should measure 'change'—Or should we?" *Psychological Bulletin,* 74, 68–80.

———, Gleser, G., Nanda, H. & Rajaratnam, N. (1972) *The dependability of behavioral measurements: Theory of generalizability for scores and profiles.* New York: Wiley.

——— & Meehl, P. E. (1955) "Construct validity in psychological tests." *Psychological Bulletin,* 52, 281–302.

———, Rajaratnam, W. & Gleser, G. (1963) "Theory of generalizability: A liberalization of reliability theory." *British Journal of Statistical Psychology,* 16, Part 2, 137–163.

Cummings, L. L. & El Salmi, A. M. (1968) "Empirical research on the bases and correlates of managerial motivation: A review of the literature." *Psychological Bulletin,* 70, 127–144.

——— & El Salmi, A. M. (1970) "The impact of role diversity, job level, and organizational size on managerial satisfaction." *Administrative Science Quarterly,* 15, 1–10.

——— & Schwab, D. P. (1973) *Performance in organizations: Determinants and appraisal.* Glenview, Ill.: Scott, Foresman.

Dansereau, F., Jr., Graen, G. & Haga, W. S. (1975) "A vertical dyad linkage approach to leadership within formal organizations: A longitudinal investigation of the role making process." *Organizational Behavior and Human Performance,* 13, 46–78.

Dunham, R. B. (1975) "Affective responses to task characteristics: The role of organizational function." Unpublished doctoral dissertation. University of Illinois.

———. (1976) "The measurement and dimensionality of job characteristics." *Journal of Applied Psychology,* 61, 404–409.

———, Aldag, R. J. & Brief, A. P. (1977) "Dimensionality of task design as measured by the Job Diagnostic Survey." *Academy of Management Journal,* 20, 209–223.

Dunnette, M. D. (1966) *Personnel selection and placement.* Belmont, Calif.: Wadsworth.

———, Campbell, J. P. & Hakel, M. D. (1967) "Factors contributing to job satisfaction and job dissatisfaction in six occupational groups." *Organizational Behavior and Human Performance,* 2, 143–174.

Edel, E. C. (1966) "A study in managerial motivation." *Personnel Administration,* 29, 31–38.

El Salmi, A. M. & Cummings, L. L. (1968) "Managers' perceptions of needs and need satisfaction as a function of interactions among organizational variables." *Personnel Psychology,* 21, 465–477.

Eran, M. (1966) "Relationships between self-perceived personality traits and job attitudes in middle management." *Journal of Applied Psychology,* 50, 424–430.

Evan, W. M. & Simmons, R. G. (1969) "Organizational effects of inequitable rewards: Two experiments in status inconsistency." *Administrative Science Quarterly,* 14, 224–237.

Evans, M. G. & Molinari, L. (1970) "Equity, piece-rate overpayment, and job security: Some effects on performance." *Journal of Applied Psychology,* 54, 105–114.

Ewen, R. B. (1967) "Weighting components of job satisfaction." *Journal of Applied Psychology,* 51, 68–73.

Farris, G. F. & Lim, F. G., Jr. (1969) "Effects of performance on leadership, cohesiveness, influence, satisfaction, and subsequent performance." *Journal of Applied Psychology,* 53, 490–497.

Forehand, G. A. & Gilmer, B. (1964) "Environmental variation in studies of organizational behavior." *Psychological Bulletin,* 62, 361–382.

Friedman, A. & Goodman, P. (1967) "Wage inequity, self-qualifications, and productivity." *Organizational Behavior and Human Performance,* 2, 406–417.

Gillet, B. & Schwab, D. P. (1975) "Convergent and discriminant validities of corresponding Job Descriptive Index and Minnesota Satisfaction Questionnaire scales." *Journal of Applied Psychology,* 60, 313–317.

Goodman, P. & Friedman, A. (1968) "An examination of the effect of wage inequity in the hourly condition." *Organizational Behavior and Human Performance,* 3, 340–352.

——— & Friedman, A. (1969) "An examination of quantity and quality of performance under conditions of overpayment in piece rate." *Organizational Behavior and Human Performance,* 4, 365–374.

——— & Friedman, A. (1971) "An examination of Adams' theory of inequity." *Administrative Science Quarterly,* 16, 271–288.

Gordon, B. F. & Lowin, A. (1965) "Qualifications, coworker characteristics and productivity: An extension and critique of Adams' inequity theory." Technical report, Bell Telephone Laboratories.

Greene, C. N. (1975) "The reciprocal nature of influence between leader and subordinate." *Journal of Applied Psychology*, 60, 187–193.
Grigaliunas, B. & Wiener, Y. (1974) "Has the research challenge to motivation-hygiene theory been conclusive? An analysis of critical studies." *Human Relations*, 27, 839–871.
Guion, R. M. (1965), *Personnel testing*. New York: McGraw-Hill.
———, (1973) "A note on organizational climate." *Organizational Behavior and Human Performance*, 9, 120–125.
Gullicksen, H. (1950) *Theory of mental tests*. New York: Wiley.
Hackman, J. R. & Oldham, G. R. (1975) "Development of the Job Diagnostic Survey." *Journal of Applied Psychology*, 60, 159–170.
Hall, D. T. & Lawler, E. E., III. (1970) "Job characteristics and pressures and the organizational integration of professionals." *Administrative Science Quarterly*, 15, 271–281.
Heller, F. A. & Porter, L. W. (1966) "Perceptions of managerial needs and skills in two national samples." *Occupational Psychology*, 40, 1–15.
Hellriegel, D. & Slocum, J. W., Jr. (1974) "Organizational climate: Measures, research and contingencies." *Academy of Management Journal*, 17, 255–280.
Herman, J. B. & Hulin, C. L. (1973) "Managerial satisfactions and organizational roles: An investigation of Porter's need deficiency scales." *Journal of Applied Psychology*, 57, 118–124.
Herzberg, F. (1966) *Work and the nature of man*. New York: World.
Herzberg, F., Mausner, B., Peterson, R. O. & Capwell, D. (1957) *Job Attitudes: Review of research and opinion*. Pittsburgh: Psychological Service.
———, Mausner, B., & Snyderman, B. B. (1959) *The motivation to work*. (2nd Ed.) New York: Wiley.
Homans, G. C. (1961) *Social behavior: Its elementary forms*. New York: Harcourt, Brace & World.
House, R. J. & Wigdor, L. A. (1967) "Herzberg's dual-factor theory of job satisfaction and motivation: A review of the evidence and a criticism." *Personnel Psychology*, 20, 369–389.
Imparato, N. (1972) "Relationship between Porter's need satisfaction questionnaire and the Job Descriptive Index." *Journal of Applied Psychology*, 56, 397–405.
Jackson, D. N. (1969) "Multimethod factor analysis in the evaluation of convergent and discriminant validity." *Psychological Bulletin*, 72, 30–49.
——— & Messick, S. (1958) "Content and style in personality assessment." *Psychological Bulletin*, 55, 243–252.
Jacques, E. (1961) *Equitable payment*. New York: Wiley.
James, L. R. & Jones, A. P. (1974) "Organizational climate: A review of theory and research." *Psychological Bulletin*, 81, 1096–1112.
Jessor, R. & Hammond, K. R. (1957) "Construct validity and the Taylor Anxiety Scale." *Psychological Bulletin*, 54, 161–170.
Johannesson, R. E. (1973) "Some problems in the measurement of organizational climate." *Organizational Behavior and Human Performance*, 10, 118–144.
Kalleberg, A. L. & Kluegel, J. R. (1975) "Analysis of the multitrait-multimethod matrix: Some limitations and an alternative." *Journal of Applied Psychology*, 60, 1–9.
Katerberg, R., Smith, F. J. & Hoy, S. (1977) "Language, time, and person effects on attitude scale translations." *Journal of Applied Psychology*, 62, 385–391.
Kavanagh, M. J. (1971) "Leadership behavior as a function of subordinate competence and task complexity." *Administrative Science Quarterly*, 16, 591–600.
Kerr, S. & Schriesheim, C. (1974) "Consideration, initiating structure, and organizational criteria—An update of Korman's 1966 review." *Personnel Psychology*, 27, 555–568.
———, Schriesheim, C. A., Murphy, C. J. & Stogdill, R. M. (1974) "Toward a contingency

theory of leadership based upon the consideration and initiating structure literature." *Organizational Behavior and Human Performance,* 12, 62–82.

King, N. (1970) "Clarification and evaluation of the two-factor theory of job satisfaction." *Psychological Bulletin,* 74, 18–31.

Korman, A. K. (1966) " 'Consideration,' 'initiating structure,' and organizational criteria—a review." *Personnel Psychology,* 19, 349–362.

Lawler, E. E. III. (1966) "Ability as a moderator of the relationship between job attitudes and job performance." *Personnel Psychology,* 19, 153–164.

———. (1967) "The multitrait-multirater approach to measuring managerial job performance." *Journal of Applied Psychology,* 51, 369–381.

———. (1968a) "The effects of hourly overpayment on productivity and work quality." *Journal of Personality and Social Psychology,* 10, 306–313.

———. (1968b) "Equity theory as a predictor of productivity and work quality." *Psychological Bulletin,* 70, 596–610.

——— & Hall, D. T. (1970) "Relationship of job characteristics to job involvement, satisfaction, and intrinsic motivation." *Journal of Applied Psychology,* 305–312.

———, Koplin, C. A., Young, T. F. & Fadem, J. A. (1968) "Inequity reduction over time in an induced overpayment situation." *Organizational Behavior and Human Performance,* 3, 253–268.

——— & Porter, L. W. (1967) "The effect of performance on job satisfaction." *Industrial Relations,* 7, 20–28.

Litwin, G. H. & Stringer, R. A. (1968) *Motivation and organizational climate.* Boston: Harvard Business School.

Locke, E. A. (1968) "Toward a theory of task motivation and incentives." *Organizational Behavior and Human Performance,* 3, 157–189.

———. (1969) "What is job satisfaction?" *Organizational Behavior and Human Performance,* 4, 309–336.

———. (1976) "The nature and causes of job satisfaction." In M. D. Dunnette (Ed.) *Handbook of industrial and organizational psychology.* Chicago: Rand McNally, 1297–1349.

Lowin, A. & Craig, J. R. (1968) "The influence of level of performance on managerial style: An experimental object-lesson in the ambiguity of correlational data." *Organizational Behavior and Human Performance,* 3, 440–458.

Margenau, H. (1950) *The nature of physical reality.* New York: McGraw-Hill.

Marx, M. H. (1963) "The general nature of theory construction." In M. H. Marx (Ed.) *Theories in contemporary psychology.* New York: Macmillan, 4–46.

Maslow, A. H. (1943) "A theory of human motivation." *Psychological Review,* 50, 370–396.

———. (1954) *Motivation and personality.* New York: Harper.

McCornack, R. L. (1956) "A criticism of studies comparing item-weighting methods." *Journal of Applied Psychology,* 40, 343–344.

McNemar, Q. (1969) *Psychological statistics.* (4th Ed.) New York: Wiley.

Messick, S. (1975) "The standard problem: Meaning and values in measurement and evaluation." *American Psychologist,* 30, 955–966.

Mikes, P. S. & Hulin, C. L. (1968) "Use of importance as a weighting component of job satisfaction." *Journal of Applied Psychology,* 52, 394–398.

Miller, E. (1966) "Job satisfaction of national union officers." *Personnel Psychology,* 19, 261–274.

Miner, J. B. & Dachler, P. D. (1973) "Personnel attitudes and motivation." In P. H. Mussen & M. R. Rosenzweig (eds.) *Annual review of psychology.* Palo Alto, Calif.: Annual Reviews, 24, 379–402.

Mitchell, V. F. (1970) "Need satisfactions of military commanders and staff." *Journal of Applied Psychology,* 54, 282–287.

Mobley, W. H. & Locke, E. A. (1970) "The relationship of value importance to satisfaction." *Organizational Behavior and Human Performance,* 5, 463–483.

Nagle, B. F. (1953) "Criterion development." *Personnel Psychology,* 6, 271–289.

Newman, J. E. (1977) "Development of a measure of perceived work environment" (PWE). *Academy of Management Journal,* 20, 520–534.

Northrop, F. S. C. (1959) *The logic of modern physics.* New York: Macmillan.

Nunnally, J. C. (1967) *Psychometric theory.* New York: McGraw-Hill.

Opsahl, R. L. & Dunnette, M. D. (1966) "Role of financial compensation in industrial motivation." *Psychological Bulletin,* 66, 94–118.

Paine, F. T., Carroll, S. J., Jr., & Leete, B. A. (1966) "Need satisfactions of managerial level personnel in a government agency." *Journal of Applied Psychology,* 50, 247–249.

Patchen, M. (1961) *The choice of wage comparisons.* Englewood Cliffs, N.J.: Prentice-Hall.

Payne, D. & Pugh, D. S. (1976) "Organizational structure and climate." In M. D. Dunnette (ed.) *Handbook of industrial and organizational psychology.* Chicago: Rand McNally, 1125–1173.

Payne, R. (1970) "Factor analysis of a Maslow-type need satisfaction questionnaire." *Personnel Psychology,* 23, 251–268.

Pierce, J. L. & Dunham, R. B. (1978) "An empirical demonstration of the convergence of common macro- and micro-organization measures." *Academy of Management Journal,* 21, 410–418.

Porter, L. W. (1961) "A study of perceived need satisfactions in bottom and middle management jobs." *Journal of Applied Psychology,* 45, 1–10.

———. (1962) "Job attitudes in management: I. Perceived deficiencies in need fulfillment as a function of job level." *Journal of Applied Psychology,* 46, 375–384.

———. (1963a) "Job attitudes in management: III. Perceived deficiencies of need fulfillment as a function of line versus staff type of job." *Journal of Applied Psychology,* 47, 267–275.

———. (1963b) "Job attitudes in management: IV. Perceived deficiencies in need fulfillment as a function of size of company." *Journal of Applied Psychology,* 47, 385–397.

——— & Lawler, E. E., III. (1964) "The effects of tall vs. flat organization structures on managerial job satisfaction." *Personnel Psychology,* 17, 135–148.

——— & Lawler, E. E., III. (1965) "Properties of organization structure in relation to job attitudes and job behavior." *Psychological Bulletin,* 64, 23–51.

——— & Mitchell, V. F. (1967) "Comparative study of need satisfactions in military and business hierarchies." *Journal of Applied Psychology,* 51, 139–144.

——— & Siegel, J. (1965) "Relationships of tall and flat organization structures to the satisfaction of foreign managers." *Personnel Psychology,* 18, 379–392.

——— & Steers, R. M. (1973) "Organizational, work, and personal factors in employee turnover and absenteeism." *Psychological Bulletin,* 80, 151–176.

Pritchard, R. D. (1969) "Equity theory: A review and critique." *Organizational Behavior and Human Performance,* 4, 176–211.

———, Dunnette, M. D., & Jorgenson, D. O. (1972) "Effects of perceptions of equity and inequity on worker performance and satisfaction." *Journal of Applied Psychology Monograph,* 56, 75–94.

Quinn, R. P. & Mangione, T. W. (1973) "Evaluating weighted models of measuring job satisfaction: A Cinderella story." *Organizational Behavior and Human Performance,* 10, 1–23.

———, Staines, G. L., & McCullough, M. R. (1974) *Job satisfaction: Is there a trend?* Washington, D.C.: U.S. Government Printing Office, Department of Labor.

Rabinowitz, S. & Hall, D. T. (1977) "Organizational research on job involvement." *Psychological Bulletin,* 84, 265–288.

Rhinehart, J. B., Barrell, R. P., DeWolfe, A. S., Griffin, J. E. & Spaner, F. E. (1969)

"Comparative study of need satisfactions in governmental and business hierarchies." *Journal of Applied Psychology,* 53, 230–235.
Rizzo, J. R., House, R. J. & Lirtzman, S. I. (1970) "Role conflict and ambiguity in complex organizations." *Administrative Science Quarterly,* 15, 150–163.
Roberts, K. H., Walter, G. A. & Miles, R. E. (1971) "A factor analytic study of job satisfaction items designed to measure Maslow need categories." *Personnel Psychology,* 24, 205–220.
Schneider, B. & Alderfer, C. P. (1973) "Three studies of measures of need satisfaction in organizations." *Administrative Science Quarterly,* 18, 489–505.
——— & Hall, D. T. (1972) "Toward specifying the concept of work climate: A study of Roman Catholic diocesan priests." *Journal of Applied Psychology,* 56, 447–455.
Schöneman, P. H. (1966) "A generalized solution of the orthogonal procrustes problem." *Psychometrika,* 31, 1–10.
Schriesheim, C. A. & Kerr, S. (n.d.) "Theories and measures of leadership: A critical appraisal." Department of Administrative Sciences, Graduate School of Business Administration, Kent State University,
Schuler, R. S., Aldag, R. J. & Brief, A. P. (1977) "Role conflict and ambiguity: A scale analysis." *Organizational Behavior and Human Performance,* 20, 111–123.
Schwab, D. P. & Cummings, L. L. (1970), "Theories of performance and satisfaction: A review." *Industrial Relations,* 9, 408–430.
———, Heneman, H. G., III & DeCotiis, T. A. (1975), "Behaviorally anchored rating scales: A review of the literature." *Personnel Psychology,* 28, 549–562.
——— & Wallace, M. J., Jr. (1974) "Personal and organizational correlates of the pay satisfaction of nonexempt employees." *Industrial Relations,* 13, 78–89.
Sims, H. P., Jr. & Szilagyi, A. D. (1976) "Job characteristic relationships: Individual and structural moderators." *Organizational Behavior and Human Performance,* 17, 211–230.
———, Szilagyi, A. D., & Keller, R. T. (1976) "The measurement of job characteristics." *Academy of Management Journal,* 19, 195–212.
Smith, H. W. (1975) *Strategies of social research.* Englewood Cliffs, N.J.: Prentice Hall.
Smith, P. C. (1976) "Behaviors, results, and organizational effectiveness: The problem of criteria." In M. D. Dunnette (ed.) *Industrial and organizational psychology.* Chicago: Rand McNally, 745–775.
———, Kendall, L. M. & Hulin, C. L. (1969) *The measurement of satisfaction in work and retirement.* Chicago: Rand McNally,
Stogdill, R. M. (1974) *Handbook of leadership: A survey of theory and research.* New York: Free Press.
"Technical recommendations for psychological tests and diagnostic techniques." (1954) *Psychological Bulletin Supplement,* 51, Part 2, 1–38.
Tenopyr, M. L. (1977) "Content-construct confusion," *Personnel Psychology,* 30, 47–54.
Tesser, A. & Krauss, H. (1976) "On validating a relationship between constructs." *Educational and Psychological Measurement,* 36, 111–122.
Thorndike, R. L. (1949) *Personnel selection: Tests and measurement techniques.* New York: Wiley.
Vroom, V. H. (1964) *Work and motivation.* New York: Wiley.
Wahba, M. A. & Bridwell, L. G. (1976) "Maslow reconsidered: A review of research on the need hierarchy theory." *Organizational Behavior and Human Performance,* 15, 212–240.
Wall, T. D. & Payne, R. (1973) "Are defiiciency scores deficient?" *Journal of Applied Psychology,* 58, 322–326.
Wallace, M. J. & Berger, P. K. (1973) "The reliability of difference scores: A preliminary

investigation of a need deficiency satisfaction scale." *Academy of Management Proceedings,* 33, 421–427.
Walster, E., Berscheld, E. & Walster, G. W. (1973) "New directions in equity research." *Journal of Personality and Social Psychology,* 25, 151–176.
Wanous, J. P. & Lawler, E. E., III. (1972) "Measurement and meaning of job satisfaction." *Journal of Applied Psychology,* 56, 95–105.
Waters, L. K. & Roach, D. (1973) "A factor analysis of need-fulfillment items designed to measure Maslow need categories." *Personnel Psychology,* 26, 185–190.
Webb, E. J., Campbell, D. T., Schwartz, R. D. & Sechrest, L. (1966) *Unobtrusive measures: Nonreactive research in the social sciences.* Chicago: Rand McNally.
Weiner, Y. (1970) "The effects of 'task- and ego-oriented' performance on two kinds of overcompensation inequity." *Organizational Behavior and Human Performance,* 5, 191–208.
Weiss, D. J., Dawis, R. V., England, G. W. & Lofquist, L. H. (1967) "Manual for the Minnesota Satisfaction Questionnaire." Minneapolis: Minnesota Studies in Vocational Rehabilitation, 22, Bulletin 45,
Werts, C. E. & Linn, R. L. (1970) "A general linear model for studying growth." *Psychological Bulletin,* 73, 17–22.
Whitsett, D. A. and Winslow, E. K. (1967) "An analysis of studies critical of the motivator-hygiene theory." *Personnel Psychology,* 20, 391–415.
Wood, I. & Lawler, E. E., III. (1970) "Effects of piece-rate overpayment on productivity." *Journal of Applied Psychology,* 54, 234–238.

# RATIONALITY AND JUSTIFICATION IN ORGANIZATIONAL LIFE

Barry M. Staw,

NORTHWESTERN UNIVERSITY

## ABSTRACT

In this essay, rationality and justification are studied as pervasive images of both human and organizational systems. A number of outcroppings of rationality and justification are illustrated at both the individual and organizational levels of analysis, and a cross-level effect is discussed in some detail. However, the overall goal of the essay is pretheoretic. Rather than providing an exact theory or set of hypotheses, the essay addresses the rationality-justification distinction as a central element of organizing from which new hypotheses may be shaped in the future.

**Research in Organizational Behavior, Volume 2, pages 45–80**

**ISBN: 0-89232-099-0**

## INTRODUCTION

In both research and theorizing, the field of organizational behavior has made heavy use of rational models. The individual has been conceived as a rational, goal-seeking entity which processes information and makes decisions in his or her own self-interest (e.g., Vroom, 1964; Porter & Lawler, 1968; Campbell and Pritchard, 1976). The organization has also been conceived as a goal-seeking entity which behaves in ways to protect and expand its domains of interest (e.g., Thompson, 1967; Pfeffer & Salancik, 1978). However, cutting across images of rationality are theories and data on justification processes. Individuals are sometimes conceived as rationalizing as opposed to rational decision makers (e.g., Festinger, 1957; Aronson, 1976). Likewise, organizations have been conceived as vehicles by which ambiguous goals, preferences, and decision plans get put into action and justified (e.g., Cohen, March, & Olson, 1972; March & Olson, 1976). Thus, we have been beset by images of both rationality and justification on each of two levels of analysis.

The goal of this chapter is to dissect and analyze the rationality-justification distinction, but not to solve it. The interrelationship of these two forces is viewed as central to the issues of organizational behavior, yet this essay will not advocate one factor over another or propose a new contingency theory based on these factors. In contrast, this chapter represents a pretheoretic effort to focus attention upon rationality and justification as pervasive elements in organizational life. It is believed these forces are prevalent at multiple levels in organizational systems and influence each other across such levels. Thus, the chapter will first outline and defend the existence of rationality and justification at both the individual and organizational levels of analysis. In the second section of this essay, examples of how these two forces may affect the behavior of both individuals and organizations will be provided. Finally, this chapter will illustrate what is believed to be an important cross-level effect; that is, how organizational rationality can trigger individual forces for justification.

Before advancing further with a task that spans two levels of analysis it is appropriate to admit to certain theoretical preferences and prejudices. In the author's opinion, the parallels between individual and organizational analysis are not by any means a coincidence. Perhaps the similarity stems from the fact that individual scholars working in the organizational field extrapolate from their own cognitive images and metaphors to explain the workings of organizations. Or perhaps large collectivities do, in fact, behave in similar ways to individuals. As suggested by Miller (1978), there may be some principles or laws of behavior which are quite generalizable across levels of analysis. However, probably the simplest and most convincing reason for the existence of cross-level parallels is that organizations are populated by individuals, and thus if there exist any generalizable tendencies within individuals, these tendencies will likely be manifested in organizational actions. Admittedly, organizational actions are not a direct by-product of individual behavior; political maneuvering, coalition formation, and hierarchical systems generally separate the individual from organizational behavior. Yet, if individual administrators do possess strong tendencies for both rationality and justification, many actions undertaken by organizations are likely to exhibit these same general characteristics.

## EXISTENCE OF RATIONALITY

*Sources of Individual Rationality*

At the individual level of analysis, the most common model of rationality has been that of economic decision making. Individuals have been conceived as maximizing their own subjective expected utility (Edwards, 1954). Limits to the ability of man in inputting and processing information have certainly been acknowledged, and so have some limitations to ways data are interpreted into final decisions. Yet the basic conception of individuals as rational, goal-seeking entities has remained relatively intact. We still expect job applicants to choose their most highly rated organization for employment (e.g., Vroom, 1966; Wanous, 1972; Mitchell & Knudsen, 1973; Lawler, Kuleck, Rhode & Sorensen, 1975), and for individuals to perform that behavior which is viewed as subjectively most desirable (Vroom, 1964; Graen, 1969; Campbell & Pritchard, 1976).

A highly elaborate model of the cognitively rational individual is outlined in Figure 1. Several assumptions are implicit in this figure. It is hypothesized that the individual, when faced with a choice to exert a high or low amount of effort on the job, will follow an economic decision-making model. That is, based upon a set of well-articulated expectancies and valences, the individual would be expected to choose that course of action with the greatest subjective utility. For simplicity, only two levels of effort are illustrated in the figure. However, given most individuals' limit to cognitive information processing (cf., Miller, 1956; Simon, 1957; Slovic, 1972), the two-level representation may depart less from empirical reality than Vroom's (1964) or Lawler's (1971) expectancy models (these latter models implicitly assume assessment of "all" levels of effort and performance).

The rational model of individual performance illustrated in Figure 1 could be operationalized by asking individuals to think of the consequences of working at two or more levels of effort. Items could then be designed to assess the likelihood that each type of task behavior would lead to various intrinsic and extrinsic outcomes. The importance or

*Figure 1.* A Cognitively Rational Model of Task Performance (from Staw 1977a)

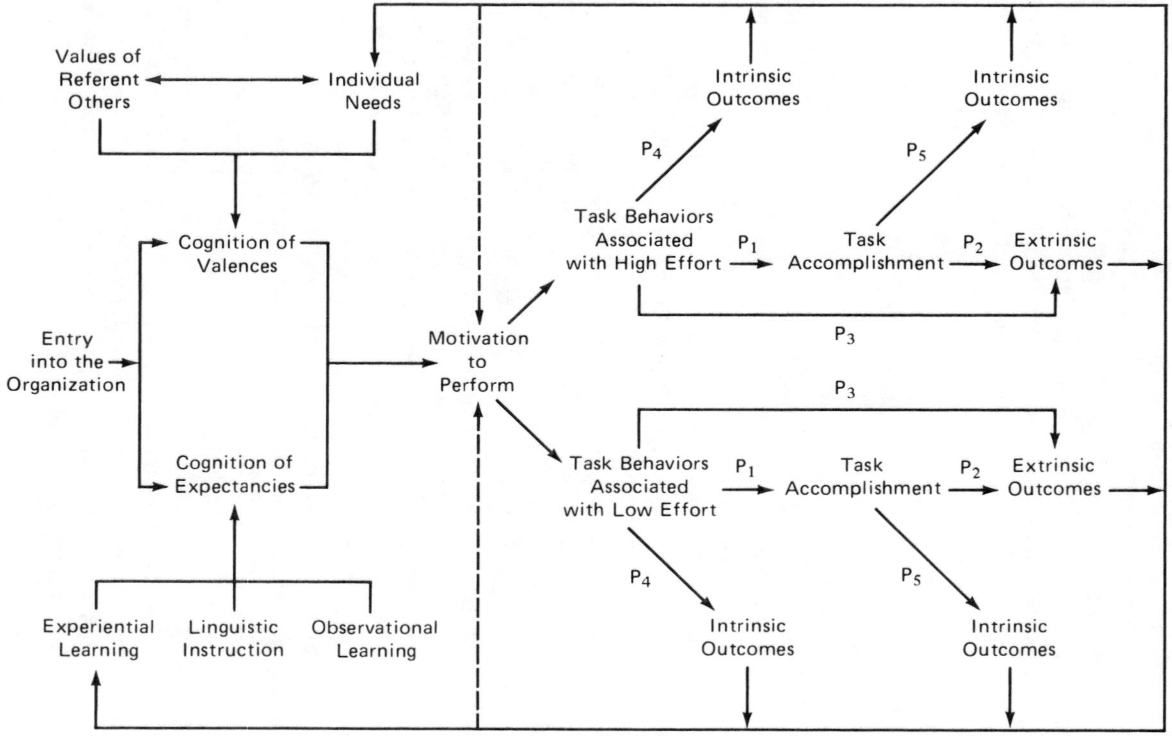

relative valence of each of these outcomes could be assessed and a weighted average of expected value could then be computed for both "working hard" and "not so hard." An operationalization of this formulation of "motivation to perform" is presented below:

$$\text{Motivation to Perform} = \left[\begin{array}{l}\text{Expected}\\ \text{Value of}\\ \text{Working}\\ \text{Hard}\end{array}\right] - \left[\begin{array}{l}\text{Expected}\\ \text{Value of}\\ \text{Working at}\\ \text{Leisurely}\\ \text{Pace}\end{array}\right],$$

where:

$$\text{Expected Value} = IV_{beh} + P_1\,(IV_{acc}) + \sum_{i=1}^{n}\left[P_1P_2(EV_i)\right] + \sum_{i=1}^{n}\left[P_3EV_i\right]$$

and

$IV_{beh}$ = the intrinsic valence associated with task behavior (i.e., working 5 vs. 12 hours);
$IV_{acc}$ = the intrinsic valence associated with task accomplishment;
$EV_i$ = the extrinsic valences associated with extrinsic rewards;
$P_1$ = the probability that task behavior will lead to accomplishment;
$P_2$ = the probability that task accomplishment will lead to extrinsic rewards;
$P_3$ = the probability that task behavior will lead directly to extrinsic rewards.

By spelling out the cognitively rational model of individual behavior in such detailed form, its implausibility becomes evident. If individuals (even in the two-level of effort case) underwent such computational gymnastics before each action alternative, they would cease to behave at all. Thus, we must acknowledge that much of our behavior has been routinized and is subject to reinforcing cycles (illustrated by broken line in Figure 1). That is, once certain behaviors have been emitted and have been rewarded, they may occur repeatedly without the need for future cognitive decision making. Similarly, individuals frequently utilize behavioral models as guides to their own behavior. They may pattern their behavior after another individual or simply accept culturally prescribed actions without undergoing any active decision making. Such reinforcement and behavioral modeling do not, however, conflict with the basic orientation of individual rationality. They are simply more abbreviated versions of rational goal-seeking than the fully cognitive utility model.

In an earlier paper, this author (Staw, 1977a) proposed that individuals do not merely react to the contingencies of their environment as a cognitive processer of events and outcomes. They are also proactive in the sense that they attempt to control the contingencies of their surroundings. Within the organizational context, control may involve the preservation of expertise, hoarding of essential information, or otherwise making oneself so valuable to the organization that one can dictate the terms of his or her involvement (Martin & Sims, 1956; Pfeffer, 1977). These are obviously the strategies of the highly skilled and centrally located participant who already has a good deal of power. However, for those at the organization's periphery and for those with less-valued skills, other strategies are possible. One method by which a low-power person may improve his situation in the organization is by ingratiation (Wortman & Linsenmeier, 1977). If the individual can manipulate his supervisor's attitudes and opinions, he can improve his share of the resources allocated by the supervisor. Through ingratiation, the individual may receive more than he deserves for a given level of work or at least assure himself of a positive evaluation of his task output. While the ingratiation strategy is highly individualistic, the organization of many low-power individuals into a coalition or union will also yield greater influence within the organization. Unionization of the workplace frequently brings increased worker control over the methods of resource allocation (e.g., weekly wage supplanting a piece-rate system of incentives) and the procedures by which the work is accomplished (e.g., highly restricted work rules replacing employer-designed job descriptions).

The rational individual, as we have described him, is both an adapting and controlling creature. He has the capacity to anticipate, evaluate, and choose that course of action which will satisfy certain needs or valued outcomes. But, he also possesses the capacity to change the very contingencies or "rules of the game" to which he is subjected. This dual nature of man which can be both proactive and reactive is implicity captured in two quotes from White's (1959) work on competence.

> As used here, competence will refer to an organism's capacity to interact with its environment. . . . In the case of man, where so little is provided innately and so much has to be learned through experience, we should expect to find highly advantageous arrangements for securing a steady cumulative learning about the properties of the environment and the extent of possible transactions. Under these circumstances we might expect to find a very powerful drive operating to insure progress toward competence, just as vital goals of nutrition and reproduction are secured by powerful drives (White, p. 297).

> [Human] organisms differ from other things in nature in that they are 'self-governing entities' which are to some extent 'autonomous'. . . . The human being has a characteristic tendency toward self-determination, that is, a tendency to resist external influences and to subordinate the heteronomous forces of the physical and social environment to its own sphere of influence. . . . Of all living creatures it is man who takes the longest strides toward autonomy. This is not because of any unusual

> tendency toward bodily expansion at the expense of the environment. . . . Man as a species has developed a tremendous power of bringing the environment into his service, and each individual member of the species must attain what is really quite an impressive level of competence if he is to take part in the life around him (p. 324).

The image projected by the quotations from White (1959) and our own discussion of rationality should convey the impression that individuals are highly motivated to predict and control their environments. Some specific hypotheses concerning the sequence by which individuals might act to increase predictability and/or control have been elaborated elsewhere (Staw, 1977a). As shown in Figure 2, individuals are postulated, first to attempt control over their environments. However, if control is not possible, then individuals are hypothesized to make their environments more predictable. As proposed, this general sequence is moderated by the individual's expectations of control. Individuals who have had a prior history of self-control are most likely to seek control over the allocation of resources. However, individuals with little history of self-control are predicted to primarily seek predictive power. Finally, if neither prediction nor control is possible, research has shown (Seligman, 1975) that individuals are likely either to become psychologically depressed (learned helplessness) or exit the situation altogether.

*Figure 2.* Flow Diagram of Upward Control (From Staw, 1977a)

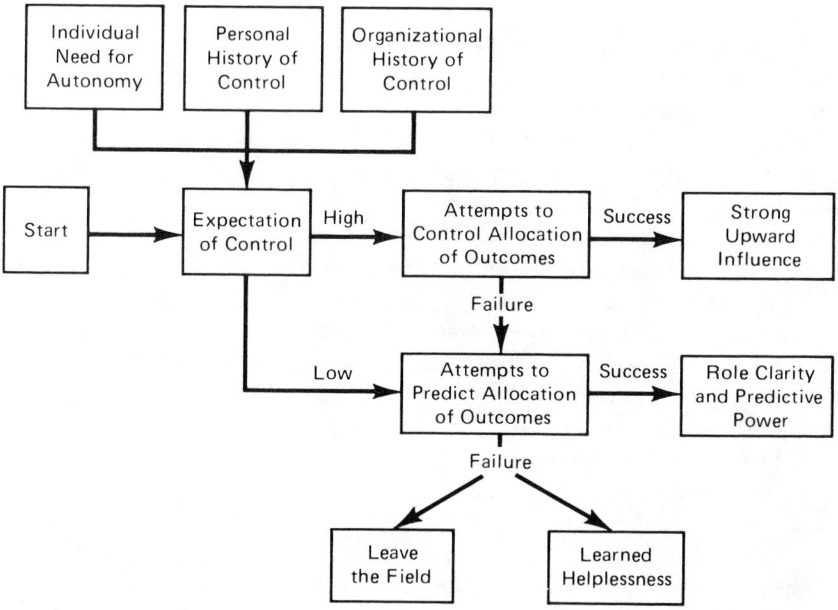

Individual Need for Autonomy
Personal History of Control
Organizational History of Control
Start
Expectation of Control
High
Attempts to Control Allocation of Outcomes
Success
Strong Upward Influence
Failure
Low
Attempts to Predict Allocation of Outcomes
Success
Role Clarity and Predictive Power
Failure
Leave the Field
Learned Helplessness

*Sources of Organizational Rationality*

At the organizational level of analysis there has also been widespread concern with rationality. The managerial literature on organization design and policy (e.g., Lawrence & Lorsch, 1967; Galbraith, 1977; Chandler, 1962) is concerned with fitting the proper organizational structure to varying environmental and task characteristics. The orientation of this body of work is clearly normative, advocating that administrators follow certain design guidelines in increasing the effectiveness of their organizations.

Other more descriptive work on organizational structure (e.g., Blau & Schoenherr, 1971; Pugh, Hickson, Hennings & Turner, 1968) has been directed toward finding verifiable relationships between structural, task, and environmental conditions. However, even the most descriptive research on organizational variables contains the tacit assumption of rationality. It is implicitly assumed in most empirical studies that some functional relationships underlie the observed correlations. Either by natural selection (Hannan & Freeman, 1977) or by conscious administrative decision (Simon, 1957) certain covariations of macrovariables come into being and are maintained over time. Although the soundness of drawing normative implications from descriptive research has been questioned (Argyris, 1972), the underlying theme of rationality is indeed very strong within the empirical literature on design and structure.

In terms of formulating *explicit* hypotheses on the rational behavior of organizations, Thompson's (1967) work is preeminent. Thompson outlined a host of propositions which encompass methods by which organizations make their environments more predictable and controllable. Figure 3 outlines the basic model of an industrial organization within Thompson's framework. As one can see, there is a technical core within which goods or services are produced by an organization. There exist sources of uncertainty in transforming raw materials to finished products and internal organizational structure is designed to reduce this type of uncertainty. Generally more problematic, however, are sources of uncertainty in the organization's environment on both the input and output sides of the system. According to Thompson, organizations reduce environmental uncertainty by various buffering activities such as material stockpiling, forecasting, and contracting. Organizations also attempt to acquire power relative to other forces within their environments, and to do this they may attempt to coopt opposing elements into the organization, form coalitions with other organizations, or seek support from other more powerful entities. In recent years there has been a great deal of empirical research supporting many of these propositions (see Pfeffer & Salancik, 1978).

The Thompsonian view of organizational rationality has dominated the

*Figure 3.* A Thompsonian View of Industrial Organizations

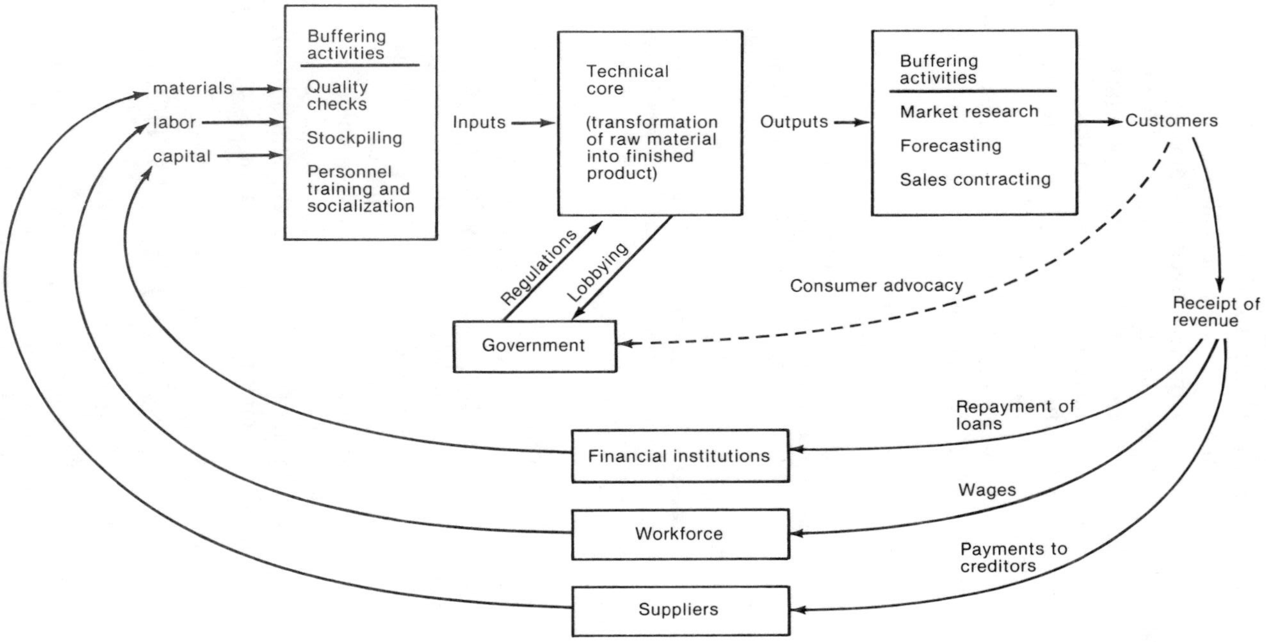

last decade of theory and research at the organizational level of analysis. According to this approach, organizations attempt to fit internal design to task and environmental contingencies, and also to buffer themselves from environmental uncertainty by selectively bringing important sources of uncertainty under greater control. Thus, like the individual who seeks to make his world both more predictable and controllable, organizations seek to reduce both their uncertainty and dependence upon the environments within which they function. Although organizations are integrally tied to the larger social system (Parsons, 1960), they seek to establish autonomy of action and domain. As a result, there are strong parallels between White's (1959) conception of man and Thompson's view of organizations. These parallels should be evident in the following excerpts from Thompson's work on organizational action.

> The domain claimed by the organization and recognized by its environment determines the points at which the organization is dependent, facing both constraints and contingencies. To attain any significant measure of self-control, the organization must manage its dependency. . . . Under norms of rationality, therefore, organizations seek to minimize the power of task-environmental elements over them. . . .
>
> The more an organization is constrained in some sectors of its task environment, the more power it will seek over remaining elements of its task environment. When the organization is unable to achieve such a balance, it will seek to enlarge its task environment. (Thompson, 1967, pp. 37–38.)

As we can see from both the Thompson and White quotes, organizations and individuals can each be viewed as reducing sources of external dependence and seeking relative autonomy. Neither approach denies the adjustive nature of individuals or organizations, however. Thompson deals extensively with organizational design as an adjustive mechanism which is less costly and disruptive than altering the environment. White, likewise, views the autonomy motive as something which does not supplant but goes beyond traditional learning mechanisms of individuals. Thus, we are left with remarkably parallel conceptions of organizations and individuals, both exhibiting rationality in a proactive and reactive sense.

## EXISTENCE OF JUSTIFICATION

### *Sources of Justification in Individuals*

Individuals, as we have noted, are motivated by a competence drive. They seek to predict and control their immediate environments and to attain particular goals they have set for themselves. However, paradoxically, the greater are individuals subject to a competence motive, the more they are also susceptible to forces that run counter to rationality. If an individual has a strong need to be correct or accurate, he is also likely to feel the need to justify his decisions—to prove to himself and others that he is indeed competent and rational as a decision maker. Unfortunately, it is precisely the need to demonstrate rationality that can lead to justification processes which run counter to our conception of rational man.

The most extensive body of research relating to self-justification processes are studies originally designed to test the theory of cognitive dissonance. As defined by Festinger (1957), dissonance theory is an extremely simple yet provocative idea. Festinger posited that any two cognitive elements which are inconsistent with each other will lead to a psychologically uncomfortable state of cognitive dissonance. According to Festinger's formulation, cognitive elements are defined broadly as any knowledge, opinion, or belief about oneself or environment. In this form, dissonance theory is almost indistinguishable from other models of cognitive consistency (e.g., Newcomb, 1953; Heider, 1958; Osgood, 1960; Rosenberg, 1960). However, dissonance theory has had a much greater influence on social science theory and research than any other consistency model.

Many reviewers have argued that the disportionate amount of influence of dissonance over other cognitive consistency theories has been due to the highly general nature of Festinger's model or the creativity of the empirical research supporting dissonance theory. In this author's opinion, however, the historical impact of Festinger's theory is at least partly due to the fact that most dissonance studies have implicitly tapped self-justification processes. Dissonance research was given credit for discovering many aspects of human justification and for finding forms of behavior which were more rationalizing than rational. It was these counterintuitive (and seemingly nonrational) findings which largely created the research interest and controversy surrounding dissonance theory.

The most controversial and heavily researched subarea of dissonance theory tests what is known as the "insufficient justification paradigm." The paradigm deals with the consequences of acting in a manner inconsistent with one's beliefs or attitudes in the absence of external justification. According to the theory, dissonance should be aroused by inconsistency between one's attitude toward an activity and knowledge of its enactment. However, as external pressure (e.g., promised reward or threatened punishment) on the individual to perform the activity is decreased, the dissonance aroused by its enactment is increased. The crucial point is that the level of dissonance aroused should be at its maximum if the extrinsic reward or punishment is just barely enough to elicit the behavior; any incentive greater than this minimal amount (being consistent with the behavior) should, theoretically, produce less cognitive dissonance (Festinger, 1957).

In reviewing much of the literature on dissonance theory, Aronson (1968) made a very important point which helps clarify the relation between cognitive dissonance and self-justification processes. Aronson noted that in most dissonance studies, it is not the two inconsistent cognitions about an activity (e.g., "I believe a task is dull," and "I am not getting paid or otherwise compensated for my efforts") that really motivates individuals to change their attitudes. Instead, Aronson noted that it is the inconsistency between behavior or attitudes and one's own self-concept that is dissonance-arousing. If individuals believe they are generally incompetent, discovering that they are undercompensated or have made a behavioral error should not be disturbing and, thus, they would not be motivated to reduce dissonance. However, if individuals possess some degree of self-confidence, being placed into a situation with insufficient rewards should be disturbing. Aronson therefore posits that positive self-concept is an important boundary condition for dissonance theory effects.

We would go beyond Aronson and state that in many dissonance experiments a need for self-justification has been tapped rather than cognitive consistency. Because researchers have typically pitted a dissonance-predicted effect versus self-interest or reinforcement effects, the impact of the theory is due largely to the ability of researchers to find counterhedonic data in a variety of situations. However, a rather parsimonious explanation of most of the dissonance findings is that individuals seek to justify their own actions, decisions, and attitudes so as to protect their self-concepts. If individuals want to demonstrate competence, they will tend to re-evaluate alternatives after a choice in order to "prove" they were right (e.g., Brehm, 1956; Vroom, 1966; Knox & Inkster, 1968). Individuals will also tend to re-evaluate their attitudes toward activities to "prove" that they did not make a mistake in committing themselves to a dull task (Brehm & Cohen, 1959; Freedman, 1963; Weick, 1964; Pallak, Sogin & Van Zante, 1974), boring discussion group (Aronson & Mills, 1959; Gerard & Mathewson, 1966) or in writing an essay against their own position on an issue (Festinger and Carlsmith, 1957; Cohen, 1962; Carlsmith, Collins & Helmreich, 1966; Linder, Cooper, & Jones, 1967).

The case for self-justification becomes even stronger when we examine some of the moderating conditions recently discovered for "dissonance effects." After intensive research using counterattitudinal advocacy as a dissonance-arousing treatment (i.e., writing an essay against one's own position on an issue as in the classic Festinger and Carlsmith procedure) it is now generally agreed that the following conditions are necessary for dissonance to be induced: the essay writing must entail strong negative consequences for the subject, there should be a high degree of choice in writing the essay, and a minimal level or absolutely no external inducement such as money should be involved (Collins & Hoyt, 1972; Calder, Ross & Insko, 1973). What these moderating conditions amount to is that the subject has made a clear-cut mistake in deciding to undertake a negative task without extenuating circumstances or the promise of other rewards. What's more, there is also research that shows that subjects must feel personally responsible for the negative consequences in order to obtain a dissonance effect (Cooper, 1971). That is, the prior decision to undertake the task must not only have been free of duress or outside pressure, but the negative consequences should have been foreseeable at the time of the decision. In short, the subject, despite his general need for competence, is typically induced into behaving irrationally in most dissonance experiments. Therefore, we would interpret most dissonance effects as products of ego-defensiveness. More specifically, we would predict that the need to justify one's actions increases as the irrationality of one's actions are exposed to both self and others.

*Prospective and Retrospective Rationality.* In moving away from dissonance theory, we have simply affirmed that individuals exhibit certain ego-defensive processes and that ego-defensiveness may often interfere with rationality. Still, if we define rationality in a totally subjective sense, any behavior, including outcroppings of ego-defensiveness, can be viewed as rational. However, we must define rationality in a more limited sense if it is to have any explanatory power. Thus, rationality is usually viewed as decision making approximating economic utility theory or expectancy theory models of behavior. We will refer to this traditional image of man as being *prospectively* rational. Under prospective rationality, the individual will process information and make decisions to attain a high level of outcomes. However, when ego-defensiveness is dominant, individuals will often behave in a *retrospectively* rational manner. They will re-evaluate alternatives and outcomes to make it *appear* that they have acted in a competent or intelligent manner. This type of behavior may be rational in a totally subjective sense, but clinicians would label it as "rationalizing" based on intersubjective criteria or standards used by outside observers of the behavior (Haan, 1977).

Probably the most critical element separating prospective from retrospective rationality is the individual's treatment of *sunk costs*. Under prospective rationality, resources are allocated and decisions entered into when future benefits are greater than future costs. Previous losses or costs, which have been suffered but which are not expected to recur in the future, should not enter into decision calculations. However, under retrospective rationality, there may be motivation to rectify past losses as well as to seek future gain. The individual, in order to appear rational in his decision making, is likely to keep sunk costs as an active part of decision making under retrospective rationality. This desire to recoup sunk costs is probably what underlies much of the behavior that we commonly label as self-justification.

Figure 4 outlines a simple model of some of the determinants of retrospective rationality. As shown in the figure, when action leads to negative consequences ego-defensiveness may occur on the part of individuals. As we have noted, there is evidence showing that the degree of defensiveness is moderated by the amount of personal responsibility for negative consequences. Two previously researched aspects of personal responsibility are *prior choice* and *foreseeability of outcomes* (Cooper, 1971); however, to these variables we have added norms for rationality. If an organization (or any subculture) highly values rationality, individuals are likely to assume greater personal responsibility for negative consequences and therefore suffer greater ego-defensiveness. If ego-defensiveness is indeed high, the individual is predicted to follow one of several retrospectively rational strategies. He may search for exogenerating explanations of the negative consequences to "take himself off the hook." There is a substantial body of research (see Stevens & Jones, 1976) that indeed shows that individuals do engage in self-serving biases. Alternatively, an individual may simply re-evaluate the outcomes he has received and conclude that they are not so bad after all or even "blessings in disguise." However, this is a difficult path to follow when outside observers can easily corroborate outcomes, as in most organizational settings. Finally, the individual can attempt to recoup his losses by committing new resources to the same, negative course of action (Staw, 1976). By throwing good money after bad, individuals sometimes attempt to "prove" that they never really made a mistake after all.

Each of the retrospective strategies shown in Figure 4 are focused upon erasing or alleviating behavioral or decisional error. However, if positive consequences had resulted from one's action, or if negative consequences did not produce ego-defensiveness, then individuals are predicted to follow a more prospectively rational strategy. They will be more likely to search for accurate or plausible explanations of events than to engage in self-serving biases (cf., Fox, 1980). Also, individual commitment to future action is shown to depend on the anticipation of future benefits and costs, rather than on an effort to recoup previous losses.

*Internal and External Justification.* Most writings on self-justification or dissonance theory have characterized rationalization as a purely internal

*Figure 4.* Some Determinants of Retrospective and Prospective Rationality

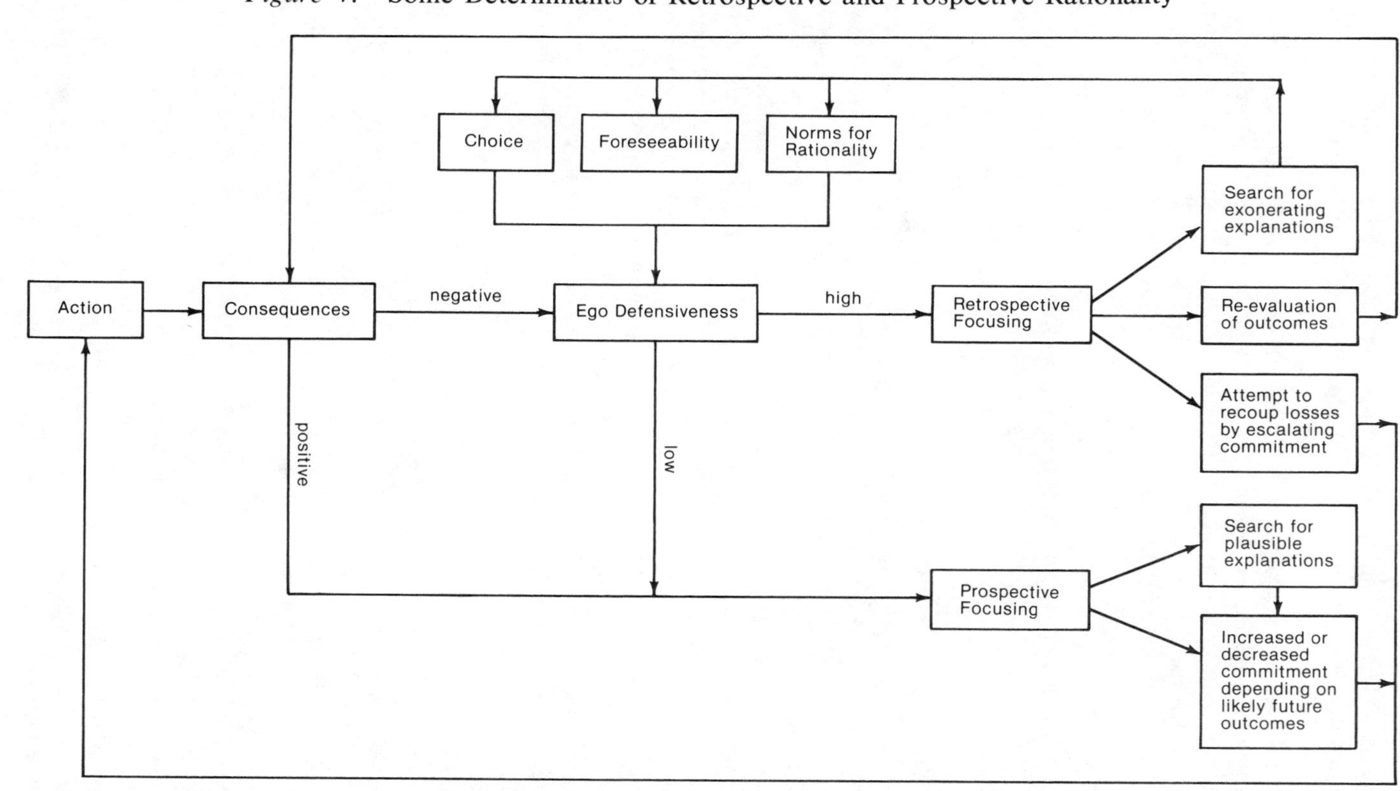

process. That is, individuals are assumed to psychologically reconstruct outcomes, events, or values to appear rational or competent *to themselves*. Like cognitive consistency, this type of justification helps protect one's ego in times of stress and failure. It can be functional or dysfunctional, depending on its extent and whether psychological adaptation rather than concrete action is required of the individual. In times of extreme stress, self-justification may, like a safety valve, reduce tension to the level that effective decision making may again be possible (cf., Janis & Mann, 1977). However, extensive and elaborate forms of justification may, by themselves, forestall necessary information processing and decision making, thus reducing individual effectiveness.

When justification is a largely internal process, individuals attend to events and reconstruct them to protect their own self-images and identities. However, justification may also be directed externally. When faced with an external threat or evaluation, individuals may resort to retrospective forms of rationality or justification. Like Richard Nixon in the case of Watergate or Lyndon Johnson in the case of the Vietnam War, individuals may go to extreme lengths to prove that they were not wrong in an earlier decision. This type of justification is externally directed, since it is designed to prove to others rather than to oneself that he or she is competent. However, as we have seen in these two widely publicized cases, the behaviors individuals can undertake in search of external justification can also be more retrospectively than prospectively rational. The individual may seal himself off from relevant information, misinterpret the facts, or simply take improper risks to save an earlier policy decision. Thus, external as well as internal justification may be viewed as "irrational" by outside observers, especially if the outcroppings of justification lead to further negative consequences.

*Sources of Justification in Organizations*

In our earlier discussion we noted that the Thompsonian perspective on organizational behavior posits that organizations are intendedly rational (not that they always act rationally). Organizations, from this viewpoint, strive to control their internal operations and external environment, but never totally achieve a closed, rational system (Hall, 1977).

The limitations to rationality in organizations are multifaceted and have been subject to much debate. Pfeffer (1977) has argued that organizations do not make choices about internal allocations of resources based upon rational criteria, but upon political influence. Internal coalitions among organizational actors may thus determine organizational actions as much as internal task demands or environmental exigencies. Hall (1977) has argued that organizations have difficulty behaving rationally since there is little consensus over goal-criteria. Educational organizations, for example, may seek to satisfy mutually inconsistent criteria (e.g., research, teaching, and service), and the process of attempting to reach one of these goals can preclude the attainment of others. Gouldner (1959) and Katz & Kahn (1978) have argued that organizations are natural systems and that organizational actions are relatively unplanned adaptive responses to threats to equilibrium. The open-systems approach stresses the interrelationship of the organization to the environment as elements of the larger social system, rather than the subjugation of the environment *by* the organization as viewed under the "rationality" perspective. Finally, March & Olson (1976) view organizations as rather irrational "garbage cans" into which goals, plans, and actions are paired together. They view organizations as possessing very imperfect sets of goals or utility functions, very limited capacities for communication and information processing, and very poor abilities to resolve conflicts among choice alternatives. From the March and Olson perspective, organizations are fortunate to survive at all, let alone function in a rational manner.

Many of the limitations to rationality can be captured effectively by Thompson & Tuden's (1959) classification of organizational decision making. Using their two-by-two matrix, decisions can be classified by certainty of cause-effect relations and clarity of preferences. According to these authors, when beliefs about cause-effect relations are relatively certain and preferences about desired outcomes are clear, then organizations can approach the rationality of a closed system. However, most organizational decisions lie in the other three cells. Thompson and Tuden have labeled the uncertain cause-effect/clear preferences cell as requiring judgmental decision making, and the certain cause-effect/unclear preferences case as requiring compromising and conflict resolution. Finally, the unclear cause-effect/unclear preferences situation is described as necessitating nothing short of inspiration. This last case brings to mind images of organizations "galumphing" along, not knowing exactly where they want to go or how to get there. Unfortunately, this last case is probably the one that best describes how many complex organizations operate.

Figure 5 attempts to graphically show why organizations are highly limited in rationality. This figure is based partially on the ideas of March & Olson (1976), but is presented in a decision format that is analogous to Figure 1. Organizational actions are shown here to be a product of organizational belief systems and preferences in a manner parallel to individual choice and action. However, organizational beliefs and preferences are not simple variables which can be assumed away or taken as given. Organizational beliefs are a complex product of internal communication, information processing, and storage systems in organizations; while organizational preferences result from complex political processes and coalition formation. We will not attempt in this paper to specify the exact linkages among these variables but must simply state that the process is not at all clear-cut and that blockages can readily occur between these factors. For example, information processing and communication may break down and not accurately affect organizational belief systems. Excess turnover of personnel may cause the loss of a large amount of the informal information on the workings of the organization. Likewise, internal organizational politics may not lead to clear preferences but may instead deadlock on mutually inconsistent goals. Finally, exogenous influences from outside organizations (e.g., consumer and government groups) as well as important changes in the general culture can affect prevailing beliefs and preferences within organizations.

If we do assume that organizations at a given point in time have arrived at some set of beliefs and preferences, this consensus will generally lead to certain organizational actions. However, as March and Olson have noted, even this rather obvious linkage is not at all assured. There may be standard procedures, work rules, or government regulations which may block even the clearest intentions. Thus, consensus that a given action will move the organization toward a valued goal will only lead to organizational behavior in the general case; it does not assure it.

Once organizations emit a particular action (e.g., reduce prices, enlarge the board of directors, revise a social action program) there is also no assurance that it will create the intended response from the organization's environment. Organizational actions are often only one of many influences within a complex environment and such actions must overwhelm exogenous influences from other organizations, institutions, and the general environment. Frequently the organization's actions are not the cause of environmental responses and, as a result, the perception of cause-effect relations can often be more superstition than fact. Indeed, it is almost rare when an organization emits a behavior which leads to an environmental response that is interpretable in a clear-cut way. Most often, organizational action constitutes a highly complex treatment or change which results in barely discernible responses from a complex set of environmental variables. This situation obviously does not lend itself to a highly valid set of causal inferences (Campbell & Stanley, 1966), nor does it provide the feedback on action which most organizations demand.

As a result of all the obstructions in the learning process (shown in Figure 5) it is no wonder that organizations fail to act rationally. Organizations, as depicted here, have difficulty translating beliefs and preferences into actions, and once the actions do take place, organizations find even more difficulty in understanding the causal nature of its environment and building a knowledge base. These imperfections in organizational learning are additionally compounded by internal political processes which may dictate goals and preferences in conflict with even the limited amount of

*Figure 5.* Organizational Learning as an Obstructed Cycle

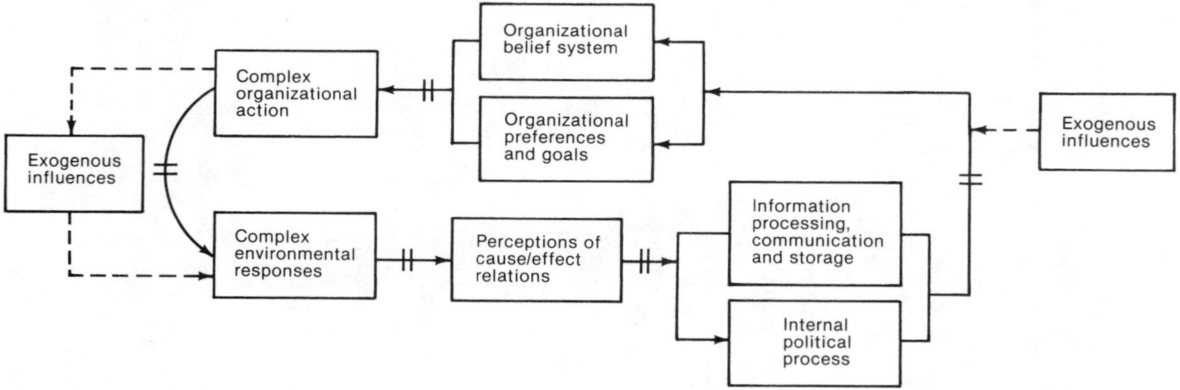

Exogenous influences
Complex organizational action
Complex environmental responses
Organizational belief system
Organizational preferences and goals
Perceptions of cause/effect relations
Information processing, communication and storage
Internal political process
Exogenous influences

valid knowledge accumulated by a particular organization. This is indeed a bleak picture that we have painted for organizational rationality, and it appears to directly conflict with Thompson's hypotheses on rationality-seeking behavior.

The major point we wish to make is that organizations, like individuals, are *intendedly* rational and operate under norms of rationality. However, organizations, even more than individuals, possess very limited information processing and learning capabilities. Thus, if the parallel still holds, organizations like individuals are often faced with inconsistency between their actions and their expectations of rationality. Just like individuals who are strongly motivated toward rationality but fall short of it, organizations are intentionally rational but do not accomplish their goals. We might therefore predict that organizations will also exhibit signs of rationalization or justification. Earlier, we had posited that the stronger are cultural norms for individual competence and rationality, the more likely it is that individuals will seek to justify their own behavior. In a parallel manner, we will therefore hypothesize that where strong norms for organizational rationality prevail (as in Western society), organizations will be prone to exhibit signs of justification.

*Types of Organizational Justification: Internal and External.* We can characterize organizational justification in much the same way as we have done with individual justification, dividing it into both internal and external varieties. However, whereas most of existing work on individual rationalization concerns internal justification, exactly the opposite is true for the organizational level of analysis.

Internal justification by organizations consists of relatively uncharted ground. Still, organizations under strong norms of rationality no doubt do exhibit many actions which attempt to create the illusion of rationality for their internal membership, although these behaviors may *not* in fact contribute to effectiveness. For example, standard operating procedures may have been evolved by the organization to deal with old environmental conditions and may now represent nothing more than vestigial characteristics—yet such procedures may still be strictly adhered to. Likewise, a great deal of attention may be paid to making internal rules and regulations consistent although such consistency may not in itself be related to effectiveness. Finally, chain of command and a logically consistent hierarchy system may be publicly defended although actual influence processes in the organization may work quite differently. All of these actions could be classified as outcroppings of internal justification, providing organizational constituents with the illusion of rationality.

We would predict, in line with our reasoning on individual justification, that an organizational failure experience (e.g., low profit or negative review by an outside agency) will increase the drive toward internal justification. That is, poor performance may lead organizations to increase or tighten the internal trappings of rationality such as rule consistency, unity of command, and hierarchical control systems. This hypothesis obviously differs from one that could be derived from an organizational learning theory (Duncan & Weiss, 1978) in which change or adaptation might be predicted to be a product of organizational crises.

In terms of external justification, the analogy between organizations and individuals is quite close. Organizations not only act under norms of rationality, but as Thompson (1967) argues, they also seek to demonstrate such rationality to outside constituents, evaluators, and potential providers of resources. Regardless of whether goals are actually fulfilled or products produced efficiently, organizations seek to demonstrate evidences of rationality. Thus, the organization will place its greatest efforts upon those areas which are most easily measured and quantified. Also, those measures upon which the organization scores well will be most publicized by the organization.

In Thompson's view, organizations act rationally to increase their evaluations or ratings by others on whom they are dependent. This type of behavior could be classified under prospective rationality, since the organization's emphasis may be upon seeking resources and minimizing future dependencies. However, Thompson's hypotheses can also be interpreted as outcroppings of retrospective rationality. Under strict evaluation or scrutiny, organizations may concentrate so heavily on justifying the rationality of past behavior that they fail to act rationally in a prospective sense. Scarce resources may be so heavily allocated to compiling hard data on a program (e.g., number of cases seen by social workers, test scores of entering students, number of new orders filled) that other relevant aspects of performance are omitted. Thus, future opportunities may be lost when attention is unduly concentrated upon rationalizing past performance. Although the Thompsonian strategies for meeting environmental demands are rational from the organization's point of view, some are rational only in the most short-term sense and are more retrospectively than prospectively focused.

## THE CONFLUENCE OF RATIONALITY AND JUSTIFICATION IN ORGANIZATIONAL SETTINGS

In this section we will address the issue of how rationality and justification combine in organizational settings. From a simple theoretical perspective these two forces appear in direct conflict with one another. However, we will argue that justification as well as rationality are two central themes of organizational life which are in constant ebb and flow. From the perspective of both individuals and organizations, the merits of rationality are rather obvious and we will not concentrate on them here. Instead, we will argue that some measure of justification can also aid the functioning of both individuals and organizational systems. Although the balance of forces most appropriate for effective functioning is not yet clear, we will emphasize the functional significance of both of these factors.

*Some Individual Effects*

Individual behavior, even as viewed from the cognitive motivation model of Figure 1, is not complete without justification processes. As shown in the motivation model, action does not simply flow from cognition, but feeds back onto the cognitions of valences and instrumentalities. The individual adjusts his values and probabilities of attainment to the reality around him, and this permeability can constitute a form of justification.

The subjectivity of valued outcomes has already been evidenced by two streams of research. First, research on level of aspiration (e.g., Lewin, Dembo, Festinger, & Sears, 1944; Myers & Fort, 1963; Brickman & Campbell, 1971) shows that individuals generally adjust their aspirations downward when they do not accomplish their goals. Second, research on adaptation level theory (Helson, 1964) has demonstrated that what is viewed as a positive outcome depends greatly upon its comparison level, and that the level of past outcomes provides one important guideline for satisfaction (see Beebe-Center, 1932; Brickman, Coats, & Bulman, 1978). Therefore, persons who must endure physically or psychologically hostile environments usually come to terms with these settings. Through cognitive realignment of affective baselines and aspirations, what appear to outsiders as highly dissatisfying conditions are often not perceived by individuals as so aversive.

In essence, there are almost no worldwide "objective" physical conditions (e.g., temperature, terrain, abundance of raw materials) which lead to positive or negative attitudes of peoples. For example, individuals generally prefer their home regions regardless of how hostile they appear to outsiders (Gould & White, 1974) and migrants to a new geographical area judge its characteristics (e.g., noise, pollution, friendliness) in relation to their previous residences rather than on an absolute standard (Wahlwill & Kohn, 1973). Even "objective" quality of life indicators such as per capita income, life expectancy, and average literacy do not relate strongly to life satisfaction across nations. For instance, Cantril's (1965) thirteen-nation survey showed that, although countries in which individuals were most satisfied appeared to be those in which there had been some recent improvement, the absolute level of a country on objective indicators did not always predict satisfaction. Interestingly, highest levels of satisfaction were scored in countries where there were strong national goals and ideologies (e.g., Cuba and Yugoslavia). Political leadership and ideology may thus provide the justifications and rationale needed by individuals to maintain positive attitudes when objective indicators are still low.

Salancik & Pfeffer (1978) have recently made a similar argument in the case of job attitudes in organizations. Although job characteristics as observed by outsiders do relate to individual satisfaction (Hackman & Oldham, 1976), few of these job characteristics are really objective. Factors such as task variety, significance, and autonomy are frequently the product of individual and social labeling. Therefore, what is a significant and meaningful task may depend greatly on one's definition of the situation, and may be highly subject to cognitive manipulation and realignment (see O'Reilly & Caldwell, in press, for a recent test of this notion).

Although satisfaction appears to be a highly subjective variable, one factor that is central to any tendency of individuals to justify their fates is behavioral commitment. Recent research on commitment (see Zimbardo, Ebbeson & Maslach, 1977; Salancik, 1977) shows that, once individuals are committed to a stream of action, their attitudes will generally become aligned with the behavior. Moreover, the ability of individuals to justify their fates can be heightened when there has been a marked reduction in outcomes. If individuals can easily reverse or change the situation to improve outcomes, this obviously will be the first choice of action for many individuals. However, if the decrease in outcomes in long-lived or irrevocable, readjustment will occur.

The moderating effect of commitment was observed by Staw (1974) in his study of ROTC cadets before and after the draft lottery. Most members of ROTC had joined the military organization to avoid being drafted and were not attracted to the organization by the intrinsic characteristics of ROTC (i.e., drills, summer camp, coursework). However, after the draft lottery occurred some individuals were informed that they were no longer vulnerable to the draft, while others were told that they would almost certainly be drafted if they left ROTC. As expected, most persons who were safe from the draft disenrolled from ROTC since they no longer needed the inducement of a draft deferment. However, a large number of individuals could not easily leave ROTC since they had already signed a binding contract with the Army before the lottery had occurred. Clearly, this group had made a mistake in joining ROTC and did not receive many valued outcomes from the organization after the lottery. They no longer really needed the deferment provided by ROTC and thus they were left only with drills, coursework, and summer camp as organizational outcomes. Yet, because of their binding commitment, this group was not allowed to disenroll. Rather remarkably, these individuals not only maintained their attitudes toward ROTC but actually increased them.

As shown in the ROTC study, the inevitability of a set of outcomes may moderate whether individuals justify their fates or not. When individuals can change their behavior to gain more positive outcomes, they generally will do this, as expected by most rational models in psychology (e.g., expectancy theory, exchange theory, decision theory). However, if individuals are locked into a course of action, they will tend to adjust to the level of outcomes received. Validation of adjustment effects have also been found in the interpersonal attraction area. When individuals are forced to work with another person who is disliked, opinions toward the unattractive person increase if the rater believes the work relationship will be long-lasting (Darley & Bersheid, 1967). No readjustment occurs when the relationship is believed to be temporary.

Self-justification, as evidenced in the ROTC study and the interpersonal attraction literature, does not fit well with a purely rational view of man in which both actions and attitudes are simply determined by the expectancy and receipt of outcomes. However, it is likely that justification is psychologically functional for individuals. Negative attitudes and persistent but futile efforts to avoid aversive outcomes may only lead to psychological stress if the situation is not tractable. In these situations justification is far better than rationality, at least in terms of individual mental health.

The conflict between rationality and justification can be most intense when there are norms for high performance and also a very difficult task to perform. When individuals are confronted with impossible demands, they will tend to lower their aspirations and adjust to lower goals and outcomes. However, there may be strong organizational pressure to persist at what appears to be an insurmountable task. In these cases, self-justification processes may be a less-active force and, as a result, persistence without success can lead to undo stress, depression, or learned helplessness (Seligman, 1975). This problem may especially exist when organizations and individuals do not agree upon what are attainable goals.

The recent upsurge in self-help techniques and career management may also threaten the balance between rationality and justification. The general theme of these efforts is for the individual to take responsibility for his own life and the outcomes produced by that responsibility. Guidelines and exercises are frequently posed for the individual to think of long-term goals and how his present job is or is not instrumental to achieving these goals. Organizations frequently sponsor career management sessions and generally endorse any increase in self-initiated improvement and goal-directedness. Unfortunately, however, career management techniques may actually decrease rather than increase job attitudes. By making careers and long-term goals salient, enormous demands are placed upon the current job one holds. It must not only be pleasant or not too dissatisfying, but also instrumental to important career objectives. Without such pressures for individual rationality, people may instead base satisfaction on more immediate characteristics of the job; and, if the job possesses few positive features, they may still justify it on some very subjective long-term considerations. These long-term factors may, however, be quite unrealistic idealizations of the future—closer to "pipe dreams" than realistic career goals. As Studs Terkel (1974) found in several open-ended interviews, a person performing a basically dissatisfying job may justify it as a temporary shelter or place to bide one's time until the "eventual" satisfying position opens up (e.g., opening own bar, becoming a race car driver, etc.). Obviously, few of these dreams ever materialize and would be discouraged by most realistic career counselors. Yet, these irrational hopes and justifications may actually be what psychologically sustains many individuals in their current lines of work.

*Some Organizational Effects*

There is literally no empirical research to describe the confluence of rationality and justification on organizations. Yet, we will speculate about some evidences of these cross-cutting forces and attempt to describe the functional interplay of these two factors.

As we have noted, although organizations are intendedly rational their ability to achieve any measure of rationality is quite limited. Only very imperfectly does the organization input and process information, and it has difficulty with even the most primitive learning mechanisms. Thus, we would argue that difficulty in rationally adapting to the environment is most marked for organizations when there is substantial ambiguity on both cause-effect relations and outcome criteria. A university, for example, has difficulty defining what "education" is as an outcome variable, let alone discovering the best way to increase its level. A church has a similar problem with faith, morality, or other outcome criteria by which to measure success, and even a hospital has difficulty defining "health" between the extremes of total malfunctioning and relative survival. In each of these cases, rational learning is made difficult by ambiguity of both purpose and technology.

If we order a variety of organizations along the two dimensions of ambiguity of goals and cause/effect relations, the lowest levels of ambiguity would be associated with a single proprietorship operating within a technologically stable area of the economy. Here the primary methods of production and sales would be known and the chief outcome variable would be clear: economic benefit (i.e., return on investment) for the owner of the business. However, once the private firm grows into a publicly-owned corporation outcome variables become more ambiguous. Although Becker & Gordon (1966) argue that all private corporations have maximization of stockholder wealth as their primary goals, others would disagree. Once a single proprietorship is replaced by professional management, organizational survival and growth often replace return on equity as operating goals (Hall, 1977). Serving consumer interests, providing jobs, contributing to the community tax base, and public responsibility also become more important as corporations grow in size and importance. This multivariate goal structure (Friedlander & Pickle, 1968; Pickle & Friedlander, 1967) leads to ambiguity and uncertainty over what goals to seek or satisfy. In a sense, the larger become private corporations the more like public interest organizations they become.

With most public organizations there are two sources of ambiguity about goals. First, it is not at all clear whose interests a public organization is designed to serve and there is frequently conflict among disparate goals held by each constituent group. In addition, most social service goals (e.g., health, welfare, education) are highly subjective in nature. Each administration has its own operational definition of these terms and changes in health care delivery, social welfare programs and education policy often shift with each change in organizational hierarchy. Finally, as we have noted, most social programs work with a highly ambiguous technology. Instead of hard technical facts, ideas about cause-effect relations are the subject of intuition and custom within many nonprofit organizations—theories of education and child-rearing, for example, run in historical cycles and are based only tangentially on any empirical data.

Against this background of ambiguity, Weick (1977) has developed the notion of enactment in organizations. Organizations, according to Weick, attempt to make sense out of even the most ambiguous environments. However, Weick argues that the terms, constructs, and metaphors by which organizations perceive the events around them can literally *become* the organization's environment. One can best illustrate this enactment notion by carrying it to its extreme. For example, a powerful religious organization has the ability to define religious terms in nearly any way it chooses and, in turn, it can operate upon its own definition of reality. Because outcomes and cause-effect relations are so ambiguous for this type of organization, norms and operating procedures could easily be based on an enacted environment for very long periods of time (this may be why major changes in religious organizations occur so infrequently). In contrast, a highly subjective view of reality does not last long within the competitive private economy. A company with a highly divergent view of its goals and the environment is likely to become economically extinct, with the surviving firms conforming rather well to the more objective rules of the marketplace.

We would argue that for organizations which must operate under highly ambiguous goals and cause-effect relations, justification is of vital importance. Because there is so much uncertainty, it is essential to justify whatever goals are chosen and whatever technology is followed. Under high levels of ambiguity, justification is necessary to both provide purpose for an organization's membership and rationale for parties' external to the organization. In fact, organizations which face a great deal of ambiguity are frequently perceived as more effective when they have developed an elaborate or persuasive set of justifications for their particular goals and technology. For example, the most successful educational organizations are those which have developed the *x* school of thought or approach to education. Likewise, successful governmental units are frequently those which are most eloquent in conveying their own subgoals, philosophy, and procedures. Thus, in many organizational settings justification *can* become the reality; through justification, perceived sources of ambiguity can be explained away or replaced by shared meaning as opposed to economic or technical "facts."

It should be noted that Berlew (1974) has also touched upon the essential role of justification in his discussion of organizational leadership. To Berlew the effective leader is one who clarifies organizational purposes and helps to articulate collective goals. He provides an organizational mission and a path for others to follow. In essence, the leader is one who formulates justifications of goals and technology which are believable and supportable by organizational members. Along these lines, we would predict that the more ambiguous are organizational goals and technology, the more important it is to have a leader who can provide justifications for the organization. This may be why charismatic leaders are frequently demanded by religious and governmental organizations, while less articulate but technically competent managers can successfully fulfill many other executive roles.

## A CROSS-LEVEL EFFECT: THE CASE OF ORGANIZATIONAL EVALUATION

An important cross-level effect of rationality and justification can be illustrated with the application of evaluation research techniques in organizations. We will show that evaluation research, as an outcropping of organizational rationality, is very likely to foster individual forces for justification. We will elaborate on this cross-level effect in some depth because of the increasing use of evaluation techniques in organizational settings.

Evaluation activities are representative of the core of rational decision making and can be applied to almost any level of a social system. As originally envisioned by Campbell (1969), evaluation research would aid public officials in allocating scarce resources among a large variety of social programs. The principle was simple. If data from various programs could be collected and analyzed in such a way as to increase its internal validity, public officials would have much greater knowledge of which programs do accomplish societal goals and which do not. Because of the increasing scarcity of public funds and strong cultural norms for rationality, evaluation research has become increasingly popular within the United States (see Riecken & Boruch, 1974; Wortman, 1975).

Recently, Staw (1977b) attempted to extend the principles of evaluation research from societal to organizational decision making. If organizations do operate under norms of rationality, it is strongly in their interest to clarify cause-effect relations. However, gaining greater understanding of the effects of particular organizational actions often requires entirely different modes of data collection, feedback, and analysis than those currently used by organizations. Extrapolating from the principles of evaluation research, organizations would need to systematically experiment with its environment (cf., March, 1971), intentionally trying out many divergent strategies, knowing full well that many of these new alternatives would eventually be dropped from use. In addition, a truly experimenting organization would seek to strengthen causal inference about internal policies through experimental and quasi-experimental design. For example, procedures for personnel selection, job design, pay systems, and work schedules are all variables which could be systematically altered to find the best available method. Currently, only vague knowledge, tradition, and precedent dictate many of the internal actions and policies of organizations.

In its ideal state, evaluation research would be devoted toward improving causal inference and reducing many of the sources of ambiguity which limit rationality in organizations. However, despite all of its potential to increase rationality, the application of evaluation research faces many problems in organizational settings. Many of these problems result from the conflict between organizational rationality and individual justification.

If organizations were to assiduously apply the principles of evaluation research, they would require administrators to experiment sequentially with many policy alternatives, assess each of their effects, and change to new policies when outcomes were not satisfactory. Such an experimenting organization may be difficult to achieve, however, because of individual tendencies of justify their behavior. Contrary to the principles of experimentation, administrators may often hesitate to make major changes in policy following the receipt of negative consequences. As noted earlier, some studies have shown that when people's behavior leads to negative consequences they may, instead of changing behavior, cognitively distort the negative consequences to more positively valued outcomes (e.g., Freedman, 1963; Pallak, Sogin & Van Zante, 1974; Staw, 1974; Weick, 1964). By biasing behavioral outcomes individuals can rationalize their previous actions or psychologically defend themselves against a perceived error in judgment. In addition, it is also possible in many evaluation situations for administrators to go beyond the passive distortion of adverse consequences in an effort to rationalize a behavioral error. When negative consequences are incurred, it is frequently possible for administrators to enlarge the commitment of resources and undergo the risk of additional negative outcomes in order to justify previous behavior or to demonstrate the ultimate rationality of an original course of action.

Using a simulated business decision case, Staw (1976) experimentally tested for the tendency to escalate following the receipt of negative consequences. In his study, business school students were asked to allocate research and development funds to one of two operating divisions of a company. They were then given the results of their initial decisions and asked to make a second allocation of R&D funds. In this study, some subjects also were assigned to a condition in which they did not make the initial allocation decision themselves, but were told that it was made earlier by another financial officer of the firm. The results of the experiment were as follows: (1) there was a main effect of responsibility such that subjects allocated more money when they, rather than another financial officer, had made the initial decision; (2) there was a main effect of consequences such that subjects allocated more money to the declining rather than improving division; and (3) there was a significant interaction of responsibility and consequences. That is, subjects allocated even more money when they were responsible for negative consequences than would be expected by the two main effects acting alone. Personal responsibility for negative consequence, therefore, may lead to the greatest likelihood of escalation behavior.

From these experimental results and observational evidence of some real-world escalations (see, e.g., "Pentagon Papers" for a description of policy making during the Vietnam War), it seems possible that administrators can become trapped by their own previous mistakes. They may, counter to principles of evaluation research, refuse to admit the failure of a policy or procedure and forge ahead in spite of negative consequences.

*Internal and External Justification.* The conflict we have posed in applying evaluation research has been between organizational demands for rationality and the individual's tendency for self-justification. We have characterized the self-justification effect as one of denial of error and internal rationalization. However, within organizational settings, ego-gratification and the protection of self-esteem (cf., Aronson, 1968, 1976) are often only secondary to simple administrative survival. If a policy or program fails, administrators are often replaced by the next most promising candidate rather than the next best policy alternative! Thus, the administrator may often be forced to make a policy work at almost any cost. The commitment of new and additional resources can, as a result, stem from self-protective actions of the administrator in addition to, or even in lieu of, the need to bolster individual self-esteem.

As we noted in our earlier discussion of individual justification, it is useful to draw the distinction between internal and external forms of justification. Whereas the internal form of justification refers to private monitoring and self-inflicted costs if a decision fails, the external form of justification refers to the public surveillance of one's decisions and the imposition of sanctions by others if errors are detected. Both forms of justification may lead individuals to focus retrospectively on those events and outcomes which might "save" a previous policy and to protect oneself from the exposure of a previous error. However, evaluation research activities within organizations are especially likely to activate forces for external justification. Because evaluation is an external source of control, administrators will frequently go to great length to demonstrate the rationality of the program and services they manage.

An empirical demonstration of the effect of external justification was recently conducted by Fox & Staw (1979). Two face-valid antecedents of external justification were manipulated. It was predicted that both job insecurity and policy resistance would increase administrative inflexibility to change. According to their reasoning, if an administrator is vulnerable to job loss or demotion, he would likely be highly motivated to protect his position in the organization. Thus, one would predict that a highly insecure administrator would be most likely to attempt to save a policy failure by enlarging the commitment of resources. Likewise, one would predict that resistance in the organization to one's policies might also serve to heighten an escalation effect. If an administrator implements a policy that he knows is unpopular within the organization, he may be especially concerned to protect himself against failure. The experimental results confirmed these two hypotheses. The administrator who was both insecure in his job and who faced stiff policy resistance was most likely to escalate his commitment and become locked-in to a course of action.

The situation of high insecurity and high resistance actually represents an operationalization of Campbell's (1969) notion of the "trapped administrator" who has little choice but to forge ahead in his commitment to a losing policy. The trapped administrator is one who stands only to lose if a particular program does not work and who has literally no choice but to remain fully committed to it, even in the face of failure. The trapped administrator is, of course, acting rationally from his own individual perspective. However, this form of rationality is retrospectively rather than prospectively focused. In this case, the issue is not how to maximize future outcomes but to recoup previous losses.

*Rationality versus Justification.* The notion of the trapped administrator highlights the confluence of rationality and justification. From the organization's point of view, resources should be allocated only to those programs which yield the highest future return. Yet, from the individual administrator's point of view, it is necessary to defend the usefulness of past and current projects so as to justify or demonstrate the rationality of *previous* allocations of resources. Thus, while a truly experimenting organization might wish to use prospective rationality in evaluating the use of resources, each suborganizational unit or administrator may resort to a retrospective form of rationality in their own decision making.

The conflict between rationality and justification can be manifested at multiple levels in an organizational system. Recall that organizations attempt to demonstrate their performance to outside agencies and may become retrospectively focused in order to defend its previous actions. Thus, at the most macroscopic level, one can envisage an experimenting society functioning under prospective rationality, while those individual organizations actually providing the services might still be manifesting strong outcroppings of justification. Likewise, at a much more microscopic level of analysis, the evaluation of individual role performance can involve the same conflict of forces. As Lawler (1971) has noted, the stronger is the evaluation system used for performance appraisal and salary administration, the more defensive are likely to be individuals participating in the program. Therefore, it appears that almost any system which seeks rationality must also suffer some of the costs of justification.

The conflict between rationality and justification seems to exist simply because goals and interests change across each level in an organizational system. Goal seeking and prospective rationality at one level of the system generally translate into self-protecting, retrospective rationality at lower levels. While some authors (e.g., Argyris & Schon, 1978) would characterize this transformation process as a malfunction within an organizational system, we would view it as both inevitable and necessary. Each level in an organizational system faces its own ecology of forces. Thus, whether one views behavior as an outcropping of rationality or justification depends greatly upon one's perspective. From the top of an organizational system, self-protective behaviors emitted by lower levels are frequently viewed as defensive reactions. However, the same behavior may be viewed by its source as highly rational and prospectively focused. This shifting focus of decision making is one reason why organizations can be characterized as loosely coupled systems (Weick, 1976).

Uniformity of purpose and action are difficult to effectuate when perspectives shift throughout an organizational system and when rationality can easily translate into justification across organizational levels. Yet, this loose coupling may be precisely what allows an organization to survive when it faces radically changing conditions. Our point is not to advocate justification over rationality, but again to emphasize the functional interplay of these two forces, even across levels in an organizational system.

## CONCLUSION

In this essay we have described forces for rationality and justificaiton at both the individual and organizational levels. In the first section we noted that rationality and justification comprise pervasive images of both human and organizational systems, yet these images often appear to conflict sharply. The second section of this essay dealt with several examples of the confluence of rationality and justification within organizational systems. The purpose of these examples was not simply to review established research areas or to "solve" the conflict between rationality and justification, but to illustrate the functional interrelationship of these two forces. It is this author's contention that the confluence of these two general factors underlie much of what is now known as organizational behavior. By reconsidering this conceptual underpinning, it may thus be possible to generate many new hypotheses and programs of research. Also, because rationality and justification concern central issues of organizing, further research on this issue could help to stem the present trend toward developing greater numbers of mini theories within rather isolated areas of research.

## REFERENCES

Appley, M. H. (1971) *Adaptation level theory*. New York: Academic Press.

Argyris, C. (1972) *The applicability of organizational sociology*. London: Cambridge University Press.

——— & Schon, D. A. (1978) *Organizational learning: A theory of action perspective*. Reading, Mass.: Addison Wesley.

Aronson, E. (1968) "Dissonance theory: Progress and problems." In R. Abelson, E. Aronson, W. McGuire, T. Newcomb, M. Rosenberg, & P. Tannebaum (eds.) *Theories of cognitive consistency: A sourcebook*. Chicago: Rand McNally.

Aronson, E. (1976) *The Social Animal* (2nd ed.), San Francisco: W. H. Freeman.

——— & Mills. (1959) "The effect of severity of initiation of liking for a group." *Journal of Abnormal and Social Psychology, 59,* 177–181.

Becker, S. W. & Gordon, G. (1966) "An entrepreneural theory of formal organizations." *Administrative Science Quarterly,* 1966, Vol. 11, No. 3.

Beebe-Center, J. G. (1932) *Pleasantness and unpleasantness*. Princeton, N.J.: Van Nostrand.

Berlew, D. E. (1974) "Leadership and organizational excitement." In D. Kolb, I. Rubin, &

F. McIntyre (eds.) *Organizational psychology: A book of readings*. Englewood Cliffs, New Jersey: Prentice-Hall.

Blau, P. & Schoenherr, R. A. (1971) *The structure of organizations*. New York: Basic Books.

Brehm, J. (1956) "Postdecision changes in the desirability of alternatives." *Journal of Abnormal and Social Psychology, 53*, 384–389.

——— & Cohen, A. R. (1959) "Choice and chance relative deprivation as determinants of cognitive dissonance." *Journal of Abnormal and Social Psychology, 58*, 383–387.

Brickman, P. & Campbell, D. T. (1971) "Hedonic relativism and planning the good society." In M. H. Appley (ed.) *Adaptation-level theory: A symposium*. New York: Academic Press.

———, Coates, D. & Bulman, R. (1978) "Lottery winners and accident victims: Is happiness relative?" *Journal of Personality and Social Psychology, 36*, 917–927.

Calder, B. J., Ross, M., & Insko, C. (1973) "Attitude change and attitude attribution: Effects of incentive, choice, and consequences." *Journal of Personality and Social Psychology, 25*, 84–100.

Campbell, D. T. (1969) "Reforms as experiments." *American Psychologist, 24*, 409–429.

——— & Stanley, J. C. (1966) *Experimental and quasi-experimental designs for research*. Chicago: Rand McNally.

Campbell, J. P. & Pritchard, R. D. (1976) "Motivation theory in industrial and organizational psychology." In M. D. Dunnette (ed.) *Handbook of industrial and organizational psychology*. Chicago: Rand-McNally.

Cantril, H. (1965) *The pattern of human concerns*. New Brunswick, New Jersey: Rutgers University Press.

Carlsmith, J. M., Collins, B. E., & Helmreich, R. L. (1966) "Studies in forced compliance: The effect of pressure for compliance on attitude change produced by face-to-face role playing and anonymous essay writing." *Journal of Personality and Social Psychology*, 1966, *1*, 1–13.

Chandler, A. D. (1962) *Strategy and structure*. Boston: MIT Press.

Child, J. (1973) "Predicting and understanding organization structure." *Administrative Science Quarterly*, 168–185.

Cohen, A. R. (1962) "An experiment on small rewards for discrepant compliance and change." In J. Brehm & A. R. Cohen (eds.) *Explorations in cognitive dissonance*. New York: Wiley, 73–78.

Cohen, M. D., March, J. G., & Olsen, J. P. (1972) "A garbage can model of organizational choice." *Administrative Science Quarterly, 17*, 1–25.

Collins, B. E. & Hoyt, M. F. (1972) "Personal responsibility for consequences: An integration and extension of the forced compliance literature." *Journal of Experimental Social Psychology, 8*, 558–594.

Cooper, J. (1971) "Personal responsibility and dissonance: The role of foreseen consequences." *Journal of Personality and Social Psychology, 8*, 354–363.

Darley, J. & Berscheid, E. (1967) "Increasing liking as the result of the anticipation of personal contact." *Human Relations 20*, 29–40.

Duncan, R. & Weiss, A. (1978) "Organizational learning: Implications for organizational design." In B. Staw (ed.) *Research in Organizational Behavior* (Volume 1). Greenwhich, Conn.: JAI Press.

Edwards, W. (1954) "The theory of decision making." *Psychological Bulletin, 51*, 380–417.

Festinger, L. (1957) *A theory of cognitive dissonance*. Stanford University Press.

——— & Carlsmith, J. M. (1959) "Cognitive consequences of forced compliance." *Journal of Abnormal and Social Psychology, 58*, 203–210.

Fox, F. *Persistence: Effects of Commitment and Justification Processes on Efforts to Succeed with a Course of Action*. Ph.D. dissertation, University of Illinois, 1980.

Fox, F. & Staw, B. M. (1979) "The trapped administrator: The effects of job insecurity and policy resistance upon commitment to a course of action." *Administrative Science Quarterly*, September, 1979.

Freedman, J. (1963) "Attitudinal effects on inadequate justification." *Journal of Personality*, 31, 371–385.

Friedlander, F. & Pickle, H. (1968) "Components of effectiveness in small organizations." *Administrative Science Quarterly*, 13, 289–304.

Galbraith, J. R. (1977) *Organizational design*. Reading, Mass.: Addison-Wesley.

Gerard, H. & Mathewson, G. (1966) "The effects of severity of initiation on liking for a group: A replication." *Journal of Experimental Social Psychology, 2,* 278–287.

Gould, P. & White, R. (1974) *Mental maps*. Middlesex, England: Penguin Books.

Gouldner, A. W. (1959) "Organizational analysis." In R. K. Merton, L. Brown, & L. S. Cothell (eds.) *Sociology today*. New York: Basic Books.

Graen, G. (1969) "Instrumentality theory of work motivation: Some experimental results and suggested modifications." *Journal of Applied Psychology Monograph, 53,* 1–25.

Haan, N. (1977) *Coping and defending: Processes of self-environmental organization*. New York: Academic Press.

Hackman, J. R. & Oldham, G. R. (1976) "Motivation through the design of work: Test of a theory." *Organizational Behavior and Human Performance, 16,* 250–279.

Hall, R. H. (1977) *Organizations: Structure and process*. Englewood Cliffs, New Jersey: Prentice-Hall.

Hannan, M. T. & Freeman, J. H. (1977) "The population ecology of organizations." *American Journal of Sociology, 82,* 929–964.

Heider, F. (1958) *The psychology of interpersonal relations*. New York: Wiley.

Helson, H. (1964) *Adaptation-level Theory: An Experimental and Systematic Approach to Behavior*. New York: Harper & Row.

Janis, I. L. & Mann, L. (1977) *Decision-making: A psychological analysis of conflict, choice, and commitment*. New York: Free Press.

Katz, D. & Kahn, R. (1978) *The social psychology of organizations*. New York: Wiley.

Knox, R. & Inkster, J. (1968) "Postdecision dissonance at post time." *Journal of Personality and Social Psychology, 8,* 319–323.

Lawler, E. E. (1971) *Pay and organizational effectiveness: A psychological view*. New York: McGraw-Hill.

———, Kuleck, W. J., Rhode, J. G., & Sorensen, J. E. (1975) "Job choice and post decision dissonance." *Organizational Behavior and Human Performance, 13,* 133–145.

Lawrence, P. & Lorsch, J. (1967) *Organization and environment*. Boston: Harvard Business School Division of Research.

Lewin, K., Dembo, T., Festinger, L., & Sears, P. (1944) "Level of aspiration." In J. M. Hunt (ed.) *Personality and the behavior disorders*. New York: Ronald.

Linder, D., Cooper, J., & Jones, E. (1967) "Decision freedom as a determinant of the role of incentive magnitude on attitude change." *Journal of Personality and Social Psychology, 6,* 245–254.

March, J. G. & Olsen, J. P. (1976) *Ambiguity and choice in organizations*. Bergen, Norway: Universitelsforlaget.

———. (1971) "The technology of foolishness." *Civilokonamen, 18,* 7–12.

Martin, N. H. & Sims, J. H. (1956) "Power Tactics." *Harvard Business Review*, pp. 25–29.

Miller, G. A. (1956) "The magical number seven, plus or minus two: Some limits on our capacity for processing information." *Psychological Review, 63,* 81–97.

Miller, J. G. (1978) *Living Systems*. New York: McGraw-Hill.

Mitchell, T. R. & Knudsen, B. W. (1973) "Instrumentality theory predictions of students' attitudes towards business and their choice of business as an occupation." *Academy of Management Journal, 16,* 41–52.

Myers, J. L. & Fort, J. G. (1963) "A sequential analysis of gambling behavior." *Journal of General Psychology, 69,* 299–309.
Newcomb, T. M. (1953) "An approach to the study of communicative acts." *Psychological Review, 60,* 393–404.
O'Reilly, C. & Caldwell, D. F. (in press) "Informational influence as a determinant of perceived task characteristics and job satisfaction." *Journal of Applied Psychology.*
Osgood, C. E. (1960) "Cognitive dynamics in human affairs." *Public Opinion Quarterly, 24,* 341–365.
Pallak, M. S., Sogin, S. R., & Van Zante, A. (1974) "Bad decisions: Effects of volition, locus of causality, and negative consequences on attitude change." *Journal of Personality and Social Psychology, 30,* 217–227.
Parsons, T. (1960) *Structure and process in modern society.* New York: The Free Press.
"Pentagon Papers, The" (1971) The *New York Times* (based on investigative reporting of Neil Sheehan). New York: Bantam Books.
Pfeffer, J. (1977) "Power and resource allocation in organizations." In B. Staw & J. Salancik (eds.) *New Directions in Organizational Behavior.* Chicago: St. Clair Press.
——— & Salancik, J. R. (1978) *The external control of organizations: A resource dependence perspective.* New York: Harper and Row.
Pickle, H. & Friedlander, F. (1967) "Seven societal criteria of organizational success." *Personnel Psychology, 20,* 165–178.
Porter, L. W. & Lawler, E. E. (1968) *Managerial Attitudes and Performance.* Homewood, Illinois: Irwin-Dorsey.
Pugh, D., Hickson, D., Hinings, C., & Turner, C. (1968) "The dimensions of organization structures." *Administrative Science Quarterly, 13,* 65–105.
Riecken, H. W. & Boruch, R. F. (1974) *Social experimentation: A method for planning and evaluating social intervention.* New York: Academic Press.
Rosenberg, M. J. (1960) "A structural theory of attitude dynamics." *Public Opinion Quarterly, 24,* 319–340.
Salancik, G. R. (1977) "Commitment and the control of organizational behavior and belief." In B. Staw & G. Salancik (eds.) *New Directions in Organizational Behavior.* Chicago: St. Clair Press.
——— & Pfeffer, J. (1978) "A social information processing approach to job attitudes and task design." *Administrative Science Quarterly,* June 1978.
Seligman, M. E. P. (1975) *Helplessness.* San Francisco: W. H. Freeman.
Simon, H. A. (1957) *Administrative behavior.* New York: Macmillan.
Slovic, P. (1972) "From Shakespeare to Simon: Speculations—and some evidence about man's ability to process information." *Oregon Research Institute Monograph, 12,* No. 2.
Staw, B. M. (1974) Attitudinal and behavioral consequences of changing a major organizational reward: A natural field experiment. *Journal of Personality and Social Psychology, 6,* 742–751.
Staw, B. M. (1976) "Knee-deep in the big muddy: A study of escalating commitment to a chosen course of action." *Organizational Behavior and Human Performance, 16,* 27–44.
——— (1977a) "Motivation in organizations: Toward synthesis and redirection." In B. Staw and J. Salancik (eds.) *New directions in organizational behavior.* Chicago: St. Clair Press.
——— (1977b) "The experimenting organization." In B. Staw (ed.) *Psychological foundations of organizational behavior.* Santa Monica, Cal.: Goodyear Publishing.
Stevens, L. & Jones. E. E. (1976) "Defensive attribution and the Kelley cube." *Journal of Personality and Social Psychology, 34,* 809–820.
Terkel, S. (1974) *Working.* New York: Pantheon Books.
Thompson, J. D. (1967) *Organizations in action.* New York: McGraw-Hill.

——— & Tuden, A. (1959) "Strategies, structures, and processes of organizational decisions." In J. D. Thompson et al. (eds.) *Comparative studies in administration*. Pittsburgh: University of Pittsburgh Press.

Vroom, V. H. (1964) "Work and motivation." New York: John Wiley.

——— (1966) "Organizational choice: A study of pre- and post-decision processes." *Organizational Behavior and Human Performance, 1,* 212–225.

Wanous, J. (1972) "Occupational preferences: Perceptions of value and instrumentality, and objective data." *Journal of Applied Psychology, 56,* 152–155.

Weick, K. E. (1964) "Reduction of cognitive dissonance through task enhancement and effort expenditure." *Journal of Abnormal and Social Psychology, 68,* 533–549.

——— (1976) "Educational organizations as loosely coupled systems." *Administrative Science Quarterly, 21,* 1–19.

——— (1977) "Enactment processes in organizations." In B. Staw & G. Salancik (eds.) *New directions in organizational behavior*. Chicago: St. Clair Press.

White, R. W. (1959) "Motivation reconsidered: The concept of competence." *Psychological Review, 66,* 297–333.

Wohlwill, J. F. & Kohn, I. (1973) "The environment as experienced by the migrant: An adaptation-level view." *Representative Research in Social Psychology, 4,* 135–164.

Wortman, C. B. & Linsenmeier, J. (1977) "Interpersonal attraction and techniques of ingratiation in organizational settings." In B. Staw & J. Salancik (eds.) *New directions in organizational behavior*. Chicago, Illinois: St. Clair Press.

Wortman, P. M. (1975) "Evaluation research: A psychological perspective." *American Psychologist, 30,* 562–575.

Zimbardo, P., Ebbeson, E., & Maslach, C. (1977) *Influencing attitudes and changing behavior*. Reading, Mass.: Addison-Wesley.

# TIME AND WORK:

## TOWARD AN INTEGRATIVE PERSPECTIVE

Ralph Katz

MASSACHUSETTS INSTITUTE OF TECHNOLOGY

### ABSTRACT

Among the more prevalent ideas associated with the study of individual attitudes and behaviors in work organizations is the rather broad viewpoint that employees are continuously negotiating and revising their perspectives toward their organizations and their roles in them. One of the basic assumptions underlying these implied changes is that employees are not especially receptive to disorder and uncertainty but will endeavor to structure, interpret, and redefine their work settings. Moreover, the process by which individuals mediate and cope with their present organizational surroundings is most likely a product of their past experiences as well as their expectations and hopes of the future. Building on these notions, the present essay begins to develop a descriptive framework to show how employees' perspectives unfold and change as they pass through their respective job environments. In particular, a three-transitional stage model of job longev-

**Research in Organizational Behavior, Volume 2, pages 81–127**

**ISBN: 0-89232-099-0**

ity is developed to illustrate the major kinds of issues and concerns that seem to preoccupy and guide employees as they work at a given job position.

## INTRODUCTION

One of the more important notions associated with the study of work organizations is the idea that employees are constantly seeking to interpret, understand, and organize the world of their experience (e.g., Schutz, 1967; McHugh, 1968; Van Maanen, 1977). Individual perceptions and responses, however, are not developed or revised in a social vacuum but evolve through successive encounters with one's work environment. Weick (1969), for example, relies on this perspective in his discussion of how individuals, groups, and organizations "enact" their environments and direct their activities toward the removal of *equivocality*—processes in which energies are devoted toward establishing a more workable and comfortable pattern of overall clarity and certainty. In a similar vein, Salancik & Pfeffer (1977) have recently argued that individuals typically cope with their job and social environments either by creating meaning that makes the context more acceptable and satisfying or by restructuring and redefining the context of their overall job situations. Such universal phenomena are related to the Meadian process whereby individuals construct situational definitions that serve to both account for and anticipate the reoccurring events of everyday life (Van Maanen & Katz, 1979). The most useful questions from such a perspective, therefore, involve not only the identification of important features that might be associated with a particular work setting (e.g., challenging work, competent supervision, promotional opportunities) but also encompass a better understanding of the various methods, strategies, and cognitive processes by which individuals acquire and revise their knowledge and views about their job environments as they experience different kinds of work and organizational events.

At the heart of this argument is the idea that one must consider the social environment in which jobs are embedded in order to determine how individuals describe and feel about their work as well as to gain a better understanding of individual behavior. Much of an employee's reactions tend to develop and change as the individual interacts with and acts upon different aspects of the job setting, not simply from efforts to fulfill psychological needs. Employee responses, therefore, cannot be viewed in total isolation, for it is the social context that provides the information and cues with which individuals define and interpret their work experiences.

Recently, a number of researchers (including Bray, Campbell, & Grant, 1974; Hall, 1976; Sofer, 1970; Van Maanen & Schein, 1977; and Levinson,

1978) have been especially cognizant of the changes which presumably occur as the result of one's passage through a particular job, career, or life cycle. In a recent book on career dynamics and cycles, for example, Schein (1978) illustrates explicitly the changing nature of the primary issues and concerns that apparently preoccupy individual employees as they proceed through their respective careers and personal lives. By adopting this kind of career-specific and time-dependent perspective, Van Maanen & Katz (1976) concluded, after carefully examining the satisfaction responses of a large and diverse respondent sample, that differences in employee reactions may be at least as great within an occupational career at various career stages as those differences in reactions that are uncovered among a wide range of occupational and/or organizational categories.

Despite this awareness, few analytical studies in organizational behavior have incorporated such ideas and possibilities into their theoretical or empirical frameworks. *The prime objective of this chapter, therefore, is to begin to develop a more temporally-based framework for conceptualizing and studying the major concerns and reactions of employees to their overall work environments particularly as they enter and proceed through their job experiences.* Much of the literature, for example, has continued to examine the relationships between individual perceptions of numerous role features of the work environment and assorted dependent measures (such as employee satisfaction and performance) without apparent regard for the substantial diversity in job, career, and life stages that might be represented within their respective samples. For the most part, such studies assume a cross-sectional view of individual workers and employees making explicit causal linkages between significantly associated variables indeterminable. And just as important, such studies invariably lump together employees who are at very different stages of their respective jobs and organizational careers; employees who may well define their present circumstances from vastly different sets of past experiences as well as in terms of different expectations and hopes.

Thus, while a whole host of work-related features and characteristics have been significantly associated with some particular subset of attitudinal or behavioral criteria in one study or another, the field is virtually silent on the question of the relative importance of these many factors for influencing employee reactions at a particular job or career-related period. It is further unclear how the influence of such factors might vary in importance with time (as one's work and life histories unfold). Consider, for instance, the hypothetical case of a young, unmarried research chemist, newly hired by a large consumer product corporation, whose primary project assignment involves the betterment of soap fragrance. Five or ten years later, this same chemist, now married with perhaps one or two children, is still working on making this soap product smell fresher. One could easily argue that the energy, excitement, and expectations—in essence, the overall meaning—by which the chemist defined and pursued his or her work has probably changed significantly over this job tenure period even though the general demands of the job may have remained the same. Yet it is the importance of such contextual information that is often absent from much of the theoretical and empirical work that is directed toward a more complete understanding of the determinants of individual attitudes and behaviors in contemporary organizations.

## ROLE DESIGN FEATURES

Considerable theorizing and empirical effort has been directed over the past 20 or so years toward analyzing how various design-relevant features (i.e., task characteristics, reporting relationships, pay systems, promotional procedures, etc.) affect specific aspects of workers' lives. In spite of their formidable methodological problems and constraints, self-administered attitude surveys and questionnaires represent by far the most common approach by which researchers have tried to understand employee perceptions of the many role design features characterizing everyday work situations. In trying to identify the most consequential aspects or properties of job settings, many studies have required respondents either to rank order or to rate a diverse collection of work-related attributes (such as good pay, interesting tasks, or pleasant surroundings) according to how much "importance" they would personally attach to each of the elements. On the other hand, most of the remaining empirical investigations have concentrated on measuring the degree to which such attributes accurately depict the present job environment and then relating such perceptions to some complementary set of affective or behavioral responses.

Although employee attitude surveys and studies can differ markedly in their specific questionnaire items, formats, scales, and underlying objectives, the broad range of role design features that can be subjected to analytical scrutiny is somewhat bounded. More importantly, it has been shown elsewhere (see Katz & Van Maanen, 1977) that this range of features can be structured around three conceptually independent and distinct aspects of organizational and occupational settings as follows. 1) *Job Properties*—characteristics of the everyday task processes involved in a particular line of work. 2) *Interaction Context*—characteristics of the day-to-day interpersonal environment in which the person carries out his or her work. 3) *Organizational Policies*—characteristics of the general rules, standards, and policies enforced managerially in the workplace regarding compensation, promotion, training, and the like. Table 1 illus-

*Table 1.* Role Design Features of Work Environments

| *Cluster* | *Representative Role Design Features*† |
|---|---|
| Job Properties | *Skill Variety:* The degree to which the job requires different skills.<br>*Task Identity:* The degree to which the job requires completing a whole and identifiable process or piece of work.<br>*Task Significance:* The degree to which the job has a perceivable impact on other people or their jobs.<br>*Autonomy:* The degree to which the job gives an employee freedom, independence, and discretion in scheduling and executing work assignments.<br>*Job Feedback:* The degree to which an employee learns the efficacy of one's work from the job itself. |
| Interaction Context | *Colleague Assistance:* The degree to which an employee receives (or provides) sufficient help in carrying out everyday responsibilities.<br>*Agent Feedback:* The degree to which an employee receives clear information about performance effectiveness from co-workers or supervisors.<br>*Participation:* The degree to which an employee participates in establishing task objectives and procedures.<br>*Positively Rewarding:* The degree to which an employee is positively rewarded by supervisors and peers for producing good work. |
| Organizational Policy | *Training:* The degree to which additional training is available to the employee for the improvement of skills and abilities.<br>*Pay Equity:* The degree to which an employee's present pay approximates what others, performing the same job elsewhere, are paid.<br>*Promotion Fairness:* The degree to which an employee's advancement procedures in the organization are known and standardized (i.e., based on known criteria).<br>*Role Conflict:* The degree to which an employee's task assignments involve conflicting demands or incompatibilities. |

† These features represent only a partial listing and are presented here simply for illustrative purposes.

trates the kinds of role design features subsumed under each of these three cluster headings.

This particular classification scheme, however, is not especially surprising or unique. Indeed, each of the clusters corresponds to some rather traditional points of view on what has come to be labelled "the quality of working life." The design features associated with the job property category, for example, can be identified directly with the kinds of variables emphasized by proponents of job enrichment or job design as the most essential determinants of employee attitudes and work performance. Advocates of this position (typically, the domain of the applied behavioral scientist) assume that the set of daily tasks performed by employees, including experienced levels of autonomy, responsibility, variety, and so on, are strongly related to employee responses (e.g., Davis, 1966; Herzberg, 1966; Hackman & Lawler, 1971).

At the same time, the design features connected with the interaction context category closely parallel the sorts of variables used by supporters of the group process or human relations school of thought (primarily the management or training specialist) to explain employee reactions in the workplace. Those favoring this position view the interpersonal context of work environments, including such characteristics as supervisory-subordinate cooperation, employee participation in decision-making, performance feedback, colleague relationships, and the like, as the most important correlates of individual attitudes and behaviors (McGregor, 1960; Blake & Mouton, 1969; Odiorne, 1965).

Finally, the so-called structuralist approach can be detected within the representative design features associated with the organization policy cluster. Supporters of this third perspective (frequently the labor relations expert or union leader) argue persuasively that the organization's (or industry's) policies regarding rewards, mobility, scheduling, training, and so forth are the principal factors influencing worker reactions and perceptions (e.g., Perrow, 1972, Gomberg, 1973; Strauss, 1977).

*Need for Multiple Perspective*

These three points of view, which one might also label "human resources," "human relations," and "human rewards," respectively, seem to capture the major research and applied paradigms popularized within management circles over the past 40 or so years. Evolutionary progress toward a more bona fide understanding of employee attitudes and responses seems dubious, however, for no single model or viewpoint is applicable to all situations, nor has any particular framework eclipsed the others. Moreover, each of the three outlooks tends to propose reforms and organizational changes around a somewhat limited set of conceptual variables consistent with the overall slant of the particular paradigm.

What is important to recognize, therefore, is not the explanatory power of any specific theoretical standpoint but rather the necessity to view worker reactions from all three perspectives conjointly (Davis, 1976; Katz & Van Maanen, 1977; Strauss, 1977). Unfortunately, what we do not as yet know is how the influences of the different perspectives vary with one's job and career experiences. Are the relationships between the various role design features and individual outcomes stable and invariant or is there some vacillating pattern, either increasing or decreasing in strength, that demands further elaboration and explanation?

Part of the dilemma is that the most well-known and proselytized managerial theories of motivation, including Maslow (1954), McClelland (1961), and Herzberg (1966), concentrate almost exclusively on trying to specify exactly what it is that energizes and sustains positive employee reactions (especially high job satisfaction). These theories typically help one weigh the comparative advantages (or disadvantages) of the various kinds of role design features clustered within each of the previously described paradigms according to whether they somehow "match" the needs and motives of an organization's or unit's workforce. As pointed out by Campbell, Dunnette, Lawler, & Weick (1970), such theories fixate only on alternative content areas completely eschewing considerations of how individual perspectives and predispositions are initially formed, maintained, and even altered. What, in fact, are the most important mental and cognitive processes by which employees make sense out of their ever-changing set of daily work experiences? There are, of course, a number of motivational theories, including expectancy, equity, and reinforcement kinds of models, which try to delineate more explicitly the process of motivation. But even these theories fail to consider how one's outcome preferences, expectancies, valences, or comparative "others" are originally formulated and subsequently revised over some meaningful time horizon.

All task and job-related activities are part of some larger temporal sequence. As such, wants, actions, and employee assessments of their job situations, in general, are greatly affected by the succession of events extended over time. As a result, the notion that a few, supposedly unchanging, monolithic needs or psychological predispositions determine the course of events has been subjected to mounting criticism (Warr, 1976; Salancik & Pfeffer, 1977, 1978). All theories of motivation seem to acknowledge that certain individual needs or design components may be more salient than others in a given situation. Nevertheless, none of the theories has made much progress in predicting which particular set of needs or design features will emerge as more important at a particular point in time.

Herzberg's two-factor theory of motivation, for example, argues convincingly that the determinants of satisfaction and dissatisfaction are, for the most part, mutually exclusive. The cluster of role design features associated with the job property category being the chief components of satisfaction while the human relations and human reward kinds of features primarily reduce dissatisfaction. Furthermore, the theory implies that this distinction remains constant throughout one's entire work life. Although numerous critiques concerning Herzberg's use of retrospective data from critical incident methodology abound (for example, see House & Wigdor, 1967), what is most important to recognize is that neither the data nor the theory directly addresses the question of whether the individual was motivated *before, during,* or *after* the critical incident. In short, the relevant temporal framework is omitted. This is not meant to suggest that

Herzberg's theory is totally irrelevant but rather to emphasize that the theory, in its present form, is substantially incomplete and oversimplified. One needs to know more than simply what employees say they are looking for in their work. *We need to begin to explicate more precisely the process of motivation by building an integrative theoretical framework that deals with how employees interpret and respond to the numerous role design features embedded within job environments as they enter their organizations and proceed through their jobs and occupational careers!*

## THE INFLUENCE OF JOB LONGEVITY

In order to gain some initial insight as to how the relative influence of alternative role design features might fluctuate in importance as individuals pass through their own discrete sequence of job positions, Katz (1978a, 1978b) investigated the satisfaction and behavioral reactions of employees as a function of the actual length of time a person had been working on a given job (i.e., as a function of job longevity). After carefully examining the pattern of employee responsiveness to the job property factors developed by Hackman & Oldham (1975) along the job longevity continuum, Katz concluded that a three-transitional stage framework could meaningfully capture the "average" individual's reactions and concerns as he or she journeyed through each distinctly different job position. The three differential states begin with a relatively brief but important socialization stage followed by an innovation period that gradually shifts into an adaptation stage with extensive job longevity.

To explicate somewhat further, during the first few months on a job, each individual employee is primarily concerned with establishing and clarifying his or her own situational identity within his or her new organizational boundaries, as well as with learning all of the unfamiliar social and technical aspects peculiar to his or her recently acquired job assignments and required interactions. Such a "breaking-in" period has been commonly recognized in discussions of how socialization processes affect organizational newcomers, i.e., individuals who are beginning not only a new job but more importantly are entering a new organization (Schein, 1968; Berlew & Hall, 1966). Organizational socialization, therefore, tends to direct our attention to those introductory events and activities by which individuals come to know and make sense out of their newfound work experiences.

It is also important to recognize that this entry phenomenon also characterizes those veteran employees trying to relocate and resocialize themselves following some degree of displacement induced by a clearly defined shift in their organizational job positions. Just as newcomers have to define and interpret their new territorial domains, veteran employees leaving one job to assume another within the same organization, must also restructure and reformulate perceptions regarding their new social and task realities. Such perceptual revisions are necessary simply because work groups and other organizational subunits are often highly differentiated with respect to their idiosyncratic sets of norms, values, time perspectives, language schemes, goal orientations, etc. (Lawrence & Lorsch, 1967; Wheeler, 1966; Van Maanen & Schein, 1977). In fact, after a decade of research on technical communications, Allen (1977) reports, in a very clear and scientifically defensible manner, that these kinds of cultural and task differences often impede and distort the transfer of technological information across organizational boundaries to the extent that technical communications between professionals from different R & D organizations are extremely difficult and, on the average, significantly ineffective.[1]

Based on these ideas, Katz (1978a) demonstrated that in the initial socialization stage of job longevity, employees are not significantly responsive to all of the supposedly challenging features of their assigned tasks. In particular, it seems that employees are sufficiently absorbed with building their new situational perspectives during the first three or four months on their jobs such that job-property features encompassing the extensive use of multiple and varied skills and abilities (*skill variety*) as well as the ability to decide and influence how one's work might be accomplished (*autonomy*) are not necessarily of immediate importance or relevance.[2] At the same time, these employees appear to be highly receptive to role features that help establish and solidify feelings of personal acceptance, reassurance, and contribution within their new job environments. Such factors seem to include the job properties of *task significance* and *feedback*.

Once employees have learned how to function appropriately in their jobs and feel sufficiently secure in their perceptions of their workplace, they are able to concentrate more fully on their respective task assignments. Individual energies can then be directed more toward task performance instead of being expended on learning the previously unfamiliar social knowledge and skills necessary to assume one's organizational role in some customary and desirable fashion. The passage from the socialization stage to the innovation stage, therefore, is most likely marked by a shift in the individual's immediate job concerns, from an initial emphasis on psychological safety and identity to a concern for achievement and accomplishment (Buchanan 1974; Hall & Nougaim, 1968). As a result, it is within this innovation stage, predominantly between the sixth month and third year of job longevity, that employees are particulary responsive, in terms of both satisfaction and performance, to the richness of their jobs along *all* of the job-property dimensions. A key assumption underlying the innovation stage, therefore, is that with increasing task challenge, employees are willing to exert greater effort and involvement into their jobs as well as derive more overall satisfaction from such challenge because their primary interests and concerns can be devoted, without significant distraction or disturbance, to achievement, responsibility, and recognition kinds of factors!

As a person remains on the same job for a considerable length of time, however, it is likely that even the most challenging jobs will eventually become routine and habitual as the individual becomes increasingly proficient at and accustomed to his or her everyday task assignments. Such changes seem inevitable whether one is talking about the patrolman making the required daily rounds, the financial officer drawing up income statements or cash flow analyses, the marketing researcher conducting focus group studies or consumer preference surveys, the lawyer writing up case briefs or drawing up contracts, or the physicist measuring the Mössbauer effect in various compounds. In his classic thesis, Argyris (1957) argues persuasively that jobs perceived as highly standardized, unexciting, or no longer stimulating can help produce conditions that diametrically conflict with the most basic demands and needs of mature adults. With considerable job longevity and stability, therefore, it is likely that workers will eventually experience such kinds of imbalances along with the ensuing frustrations and possible tensions, pressuring them to either remove themselves from their settings or to adjust to their purportedly impoverished job situations. One way of adapting to one's long-term job tenure is simply by redefining what is important in the overall work situation, perhaps placing greater value on the so-called extrinsic factors and less value on the contrasting intrinsic factors. If a person stays on a job for a substantial period of time, for example, all of the human resource or job-related features may become progressively less important and less influential on subsequent work attitudes and behaviors. On the other hand, certain extrinsic or contextual role features belonging to the human relations or human rewards paradigms, such as interpersonal or collegial relationships, the nature of supervision, working conditions, or good pay and benefits policies, may become relatively more important and influential even though they may have always been somewhat important.

Based on such a developmental perspective, it is suggested that employees gradually shift from an innovation stage to an adaptation stage with increasing job longevity in the sense that individual workers ultimately succeed in adapting to their steady employment on the same job by becoming increasingly indifferent toward the challenging task features represented within their specific job duties. In fact, Katz (1978a, 1978b) suggests that, in all likelihood, most employees possessing five or more years of steadfast work at the same organizational position have slowly but methodically adjusted to their job situations by becoming virtually unresponsive to all of their task characteristics on that job. Yet, at the same time, they feel rather ensconced and content with their very familiar and, most likely, very predictable set of daily task activities!

The thrust of the preceding framework is not meant to suggest, either empirically or theoretically, that job satisfaction or performance necessarily deteriorate with increasing job longevity. On the contrary, the very meaning of the term "successful adaptation" seems to imply that employee expectations have become adequately satisfied as they perform their customary assignments in some reasonably acceptable fashion. If anything, it is more probable that employee attitudes would increase slightly with tenure on the job, rather than decline, just as many previous studies have demonstrated a positive relationship between organizational tenure and the individual attitudes of job satisfaction and organizational commitment (Quinn, Staines, & McCullough, 1974; Hall, 1976).[3]

The findings outlined in the present chapter, however, emphasize the notion that employee attitudes and behaviors may be significantly related to the different role design features, especially the various job-property or task-related characteristics, only at particular stages of one's job and organizational career. *As a result, the crux of the job longevity argument is that the sources or determinants of job satisfaction and performance change systematically with the actual length of time a person has had a given job.* In short, as employees move from socialization to the innovation stage, many job factors (and autonomy in particular) become relatively more important, while the gradual shift of employees from innovation to the adaptation stage is essentially defined around a relative decrease in the importance of job property factors, accompanied perhaps by a relative increase in the salience of the more extrinsically-oriented role design features.

*Reducing Uncertainty*

Underlying the descriptive changes represented by this job longevity model is the idea that people are not especially receptive to disorder and uncertainty but will endeavor to structure, organize, and interpret the world of their experience. Garfinkel (1967), for example, stresses the individual's strong, perennial need to unravel and define the many "constitutive rules" that help to keep and steer the workplace toward social order rather than toward social chaos. By and large, people must succeed over time in formulating situational definitions of their workplace with which they can coexist and function comfortably; else they will feel terribly strained and seek to withdraw from the given work environment.[4] To some extent, therefore, individuals and organizations are much alike, for they both seek to reduce uncertainty. As Thompson (1967) clearly emphasizes, organizations strive to structure themselves in ways that will reduce their perceived dependence upon unexpected variations in their input and output domains. In a similar vein, individuals are also motivated to make their knowledge of social and task interactions more predictable and less ambiguous.

Take, for example, the common experience of passengers temporarily diverted to a secondary airport because of bad weather conditions. It is not so much the diversion itself that eventually upsets the travellers but the lack of information and explanation about what is happening that is so frustrating. As a result, individuals cannot effectively reduce uncertainty and find it difficult to cope with the unexpected delay. Generally speaking, people feel very strongly about regularizing their social behavior to achieve "interactional order" (Goffman, 1967). Students, for example, will invariably take the same seat class after class after class, even though they are usually not permanently assigned. It is simply their way of reducing uncertainty! The crucial questions from such a perspective, therefore, surround the various mechanisms by which people establish order and reduce uncertainty, particularly within each of the job-longevity stages.

## SOCIALIZATION STAGE

During the initial socialization stage, an individual embarking upon a new job in a new organization is, to a large degree, *tabula rasa* insofar as organizationally-relevant attitudes, behaviors, and procedures are concerned. It is in this initial period, therefore, that the individual newcomer or novice typically learns not only the technical requirements of the job but also the social behaviors and attitudes that are acceptable and necessary for becoming a functioning member—both from the viewpoint of the individual and from the viewpoint of others in the organization. Using the theatrical similarities originally presented by Messinger, Sampson, & Towne (1962), Van Maanen (1977) metaphorically compares an organizational newcomer with an actor, who upon moving to a new stage (i.e., organizational workplace), must at first construct a dramaturgic interpretation of the plot's meaning, the cast of characters, the background of events, the props, the audiences, and the actor's own role in the sequence of incidents in order to direct and orient his own performance in a meaningful and contributive manner. Thus, if the newcomer is to develop a realistic understanding of the events and activities taking place around him or her, the person must actively build a situational definition within which certain ideas and assumptions can be tested and interpreted.

As pointed out by Katz & Van Maanen (1977), creating such a situational definition is analogous to building a mental or cognitive map of one's organizational surround. And it is within these individual mental maps that newcomers must eventually locate themselves by learning the practiced norms of conduct and outlook or what has been more colloquially phrased as "learning the ropes." To come to know a job situation and act within it, therefore, implies that the person has developed a scheme for interpreting and understanding the vast array of experiences associated with participating in that specific setting. In their action-research oriented projects, for example, Lou Davis and his colleagues at U.C.L.A.'s Center for Quality of Working Life have recently begun to use this notion of cognitive maps, rather than the more conventional attitude survey, for understanding more meaningfully the complex set of interactions between employee perceptions and role redesign issues.

As suggested by the descriptions of McHugh (1968) and Van Maanen (1977), to become a viable, functioning member of the organization and pursue a successful long-term relationship requires the newcomer to orient and locate himself or herself along the twin dimensions of social time and social space. Newcomers must discover when to ask questions, give advice, take a vacation, quit early, or push for a pay raise or promotion. Certain beliefs and assumptions about means-end relationships or expectancies have to be established to help guide and organize the newcomer's eventual participation in the workplace. Similarly, newcomers must locate themselves along the social space dimension by matching their own assumptions, values, and behavioral modes against the parallel perspectives held by their comparative or reference groups. Knowing where one stands in the spatial sense, therefore, connotes a set of perceptions held by the newcomer concerning what others in the organization are about, how they operate, and how he or she should perform on the job relative to these others. More figuratively, one must rely on social space awareness and understanding to determine where one "is at" and where others are "coming from" (Van Maanen and Katz, 1979). Such knowledge provides the newcomer with a way of classifying and categorizing numerous events and elements in his or her work environment. *Indeed, unless the newcomer succeeds in formulating a new perspective on his or her work situation that is commonly shared by other relevant employees, the individual's relationships with these employees in the workplace will probably be severely strained.*

*Initial Concerns*

Given the strong need of newcomers to define their organizational arenas and to develop constructs relating themselves to such a perspective, it becomes clear why a number of writers have empirically discovered that organizational newcomers tend to be more preoccupied than other employees with psychological safety and security, and with clarifying their own identities within the organization. Kahn, Wolfe, Quinn, Snoek & Rosenthal (1964) assert from their research findings, for example, that individual newcomers, faced with extremely new experiences at the beginning of their organizational careers, are predominantly concerned with feeling secure in their new situations as well as with searching for the means to integrate themselves into their settings as quickly and as effectively as possible. In their study of young A.T. & T. managers, Hall & Nougaim (1968) also found especially strong concerns for psychological safety during the managers' first year of employment, that is, getting established within and accepted by the organization. To give an additional example, Van Maanen (1975) has demonstrated, in a synthesis of ethnographic and questionnaire data, that police recruits are busily absorbed in reformulating and solidifying their own self and job-related perceptions for about the first four months following their actual introductory contacts and interactions with the realistic norms, values, assumptions, and behaviors held and demanded by their fellow patrolmen.

Because of these initial concerns surrounding integration, security, and identity, it should not be particularly surprising that Katz (1978a) presumed that newcomers, in the first three or so months of their new organizational jobs, would not be able to comfortably direct their attention, during this socialization stage, to role features involving *high* achievement, challenge, and autonomy, though of course some degree of challenge is essential (Schein, 1964). Not only were newcomer reactions unrelated to the task dimensions of skill variety and autonomy in this initial learning period, but Katz discovered that high amounts of early autonomy were perceived to be particularly disturbing, most likely because such autonony precluded a rapid and sound process for establishing a well-defined situational definition and identity.

*Establishing Role Identities*

Having crossed the organizational boundary, the newcomer cautiously enters a somewhat alien and stressful territory, confronted quickly, nevertheless, with the problem of creating order out of unfamiliar and vague surroundings. Consequently, the newcomer is primarily motivated during the socialization stage to reduce uncertainty by building a clear and acceptable identity within his or her progressively enacted work environment. In discussing his philosophy about process observation, Schein (1969) states explicitly that the foremost issue facing an individual in a new situation is the problem of developing a role identity which will be viable and suitable both from the standpoint of the individual person as well as from the standpoint of other persons within the relevant organizational area. Furthermore, it is likely that neither the newcomer nor the organizational subunit will feel comfortable in their new situation until this issue is resolved. It is precisely for this reason, for instance, that change agents or consultants, whether they be internal or external, will probably be more effective if they spend a sufficient amount of initial time clarifying and defining their role and situational identities vis-à-vis any new client body or group before attempting to tackle or deal with substantive issues.

Recognizing the numerous insights and ideas of George Herbert Mead, (1932), many sociologists have emphasized that new employees absorb the subtleties of organizational culture and climate and construct their own definitions of organizational reality—and in particular their own role identities—through *symbolic interactions* with other individuals, peers as well as supervisors. According to Mead's notions of symbolic interactionism, an individual newcomer would orient himself or herself to the new organizational world, not by responding directly and instinctively to events themselves, but by responding in terms of his or her own "interpretation" of those events. Since multiple meanings are possible for any particular event, it is the newcomer's active interpretation of the event, strongly influenced by numerous interconnected variables including past experiences, future expectations, reference groups, and educational backgrounds, that is important in constructing his or her personal view of reality (Schutz, 1967; Blumer, 1969).

Based on the present discussion, it can be argued that in the initial stage of organizational socialization, the new employee reduces uncertainty primarily through interpersonal and feedback processes and interactions. *Social relationships, therefore, are particularly important in shaping one's interpretive scheme of reality and in formulating a perspective about what is expected and acceptable in a given role.* Evan (1963), for example, demonstrated that unstructured interaction time with peers can help newcomers learn quickly the necessary and appropriate behaviors and attitudes. Moreover, Sims & Szilagyi (1976) interpreted their findings to suggest that interpersonal components of jobs can be a significantly positive influence on employee attitudes when individuals have particularly high needs for social interactions. If, in fact, interpersonal and social interactions are instrumental in reducing uncertainty and in establishing subjective order during the initial socialization stage, then one can understand more clearly how autonomy might interfere with this learning process. It seems reasonable, on the other hand, that some of the role design features subsumed under the human relations paradigm or clustered within the interaction context of Table 1 (such as colleague assistance and agent feedback) might be especially important during this same period.

One can conclude from these ideas that newcomers' initial perceptions and behaviors are significantly shaped by their conceptual schemes of their work environments acquired through social interaction. As a result, they tend to formulate their concepts and guide their activities vis-à-vis the anticipated reactions and expectations of other employees with whom they interact. And of all the concepts that the individual newcomer acquires through such social interaction, perhaps the most important is one's self-concept. Building on Cooley's (1922) notion of the "looking-glass self," it can be argued that one's fellow employees, or more specifically one's initial reference groups or role senders, help to define for the newcomers many of the diverse aspects in their new job settings by the way they act and behave toward these aspects; for example, how the more experienced employees deal with their superiors, customers, staff engineers, late reports, budget overruns, or subordinate suggestions. In a similar fashion, fellow employees also help the newcomers create a perspective on themselves as particular kinds of individuals by the way the fellow employees act and behave toward these organizational newcomers.

As a result, one's self-image is, to some extent, a social product, significantly influenced by the actions and attitudes of employees within one's organizational neighborhood—especially during one's initial socialization stage. It also follows that a newcomer's initial perceptions are significantly affected or molded by the *range* of reactions emerging from a set of people he or she knows and with whom he or she interacts. As such, the newcomer must respond to the actual views and behaviors constituting their new organizational world, regardless of their appropriateness or whether such perspectives are limited or broad, positive or negative, or open or closed. At least at the outset, therefore, the newcomer is strongly influenced by the self-concept that is gleaned and interpreted from the eyes of significant others. Hence, the newcomer who is viewed or sponsored as a high performer has an initial advantage over other newcomers not so fortunately viewed—the proverbial self-fulfilling prophecy. And as discussed later, such initial images can have a strong and lasting influence on one's future task assignments, perceived performance and abilities, and promotional success (Farris, 1969; Hall, 1976). It is partially for these reasons, perhaps, that so many studies have pinpointed one's initial supervisor or mentor as being such an important and influential factor with respect to one's successful organizational or professional career (e.g., Vicino & Bass, 1978; Thompson & Dalton, 1976; Kanter, 1977).[5]

*Goal-Setting and Feedback*

Irrespective of one's previous assumptions, motives, relationships, and abilities, as a newcomer, one strives to reduce uncertainty by locating and orienting oneself relative to the views and expectations emerging from those sources with whom one is most interactive and dependent—

including subordinates, peers, or supervisors. Perhaps the most influential experiences for new organizational employees, therefore, are those which attune them to what is expected of them (Berlew & Hall, 1966). As a result, there is a strong need for newcomers to identify more closely with those colleagues, supervisors, or other reference groups who can furnish guidance and reassurance concerning such expectations (Merton, 1957; Schacter, 1959).[6] Although employees generally express a particularly strong need for information regarding what is expected of them (Meyer, 1966), the need for information of this sort might be even more important and useful to newcomers building perceptual frameworks about their new organizations and themselves.

Goal-setting activities, as a result, are probably very beneficial to the newcomer during the initial socialization stage as long as such activities help reduce uncertainty by more clearly defining and interpreting supervisory or peer expectations. In a rather well-conceived but short-term field experiment designed to test and explicitly compare the direct attitudinal and behavioral effects of increasing job challenge and of assigning specific goals, Umstot, Bell, & Mitchell (1976) discovered that the goal-setting treatment was significantly more effective than the job enrichment treatment when applied to a random distribution of new, but *temporary,* employees.[7] Such a finding is clearly in line with the present notions that newcomers, in the midst of socialization, will be significantly responsive only to those role features that help reduce uncertainty in the new and unfamiliar work setting (in this case, the field experiment) and not necessarily to role features supposedly involving relatively high task challenge.

It is also important to note that the Umstot et al. experiment examined the direct assignment of specific goals to the working subjects and did not involve participative goal-setting. Although the knowledge of specific goals seems to outperform the more nebulous philosophy of "do your best," a recent review of the literature by Latham & Yukl (1975) comparing assigned goal-setting versus participative goal-setting interventions could not identify either of the alternatives as more advantageous. One could argue, however, that during the socialization stage, newcomers would profit more from assigned goal-setting than from participative goal-setting, since it is the definition of the supervisor's expectations that helps reduce uncertainty. Maier (1973) seems to agree with such a notion when he contends that a nonparticipative approach to appraisals would have its greatest potential with new and inexperienced workers still trying to learn and define their jobs. Following this lead, Hillery & Wexley (1974) discovered that for inexperienced people in a new training setting, nonparticipative appraisal interviews were significantly more effective than their participative counterparts in producing improvements as well as in leading to higher satisfaction. In fact, many of the trainees were noticeably disappointed when asked to participate in their own appraisals, preferring direct interpersonal feedback instead. House's (1971) path-goal theory of leadership would also support this viewpoint in that leadership of a more structuring nature should be more effective when tasks are reasonably ambiguous.

As a result, positive feedback may be particularly important during this initial socialization stage in that it helps establish a sense of contribution and worth in addition to providing a more secure and acceptable feeling. But it is *interpersonal* feedback that may be most important here—not necessarily job feedback or feedback from the job itself. For as discussed, the newcomer interprets and attaches meaning to events relative to the expectations and views of significant sources (such as one's boss) and not in an absolute sense! Knowing the organization's relevant expectations or goals may be especially important and helpful to the newcomer, for if they are defined and understood well enough, then newcomers should be able to judge pretty well for themselves how they are doing. *Clear, unambiguous expectations can help expedite knowledge-of-results!* Interpersonal feedback can then be used more broadly to provide newcomers with a more general sense of how they are being viewed and regarded, helping them to find their overall "niche" in the scheme of things.

*Facilitating Socialization*

Because the ultimate formation of the relationship between the newcomers and the organizational setting is an interactive social process, unfolding through mutual perceptions and influences, a correct understanding and view by the newcomer of the organization's expectations of himself or herself can be very functional. From his work on how socialization affects newly hired M.B.A. graduates, Schein (1964) stresses that graduates' perceptions of their newly encountered organizational realities are often misaligned and incorrect vis-à-vis organizational assumptions and expectations. In a similar vein, Wanous (1976) found that the perceptions of new graduate business students, as well new telephone operators, were reasonably accurate insofar as the more objective aspects of the environmental settings were concerned but were substantially less accurate with respect to the more subjective or intrinsic aspects. Thus, when the newcomer discovers that his or her expectations are not completely shared by one's colleagues or boss, then one experiences what Hughes (1958) has termed "reality shock." Such differences strongly push the new person to either reorient or realign his or her set of expectations and experiences or to eventually withdraw or terminate.[8] As summarized by Porter & Steers (1973), there is sufficient evidence to suggest that when a newcomer's expectations and beliefs are more in tune with organizational reality, the probability of turnover is lessened. When the individual's reward expectations are exceeded by the organizational setting, however, the likelihood of commitment can be enhanced (Grusky, 1966). Finally, in advocating the use of the psychological contract to avoid reality shock and to define quickly and accurately the organization's expectations of and for the newcomer (and vice versa), Kotter (1973) has shown that satisfaction attitudes of new employees can be directly related to smaller discrepancies in mutual expectations.

Such findings seem to imply that the more successfully a newcomer can navigate his or her way through the sometimes stressful and anxious moments of socialization and into the organization's culture, the more successful the newcomer is likely to be in the organization. The assumption underlying such a proposition is that employees who are more effective in dealing with those problems and concerns that arise during one's socialization process are likely to be viewed more positively by the organization and vice versa! And according to the law of primacy expressed by Brown (1963), such early socialization experiences may be especially important, since they influence how *later* experiences will be interpreted.

A crucial question, therefore, concerns whether or not newcomers will function more effectively within their settings when their situational perceptions are more congruent with the corresponding perceptions of their reference groups. How well the newcomer is located along each of the aforementioned dimensions of time and space can be measured by the degree of "fit" between his or her perceptual reality and the collectively held views and beliefs of other members in the relevant organizational setting—a kind of "shared awareness" measure between the newcomer and the reference group (Glaser & Strauss, 1964). To be sure, an individual's concern for or emphasis on either the temporal or spatial dimensions will most likely vary throughout the person's career. During socialization, however, it is probably more critical for newcomers to locate themselves relative to significant organizational others (i.e., the spatial mapping) than it is for them to develop an integrated temporal perspective around their organizational past, present, and future—especially since one's organizational past is virtually nonexistent.

To explore the previous question, Van Maanen and Katz (1979) investigated the attitudinal and behavioral responses of new police recruits during their first six months of "street duty" as a function of how well each recruit's perceptual space could fit or match with the collective perceptual space of their fellow veteran patrolmen. Using multidimensional scaling methodology, the researchers derived a two-dimensional configuration, along spatial and temporal axes, to capture the collective perceptions of the entire patrolmen force of a major, urban southeastern city—a kind of shared mental or cognitive group map. As expected, the spatial dimension accounted for considerably more of the total variance within the collective perceptual map than the temporal dimension. More importantly, however, there was a strong positive association between how well each individual recruit's perspective matched this shared perceptual space and the overall performance and promotability evaluations independently obtained from sergeants. Thus, it appears that recruits who were more able and willing to adopt the collectively held scheme of things during early socialization were more likely to be viewed positively by the organization and receive favorable supervisory evaluations.

On the other hand, it is extremely important to realize that the collectively held view to which a newcomer is subjected may not necessarily be a desirable or positive one. Indeed, the world of the urban police patrolman is often plagued by feelings of cynicism, isolation, and distrust or resentment of those at high bureaucratic positions (Rubenstein, 1973; Van Maanen, 1975). As a result, newcomers who most identify and perhaps assimilate such a cultural ethos during socialization may be viewed and rewarded more positively by veterans and sergeants but may, in fact, be the less motivated, creative, or innovative recruits. *In a sense, the process of organizational socialization can function as a kind of control or filtering mechanism according to the degree to which it emphasizes and rewards status quo perspectives.* Recognizing the issue and its importance, Van Maanen & Schein (1978) have been working toward a comprehensive theory of socialization by trying to isolate alternative dimensions or tactics of socialization and their potential effects vis-à-vis custodial or innovative kinds of responses.

Whether or not newcomers who are able to identify more closely with their reference groups are in the long run more favorably viewed, rewarded, or promoted remains an unsettled research issue. After reviewing the results of six experimental studies of recruitment strategies, Wanous (1977) concludes that those interventions which lead to a more realistic preview of organizational settings (via more realistic booklets, films, or simulations) tend to result in significantly less turnover. For the most part, these studies examined turnover rates only within the 3-month period following organizational entry. Hence, there really is no substantive evidence that a more realistic job preview will affect turnover across any reasonable time period. What it does suggest, however, is that a more realistic preview may help one to survive the sometimes stressful socialization period, a kind of innoculation against uncertainty and potential reality shock. Whether newcomers with more realistic previews are able to decipher or locate themselves within their particular organizational settings more easily or fare better after socialization needs further research.

*Initial Assignment*

In the process of building their situational definitions and identities, newcomers are also concerned with inclusion, that is, becoming a helpful, necessary, and important part of the overall operation. According to Schein (1971), to be accepted as a contributing member by others in the organization, "proving oneself," is one of the major obstacles with which a newcomer must struggle and eventually overcome. The importance of the initial work assignment, therefore, can be a very significant factor in the new employee's overall socialization process, for it represents a somewhat tangible and early means of achieving recognition and visibility within the workplace as well as demonstrating one's ability to contribute meaningfully to the organization. Moreover, the extent to which one's initial task demands and requirements can be seen as more central or pivotal to the organization rather than merely peripheral may also help not only in bolstering one's self-image but also in providing access to key information or permitting interactions with more important individuals. Thus, the newcomer may feel somewhat more interdependent, and may derive more easily and quickly a clearer, more perceptive understanding of the workplace. As previously discussed, therefore, it is not a matter of how much challenging work per se (e.g., skill variety and autonomy) that is important to the newcomer during the initial months of socialization but rather how much task significance characterizes the work.

What has also become clear from a large number of studies (e.g., Berlew & Hall, 1966; Bray, Campbell, & Grant, 1974; Vicino & Bass, 1978) is that the degree to which an employee perceives his or her job as important and challenging by the end of their first year will strongly influence future performance and promotional success. Using an overall index of job challenge, Berlew & Hall verified that young managers who evaluated their jobs more highly after one year of employment also had higher performance scores and rates of promotion over the next six or seven years. After carefully examining their data and possible alternative explanations, Berlew & Hall concluded that it is not the initial challenge or resulting visibility per se which leads to success but that the initial challenge must stimulate a person to perform well. *Having challenging tasks is not necessarily enough, for employees must also be responsive to the task characteristics of their jobs.* Such findings, therefore, are completely consistent with the job longevity argument, since by the end of the first year, employees have probably eased their initial security concerns and are probably more attentive to challenging kinds of activities.[9]

Moreover, the measure of job challenge in the Berlew & Hall study was an extremely broad index, aggregated across a large variety of items including communication skills, group skills, persuasiveness, sociability, supervisory skills, etc. Such an index of job challenge focuses not only on perceptions of job-property items but, more generally, on the degree to which the individual was required or asked to contribute in a highly important and central manner to the organization, which would appear to support the importance of initial task significance.

## INNOVATION STAGE

As previously discussed, individuals regularly revise and reformulate their definitions of their organizational situations. Such redefinitions occur primarily because the information or processes by which (or simply how) workers evaluate and interpret their specific job settings are not constant but vary according to the salience of specific types of concerns, issues, or problematic questions.

In particular, the movement from socialization to the innovation stage of job longevity carries the implication that the information derived from social interactions and relationships is no longer as relevant or as important to the individual in deriving situational meanings. Having adequately constructed their situational definitions and identities during the socialization period, individual employees are now free to participate within their own conceptions of organizational reality. They are freer to divert their attentions from *social* concerns to *task* concerns, that is, from concerns surrounding the issues of job safety, security, and integration to concerns for achievement, accomplishment, and challenge. As a result, employees are particularly responsive to all of their job-property features, or task characteristics, but only after the required socialization or learning phase (Katz, 1978a, 1978b).

The notion of having to achieve some reasonable semblance of situational security in order to be fully responsive in the work setting is very consistent with Pelz's (1967) model of "creative tensions." After studying 1300 scientists and engineers in 11 different research and development laboratories, Pelz concluded that scientists were more effective when they had sources of stability and security to balance sources of disruption or challenge. Thus, when *both* challenge and security are present, the creative tension between them can generate scientific achievements.

If, on the other hand, the individual has been unable to decipher sufficiently the job and/or organizational surroundings, then he or she cannot act, as yet, in a completely responsive, creative, and undistracted manner. To some extent, an individual unable or prevented from making sense out of the setting is partially in a state of what Durkheim (1964) might call "anomie", the antithesis of subjective order, characterized by confusion and unconnectedness. Various circumstances can arise in any work setting to delay or inhibit one's socialization, circumstances that typically preclude the necessary and essential set of extensive social interactions. Consider, for instance, the new employee whose boss is on vacation, out of town, or simply insists that he or she is too busy to help with introductions; or the new employee who is given an office far removed from his boss or peer group because of space limitations. Chances are that the reduction of situational uncertainty under such conditions will be a more prolonged process, perhaps interfering with the newcomer's potential success in the organization.[10]

To be responsive, however, does not imply that employees are necessarily more satisfied or more effective. Rather, it implies that such outcomes revolve around one's ability to participate fully and responsibly in meaningful work. No longer burdened or distracted by vast amounts of situational uncertainty, employees in the innovation stage are sufficiently responsive to their jobs such that they are likely to define their situations according to the degree to which they can alter, change, or modify elements within their work settings. *Situational meaning is derived not so much through interpersonal processes but through one's ability to influence and contribute to one's work situation. The greater the employee's influence, the more likely he or she would be willing to increase his or her level of effort and involvement.*

Such an explanation is consistent with Weick's (1969) contention that it is not the contents of the job per se by which employees deduce meaning from their work activities. Instead, it is the degree to which they have affected or altered the numerous, possible states of information; in short, their abilities to influence and acutally change conditions within their jobs settings. A similar kind of interpretation seems to pervade Walker & Guest's (1952) famous study of automobile assembly line workers. They reported that about 90 percent of the workers disliked their jobs chiefly because of the mechanical pacing of the assembly line which they viewed as having decreased their control over their own activities, making them dependent, subordinate, and passive. In a similar vein, the notorious Lordstown strike, heralded by many as the beginning of a worker's revolt against monotonous work, was essentially a demonstration against management's purportedly unilateral speed-up of the assembly line—the core issue, therefore, being one of control.

The intention here is not to downgrade the importance of challenging work but to emphasize the broader issue of influence. Indeed, more challenging jobs—those conferring autonomy and significance but requiring a variety of skills—typically give their holders more of a chance to influence their job environments and, consequently, result in more positive outcomes. In formulating his classic personality versus organization argument, Argyris (1957) does not claim that employees simply need more challenging assignments; but instead, he clearly emphasizes that workers should be allowed to enlarge the scope of their personal work domains such that their own decisions determine the outcomes of their efforts. Healthy, mature adults, therefore, seek more control over their current world of work while also striving to occupy a higher, more influential position than their peers. Argyris goes on to suggest that if individuals have little control over their working environments, they will tend to shorten their time perspectives mainly because they do not control the information necessary to predict their futures. Furthermore, they become dependent upon, passive toward, and subordinate to their leaders—all of which are manifestations of characteristics of less individual maturity. In his work on expectancy theory, Staw (1977) takes a similar point of view when he asserts that employees must first learn to predict and then be able to control their most relevant behavior-outcome sequences.

## ADAPTATION STAGE

As employees continue to work on the same job for a long period of time without being transferred or promoted to a new job position, they will gradually move from the innovation stage to the adaptation stage, or as previously discussed, from being highly responsive to their task characteristics to becoming progressively unresponsive. With extended job longevity, therefore, employees' perceptions of their present conditions and of their future possibilities become increasingly impoverished. It seems reasonable that, in time, even the most demanding task assignments and responsibilities would be perceived by the incumbent as progressively customary and less interesting and exciting. Without the presence of continuing job challenge and self-development, employees must begin to question warily the meaningfulness of what they are doing and where it may lead.[11]

It would appear, however, that such perceptions and conditions are in direct conflict with the normal adult development trends typically characterized by Maslow's (1954) notion of self-actualization, Erikson's (1963) concept of generativity, and Argyris's (1957) characteristics of mature adults. As a result of this imbalance, the individual must either leave the setting or adapt to the job circumstances. Adaptation can be achieved when one has brought one's personality characteristics or predispositions into equilibrium with the external information or stimuli in the work environment. Consequently, one can successfully adapt either by changing one's internal structure or by redefining the external situation. Some individuals may adapt to their long-term job tenure by appreciating over time the significance and challenge of their particular task demands and performance. However, for others, psychological defense mechanisms will be more helpful in allowing an individual to deviate from the basic development trends by reformulating and revising what he or she values or considers to be important. In a four-year longitudinal study of engineers, for example, Kopelman (1977) documented that perceived decreases in intrinsic expectations, such as "more challenging work assignments," were eventually matched by parallel decreases in the attributed value of such expectancies. Similarly, Hall and Schneider (1973) concluded from their diocese study that when individuals are deprived of opportunities for achieving psychological success through challenging work for many years, they are less willing to define future psychological success in terms of tasks that permit personal growth. Given these ideas and findings, it is likely that with considerable job longevity and stability, various psychological defenses will succeed in enabling individuals to adapt to their job environments by becoming indifferent and unresponsive to the task characteristics of their jobs, most likely, by placing less value on such intrinsic or job-property kinds of features.[12]

Although the seemingly perpetual debate on the relationship between intrinsic and extrinsic factors is as ambiguous as ever—especially with regard to whether extrinsic rewards detract from intrinsic ones (Deci, 1975; Salancik & Pfeffer, 1978)—it would seem reasonable that as employees became less responsive to the more intrinsic parts of their jobs, they would have to compensate by attaching more meaning to the more extrinsic aspects such as one's perquisites, vacation time, supervision, or colleagues. Or, they might compensate by transferring more value and importance to those interests and activities outside the work environment including one's family, leisure time, and hobbies (Bailyn, 1977).

What has often been neglected in the countless debates surrounding the extent to which workers are dissatisfied with their work lives and, consequently, their supposedly professed need for more enriched jobs is the notion that people are extremely adaptable, more than we often realize. Although a strong believer in the potential of job redesign, Hackman (1978) readily acknowledges that people show an enormous capacity to adapt gradually, almost invisibly, to their particular work environments, whether they be stimulating and exciting, routine or dull. It is the lack of knowledge about how employees have actually oriented themselves, how they attribute meaning to their work and nonwork environments, that has made the reported results around job design so difficult to synthesize. As pointed out by Hall et al. (1978), despite the apparent utility of job-design interventions implied by many correlational studies, most of the careful longitudinal studies have not been able to demonstrate a clear, positive relationship between changes in job characteristics with changes in performance, effort, or satisfaction.[13] Perhaps by knowing more about how the employees have adapted to their settings, knowing more about their current cognitive maps (especially vis-á-vis the intrinsic factors and extrinsic rewards), we can determine more effectively what combination of role design changes would be most appropriate, and just as important, devise a better intervention program for implementing them.

## CYCLES OF SOCIALIZATION

If an employee does not continue to work on the same job but, instead, is assigned to a new job position within the organization, either through promotion or transfer, then the individual reenters or cycles into the socialization stage where he or she must renegotiate the initial learning phase to restore the temporary loss of the familiar. In his 3-dimensional configuration of organizations, Schein (1971) stresses that employees must not only locate themselves within their functional departments and hierarchical levels but must also locate themselves along a radial dimension of inclusion or centrality. To move along this dimension, one must learn the norms, values, and assumptions that are shared within one's work setting and must also become accepted by others in the setting as a central and contributing member. To have full inclusion is to be an "insider" with full social and task knowledge about the organization.

As one shifts across job positions, therefore, one suffers from a temporary loss of centrality or inclusion. The degree to which one feels displaced along this radial dimension depends upon the permeability of the intraorganizational and group boundaries that one must cross to assume the new job position. Just as newcomers must proceed through their socialization stage to construct their situational realities, the newly transferred or promoted employees must also proceed through essentially a parallel process to reinterpret and reorient their new situational definitions and identities. *Socialization, therefore, is a lifelong process, continuously reoccurring throughout one's organizational career.* Thus, newcomers and veterans in the beginning months of their new jobs must struggle and contend with the same kinds of concerns and preoccupations.

There can be some important differences between newcomers and veterans, however (Wheeler, 1966). The newcomer must learn about and adjust to the organization from scratch. Such a process requires that one sift and digest large amounts of sometimes vague and ambiguous information to discover where one fits in the organization's overall scheme of things. To define and nourish one's existence in the new working environment, the newcomer must build interpersonal relationships as well as learn the technical requirements of the job.

In contrast to the newcomer, a veteran employee in a new position has already acquired sizable social and task-related knowledge about the organization, established contacts within it, and developed some sort of personal reputation or image through his or her previous organizational assignments. As a result, the veteran is probably not as concerned with or dependent upon the building of new social relationships for reducing uncertainty as the newcomer, though of course such social relationships are still essential for establishing subjective order. On the other hand, the veteran is probably most concerned with demonstrating his or her mastery and competence over the tasks contained in the new job. Hence, it is reasonable to assume that it is more essential for newcomers entering the organization than for veterans changing jobs to feel that they are becoming integrated within and accepted by the organization. Newcomers, therefore, should be especially responsive to *social issues* ("getting on board") while veterans should be very sensitive to *task performance issues* ("doing a good job"). Following this lead, Katz (1978a) found that newcomers were significantly more disturbed by high autonomy than veterans during the initial months of the socialization period, while veterans appeared to respond more positively to job feedback than newcomers.[14] *Of course, the more a veteran feels displaced during the socialization period of his or her job, the more likely he or she will feel, act, and respond like a newcomer undergoing socialization.*

Regardless of their potential differences, socialization for both the newcomer and veteran denotes a clear and discrete break in the normally smooth and continuous flow of daily events. As such, they represent distinct, interruptive transitions or what Van Maanen (1977) has conveniently labeled as *breakpoints,* that is, changes which thrust one from a state of certainty to uncertainty, from knowing to not knowing, or from the familiar to the unfamiliar. These breakpoints essentially represent disjointed junctions in the sense that one must foresake and replace one's former assumptions, relationships, responsibilities, and established patterns of behavior. Almost by definition, therefore, breakpoints require one to attend to those parts of the environment that have become equivocal.

### *Unfreezing and Changing*

The emerging condition of uncertainty subsequent to the breakpoints of moving to a new job position, either in a new or the same organization is similar to the Lewinian notion of "unfreezing" in which the individual is primed for change. The lack of confirming or disconfirming information, according to Schein (1973), around either one's self-image, one's image of others, or one's situational definition will contribute to the process of unfreezing. As long as one's self-image or situational definition needs to be balanced against the views of significant others in the environment, one will remain susceptible to change. Clearly, transitions into the socialization period represent unfreezing processes.[15]

The job longevity framework seems to suggest that as a result of such breakpoints, periodic job mobility or rotation might help prevent employees from moving into an adaptation stage. Employees can simply cycle from one innovation period into another after a brief but necessary socialization or learning stage. As long as the socialization period is positively negotiated, such transitions appear highly plausible. It would appear more difficult, however, for changes in job positions per se to shift employees from adaptation to innovation via socialization, that is, changing one from being unresponsive to being responsive. Certainly, the removal of individuals from those situations in which their task and social relationships operate to maintain and reinforce the validity of old attitudes can help unfreeze the adapted individual. But this alone provides no assurance about the direction of any resulting changes or that they will have any permanency. There is no guarantee as to how the individual will reorient or reinterpret his or her new situational reality, self-image, and image of others. What we do know, however, is that these outcomes can be highly influenced by the kind of social interactions and interpersonal relationships that are developed and utilized during socialization. What is important, therefore, is to establish strong relationships between the individual and a proper reference group in order to replace any undesirable attitudes and behaviors with more acceptable ones. *Moreover, one would posit that the more one has adapted to one's work environment, in the sense of being highly complacent and unresponsive, the greater the level of displacement and ensuing uncertainty will have to be in order to unfreeze the individual.* Behaviors and attitudes are usually deeply embedded in the relationships one has built-up over the years and as long as one can sustain these supporting relationships, changes are less likely to occur.

A few companies have already instituted programs specifically designed to rejuvenate individual employees, including engineers, by relocating each person in a carefully selected and distinctly different part of the organization where the person's skills, abilities, and aspirations can still be of value. Such interventions, however, have only dealt with small numbers of purposefully-chosen employees. Chances are also that the organizations have to instigate the idea of such movements simply because individuals who have adapted to their settings have essentially become indifferent and content, perhaps even resisting or resenting any overture that hints at relocation. *What may be the most important factor in the success of any change effort is the set of attitudes belonging to the people involved in the change process.* And one critical aspect of this attitude set is the perceived need for change. The decision to actually transfer, therefore, should be a joint decision between the individual and the organization, since there is probably no advantage to be gained, at least in terms of motivation, in forcing someone to relocate. Perhaps, an easier strategy is a preventive one, using both career planning and job rotation programs to help maintain interest and responsiveness by preventing long-term job longevity.[16] Such a comprehensive human resource planning and development model has been recently outlined by Schein (1978).

## SITUATIONAL VERSUS INDIVIDUAL CONTROL

The formation of an individual's perceptual outlook of his or her job environment is, to a large extent, a retrospective process, built and shaped through the recollection and interpretation of past interactions, experiences, and activities. To develop a meaningful perspective toward one's work setting, therefore, requires time, interaction, and familiarity with one's organizational surround. One of the critical characteristics of breakpoints, however, is that the information and knowledge gathered by the employee are suddenly *insufficient* for interpreting and understanding one's new and somewhat uncertain situation. Newcomers in the midst of socialization, for example, have virtually no personal organizational histories from which to construct their perceptual views. Their lifespaces are essentially blank vis-à-vis their new organizations. In a similar vein, the extent to which veterans can utilize past events and knowledge to reformulate individual perspectives toward new job positions can also be very limited, depending upon their degrees of displacement.

Following such unfreezing events or breakpoints, therefore, individual perspectives are especially malleable and susceptible to change and revision (Parsons, 1951; Brim, 1966). They are in a sense under *situational control* in that they must respond to and interact with external sources of information within the organization in order to help define and interpret what has become unknown and unfamiliar. The greater the uncertainty or displacement in the new setting, the more the individual employee must rely on the immediate work environment to provide the necessary information and social exchanges by which he or she can establish order and construct a more complete and realistic situational definition and identity. As previously discussed, therefore, employees are particularly attentive to social cues and information emerging from the overall work situation during the socialization stage of job longevity or, more generally, after any transitional breakpoint. The sensitivity of individuals to such environmental cues is well demonstrated by King's (1974) field experiment in which management's explicit expectations of productivity improvements from change interventions involving either job rotation or job enlargement turned out to be a significantly more influential and salient piece of information with respect to actual increases in production than the particular type of job change program.

If, on the other hand, one should fail to obtain from the relevant job environment any of the clues, information, or guidelines necessary to build a sound working perspective, then one may well resort to his or her own role models, stereotyped images, or implicit theories to complete the required situational definition (Kelly, 1971). This can typically occur when everyone is new to the setting, when social interactions have been curtailed or confined, or simply when there has been insufficient time. Exemplifying all three characteristics, Zimbardo's (1972) Stanford prison experiment poignantly illustrates the extent to which individuals can employ their own stereotyped models and images when the given situation does not provide either its own or alternative behavioral and information cues. Exploring the dynamics of causality, Staw (1975) also suggests that individuals, in a new and unfamiliar setting, are likely to ascribe certain characteristics to their groups and associated intragroup activities according to their own implicit theories about performance and group processes. As suggested by DeNisi & Pritchard (1978), however, such implicit attributions may diminish over time with increased awareness of one's surroundings and extensive interpersonal interactions.

*Generally speaking, as employees become more cognizant of and familiar with their job settings, they also become less docile in their susceptibility or obedience to situational information.* No longer feeling substantially unfrozen, they are less easily manipulated. In moving from socialization into the innovation stage, employees have developed a sufficiently robust perspective or situational definition through which they can discern and judge information and social cues more consciously and knowledgeably. They are now less constrained in their *use* of information to formulate perceptions and freer to decide *how* they will perceive the information. The individuals are simply more capable of determining for themselves the relevance and salience of the various information inputs. In particular, employees are likely to interpret and process information during the innovation stage according to their abilities to influence and participate in their overall job environments. Consider, for instance, the new secretary, engineer, or technician who faithfully arrives and leaves work at the regularly scheduled hours and who seems to perform all task assignments willingly and diligently—at least initially. With increasing awareness of one's job setting, its actual practices, procedures, and norms, the individual is soon freer to decide which tasks will be performed carefully or promptly as well as to develop his or her own interpretation of permissible working hours. During the innovation stage, those tasks which permit greater involvement and challenge are likely to elicit greater effort and interest. The passage from socialization to innovation, therefore, can be characterized by a relative shift to *more* individual and *less* situational control.

Furthermore, as employees gradually adjust to prolonged periods of job stability or longevity, they may continue to become even less susceptible to situational information and control. It is likely that in the process of adaptation, employees are becoming increasingly content or ensconced in their customary ways of doing things, their comfortable routines and interactions, and their familiar sets of task activities and responsibilities. As a result, they are probably less receptive toward any change or toward any piece of information that might threaten to disturb their developing sense of complacency. Rather than actively trying to enlarge or advance the scope of their organizational activities, adapted employees may be more intent on maintaining control over their immediate work environments, most likely by buffering or insulating themselves from higher-level managers or from other sources of possible interference. In his recent study of the British constabulary, Manning (1977) clearly recounts how British constables eventually succeed in isolating themselves from their supervisory sergeants in order to secure and protect certain behavioral practices and prerogatives. Similar kinds of evidence can also be found in the survey data analyzed by Pelz & Andrews (1966) in that the longer a scientist had been assigned to a particular group, the less frequently he reported communicating with his immediate group leader about work-related matters. Nor are such kinds of protective reactions only exercised by subordinates. For example, Dubin (1972) contends from his research on the problems of obsolescence that as managers become more complacent and fail to update themselves, they begin to fear change, new ideas, and potential rivalries from younger subordinates; and consequently, are often afraid to hire anyone with more knowledge and competence.

There is also considerable evidence to suggest that as the average length of time that members of a group have been working together increases, i.e., as group longevity increases, the more the group tends to differentiate itself from external individuals and groups both within and outside its own organization (Homans, 1961; Graen, 1976; Schein, 1979). In Well's (1962) study of 83 industrial and government R&D groups, for instance, the amount of contact outside the group was found to vary inversely with group longevity (also known as group age) while, in contrast, the level of intragroup communication was found to increase. Furthermore, it has been generally demonstrated that, over time, groups tend to both attract and select new members who are highly similar to themselves, perhaps becoming in the process more homogeneous and narrow with respect to both their range of interests, abilities, and member characteristics as well as their overall group norms, values, and problem-solving strategies (Shaw, 1971; Stogdill, 1974).

One could easily argue, therefore, that groups whose members have worked together for a long period of time tend to feel and behave more cohesively amongst themselves. Yet, at the same time, the longer they are functioning together as a cohesive group, the more competitive and less cooperative they may become toward other groups (Pelz and Andrews, 1966). As a result, the well-known Not Invented Here (NIH) factor may be especially characteristic of project teams with considerable group longevity, inducing members to bias adversely their views and evaluations of any seemingly competitive ideas, innovations, or products stemming from sources outside their own group. Perhaps it is these kinds of developmental trends that might help to explain why the studies of both Shepard (1956) and Wells (1962) uncovered comparatively lower overall performance scores for those R&D groups whose group longevity exceeded 4 or more years.[17] Thus, even though one might not necessarily detect a decline in performance with long-term job longevity when examining a random sample of individuals from a wide range of heterogeneous jobs, the problem of maintaining high performance may be particularly difficult when dealing with a team of individuals, many of whom have extensive job longevity.[18]

Ever since the Hawthorne experiments, it has been generally acknowledged that the particular conditions and interactions within a group can significantly influence the behaviors, motivations, and attitudes of its individual members. Consequently, if we ever hope to make sense out of the vast multitude of studies conducted on an individual level of analysis, we will need to know and report as much as possible about the various group contexts comprising our samples. And just as many behavioralists are now intrigued with researching the maturation process of adults throughout the life cycle, we also need considerably more research on the long-term aging phenomena occurring within and between groups.

*Some Methodological Concerns*

What is also important to recognize from the formulation of a situational versus individual control continuum is that many behavioral research findings are grounded in data that has been collected from respondents and subjects under conditions typified more by situational than by individual control. Laboratory experiments, in particular, are almost always conducted under as much situational control as possible. To the subjects, the experiment is a new and uncertain situation, for they are commonly placed in an unfamiliar setting to engage in unfamiliar behavior. *As a result, their normal modes of behavior and customary supporting mechanisms are not particularly helpful or even relevant.* Nor can they rely to any great extent on their previous experiences to construct a meaningful situational definition. The experimenter, on the other hand, is in the powerful position of controlling the meaning of what is valid information (Argyris, 1975). Moreover, the subject is forced to react to such unilateral leverage, often without the benefits of increased learning or the chance to ask for additional information or even to withdraw. The degree to which one really has any semblance of free choice is greatly restricted by such singular situational control.

What can be learned from such kinds of experiments is how people might respond when they are subjected to unfamiliar but highly salient information cues or treatments in a fairly strange environment. In fact, some social psychological experiments have clearly shown that if people are put in a rather atypical and absurd situation, they can be made to respond in a rather atypical and absurd fashion. Thus, army reservists have been situationally manipulated into eating fried grasshoppers by Smith (1961) while Comer & Laird (1975) were successful in getting naive subjects to agree to eat dead worms.

What cannot be ascertained from studying short-term experimental results is how individuals might react to particular treatments or conditions as they learn and become increasingly familiar with both task and social surroundings, that is, as they are able to rely more and more on their *own* situational definitions to formulate and direct their attitudinal and behavioral responses. In their research on communication networks, for example, Faucheux & Mackenzie (1966) discovered that as subjects became more aware of their groups and the demands of the common symbol task, they soon learned to organize and function exactly like a centralized "wheel" even though they were assigned to a decentralized, "all-channel" structure. To truly understand behavioral phenomena, therefore, one cannot focus only on the most immediate reactions of individuals in the midst of new and unfamiliar circumstances but must also determine if and how such reactions change and develop with progressive exposure and interaction. It is clear from Milgram's (1963) shock experiments that individuals with, as yet, very little pertinent knowledge or understanding about acceptable or alternative modes of behavior can be extremely obedient to authority when trying to function in a somewhat strange and alien environment. It is, however, just as important to determine how obedient individuals are likely to be as they become more cognizant of their situations. A crucial question from this perspective then is whether Milgram's experimental subjects would have continued to shock additional learners on subsequent trials once they had gained experience and become familiar with the specific demands of the setting. In a similar vein, Asch's (1955) classic experimental findings on group conformity may diminish significantly as subjects interact and become fairly friendly with their other group members. This is not meant to imply that pressure toward group conformity is an unimportant behavioral phenomenon but rather to suggest that such pressures may be particularly strong when group members are unfamiliar with each other or when members have not interacted very much as a group.[19]

The critical implication is that it may be very difficult, if not erroneous, to extrapolate results from conditions of high situational control to conditions of high individual control. An individual's attitudes and behaviors, for instance, are much easier to influence and manipulate when he or she is displaced into a novel setting characterized by high situational control (Zimbardo and Ebbeson, 1969; Argyris, 1975). One notable application which illustrates just how powerful situational control can be for influencing individual responses was recently carried out by Varela (1971) in Uruguay. Rather than trying to generalize from experimental findings on attitude change to the real world, Varela concentrated on making the real world more like the laboratory. To accomplish this, he invited retailers to visit his warehouse where he was able to bring their behavior under situational control and manipulate many of them into buying large quantities of ready-made curtains—a type of purchase that was rarely practiced in Uruguay. In short, laboratory studies may be telling us considerably more about the reactions of individuals trying to reduce uncertainty during some microcosm of a socialization process than about the responses of individuals in an innovation or adaptation stage.

One should also recognize that laboratory experiments do not possess a monopoly on problems associated with interpreting data collected under novel and uncertain conditions. Indeed, many survey-type studies have required respondents to complete elaborate questionnaires under unknown circumstances. In many cases, for example, employees are gathered together in a large conference or meeting room and asked by someone they have never seen or heard of before to answer a plethora of questions as frankly and as honestly as possible. But why should they! Often respondents do not know what the survey will be used for, how it will be used, or who will use it. Is it merely just another survey, another waste of time, or will they somehow benefit from completing the forms? Without a reasonable and acceptable perspective about why and for whom they are completing the battery of instruments, there seems to be little rationale for filling them out very truthfully or candidly.

Furthermore, questionnaires are often administered during regular working hours to large groups of respondents (though, of course, they are asked and expected to complete them individually). Some respondents, as a result, may simply view the survey as a fun, novel alternative to their regular work day, while others may see it as an unnecessary, bothersome disruption. In either case, respondents are probably not especially committed to answering the questions in a very careful or very accurate manner.[20] It should not be surprising, therefore, that in choosing their answers, respondents are often susceptible to subtle information cues both within the questionnaire itself and within the overall setting in which it is administered (Salancik & Pfeffer, 1977). Employees' answers to certain questions, for example, may be very different when the survey is sponsored or okayed by the union than when it is sanctioned or approved by management; just as students answer behavioral-type questionnaires very differently in an organizational behavior class than in a course on production and control. Thus, we need to be considerably more concerned about the specific kinds of circumstances surrounding questionnaire administration if we hope to improve the validity and credibility of data responses. And it is just as important to report in one's research methodology as much as possible about the overall situational context in which the administration process occurred as it is to report descriptive statistics of samples and psychometric properties of instruments.

## CONCLUSIONS

Central to the ideas put forth in this essay is the sociological viewpoint that employees are continuously negotiating and remodeling their perspectives toward their organizations and their roles in them. And it is the perspective that one has formulated at a specific point in time that gives meaning to one's work and one's career. In particular, we have tried to show how employees' perspectives develop and change as they enter and cycle through some or all of the three transitional stages of job longevity. Underlying such changing perspectives is the basic premise that different sets of common issues and concerns are particularly important and influential in each of the separate job tenure periods. Based on the ideas developed in the previous sections, Table 2 exemplifies the major kinds of issues that seem to preoccupy and guide employees as they move along the job longevity continuum.

Briefly summarizing from Table 2, individuals in the midst of socialization are primarily concerned with issues of reality construction. In the process of building their situational definitions they must first confront the problems of developing new task and social relationships; and of forming new or modifying old attitudes about themselves and about their co-workers, supervisors, subordinates, customers, and task-related assignments. With time, interaction, and increased familiarity, however, employees are soon able to direct most of their attention toward issues of influence and achievement. What becomes progressively more pertinent

*Table 2.* Examples of Special Issues During Each Stage of Job Longevity

| *AREAS OF CONCERN* | *Job Longevity Stages* | | |
|---|---|---|---|
| | *Socialization* | *Innovation* | *Adaptation* |
| Reality Construction | a) To build one's situational identity<br>b) To decipher situational norms and identify acceptable, rewarded behaviors<br>c) To build social relationships and become accepted by others<br>d) To learn supervisory, peer, and subordinate expectations<br>e) To prove oneself as an important, contributing member<br>•<br>•<br>• | | |
| Influence and Achievement | | a) To be assigned challenging work<br>b) To enhance one's visibility and promotional potential<br>c) To improve one's special skills and abilities<br>d) To enlarge the scope of one's participation and contribution<br>e) To influence one's organizational surroundings<br>•<br>•<br>• | |
| Maintenance and Consolidation | | | a) To routinize one's task activities<br>b) To mystify and ritualize one's task procedures and resources<br>c) To protect one's autonomy<br>d) To minimize one's vulnerability<br>e) To cultivate and solidify one's social environment<br>•<br>•<br>• |

to employees as they proceed from socialization to the innovation stage are the opportunities to participate and contribute within their job settings in a meaningful, responsible, and visibly competent manner.

On the other hand, when employees continue to work at their same job positions for extended periods of time and begin to adapt to such long-term tenure, their principal concerns may gradually shift toward the consolidation and maintenance of their work environments. Rather than focusing on trying to enlarge the scope of their activities and efforts, adapted employees may concentrate more on protecting their respective individual (and group) domains. With a growing sense of complacency, they may prefer to divest themselves of many apparently extraneous and less important activities while, at the same time, endeavoring to routinize as many of the remaining task requirements as possible. Interestingly enough, to protect themselves and secure their individual autonomy, they may also become less open and candid about what they are doing and how they are doing it! Adapted employees may even try to dismiss their supervisors, perceiving higher-level inquiries and observations as a somewhat threatening intrusion. One often hears the almost rote response of "leave us alone; we're doing just fine!" In addition, it is likely that institutional rewards and perquisites are relatively more important during the adaptation stage, for they may be the only means left by which adapted employees can obtain a somewhat clear and unambiguous measure of their self-worth and appreciation within their organizational settings.[21]

It is important to emphasize that Table 2 describes only broad categories of issues and concerns that employees are working to resolve in each of the three job longevity stages. What it does not list are the specific kinds of strategies, programs, interactions, and designs through which individuals, groups, and organizations can try to deal with each set of issues (though a number have been discussed in the previous sections). Whether a particular variable will be important in any given period will depend on how relevant it is vis-à-vis the dominant issues in that stage. Clearly, we need to be able to link our empirical variables more closely to the full range of developmental issues and concerns described in Table 2. Furthermore, even though certain kinds of activities and structures may be found especially influential during a given period of job longevity, it may be difficult to establish a very consistent pattern of consequences or outcomes for such factors without knowing considerably more about the organizational and occupational contexts. To some extent, the strength and direction of certain associations may depend on existent intragroup attitudes and norms concerning colleagues, management, and work performance; the kinds of task demands and technology with which the individuals are working; and the types of control and supervisory systems that are functioning within the situation (Katz, 1977; Stogdill, 1974).

Activities which promote the learning of group norms during socialization, for example, can be either functional or dysfunctional depending on whether such norms support high effort and performance. Thus, an assembly-line worker who is rapidly socialized into a work group with low performance norms is also likely to become a low performer. As an additional example, lots of interaction, support, and structured activities may help the new assistant professor perform effectively during his or her early career; perhaps only to discover later on, however, that this individual cannot function effectively in a highly autonomous, unstructured environment. If, on the other hand, the socialization process were highly autonomous, then such a strategy may be initially frustrating but could serve as an effective filtering mechanism for screening out professors who cannot tolerate high autonomy or who have difficulty structuring and defining their own task activities. Such a strategy is often utilized either because the department does not have adequate resources or is simply not organized to perform any kind of "professional development" function. As a result, many professors who are initially ineffective at one university go on to become very productive at another. The same can be said for many students, employees, and managers.

Most likely, it is during the innovation stage that one is apt to find the most consistent positive associations between role activities that permit responsive individuals to participate and contribute in their work environments and various outcome variables such as satisfaction and performance. Contrastingly, whether or not protective and consolidating kinds of activities eventually result in lower performance for unresponsive employees in the adaptation period may very well depend on the types of tasks and the kinds of controls operating within the setting. For example, with fairly routine, simple tasks, careful supervisory controls, and monetary incentive systems, the level of performance can probably be maintained. As tasks become more complex and require some degree of creativity or innovativeness, the effects of adaptive activities are more likely to be dysfunctional, especially perhaps for groups of adapted employees. All of this strongly suggests that we need to know a great deal about the situational contexts in which we are studying relationships if we really expect to understand any emerging pattern of results and trends. Only if we are willing to delve more deeply into the specific occupational, organizational, and group cultures and rituals, will we finally begin to overcome the problem of having almost every major literature review of an important relationship conclude with the rather equivocal and meek phrase: "the situation is considerably more complex."

*Individual and Occupational Differences*

The discussions presented so far are based on the aggregated responses of many respondents from many different studies. In a sense, the job longevity model has described the issues, concerns, and reactions of everyone, but of no one in particular. Each individual employee has his or her own abilities, needs, and prior experiences. Some workers, for example, may be ready for, or even expect, substantial autonomy and variety from the outset. Others may not. Some could start to adapt to their jobs much earlier than others, having had a more difficult time maintaining a sense of challenge and stimulation. Part of the diagnostic skills of a good manager, then, may be to understand how the job longevity model *stretches and contracts* to fit the particular issues and concerns represented within the mix of people with whom he is dealing.

It is also generally agreed that behaviors and attitudes tend to be a function of the interaction between the person and his or her environmental situation. As a result, employee reactions are most likely influenced by both their psychological and personality characteristics as well as by their definitions of and interactions with the overall work setting. What needs to be studied, therefore, is how different kinds of individuals negotiate their way along the job longevity continuum. Based on some initial, albeit cross-sectional, analyses, Katz (1978b) suggests that individuals with high growth needs might be more capable of responding sooner to more challenging tasks and responsibilities but might also start to adapt sooner than employees with low growth needs.

In looking at the predictors of managerial success, Hall (1976) concludes that there is some evidence to suggest that individuals with a higher need for achievement or a higher tolerance of uncertainty are also likely to be more successful in their managerial careers. *One possible explanation, perhaps, is that such individuals are more successful because they are more adept at handling their many socialization encounters as they cycle up the managerial ladder*. Rather than simply waiting for others to define for them the many aspects of their new environments, managers with the aforementioned characteristics and abilities may be more active in seeking such definitions or in defining their own sense of reality, including goals and expectations. As pointed out by Staw (1977), individuals high in achievement motivation do not derive pleasure from uncertainty itself but from the process of reducing it.

In addition to the importance of individual characteristics and abilities, one should also examine the job longevity model with respect to occupational differences. In situations where the jobs are fairly simple and social interaction is high, both the socialization and innovation stages may be relatively brief. In fact, if new hires are socialized by employees most of whom are in the adaptation stage (as typified perhaps by assembly-line work or police patrol jobs) then many of the new workers might quickly pass from socialization directly into an adaptation stage. Conversely, when substantial learning and task complexity are involved, both the socialization and innovation stages may be relatively longer (as is probably the case for engineers and lawyers). One might also hypothesize therefore, that certain individual characteristics and abilities might be better predictors of success in some occupations than in others. *High need for achievement or high tolerance for uncertainty, for example, may be stronger predictors of success for those occupations in which socialization is either relatively long or involves considerable uncertainty.*

Finally, the framework developed in this essay has focused almost exclusively on developmental issues associated with the span of time an individual has been working at a particular job position. There are, obviously, many other developmental issues taking place in parallel. Not only may people differ from each other across the job-longevity continuum but there may also be significant differences in individual concerns according to age, career stage, and family circumstances. Older adapted employees, for example, may be relatively more concerned about career stability and economic security than their younger counterparts.

In short, organizations are comprised of people who want many different things from their work, careers, families, and outside activities. Some may be career-oriented; others family-oriented; and still others may be dual-career oriented or may work only to enjoy their leisure time. Some may be managerially oriented; others functionally or technically oriented; while others may need to be entrepreneurial, autonomous, or highly secure (Schein, 1978). As emphasized by Bailyn (1977), the real challenge to organizations in general, and to their executives, in particular, is to develop the programs, policies, and managerial talents necessary to be effective in such a pluralistic society.

## FOOTNOTES

1. Allen (1977) goes on to emphasize, however, that although the average technologist may not communicate effectively across organizational boundaries, certain R&D professionals, defined as "gatekeepers," are able to transfer effectively external technology and ideas into their organizational laboratories or departments. More recently, Katz & Tushman (1979) have extended Allen's ideas to show that, in general, problem-solving communications across highly differentiated intraorganizational boundaries are also prone to bias and distortion. As a result, the greater use of *internal* problem-solving interactions per se by groups of staff engineers and scientists are not necessarily associated with more effective overall group performance.

2. It is important to emphasize that there is nothing sacred about the temporal ranges mentioned with respect to each of the job-longevity stages. Indeed, in any given situation, the progression of the stages is influenced by a large number of factors including anticipatory knowledge, personal characteristics, occupation, organizational structure, etc.

3. In support of this position, Ronen (1978) has recently shown that job satisfaction may vary more positively with one's job longevity, or with what he calls "job seniority," rather than with the length of one's organizational employment.

4. Based on the ideas from activation theory that individuals are responsive to "affective arousals" (Scott, 1966; Young, 1961), one might argue the inverse by claiming that people

seek uncertainty, novelty, or change. It is suggested here, however, that individuals feel comfortable with novelty, risk, or uncertainty only when the overall situation is well understood and interpretable. The person who enjoys betting on horses, for example, must develop a clear perspective of the rules of the game. Moreover, there is a strong tendency for the person, over time, to develop his or her own "betting system" for dealing with the various uncertainties.

5. As pointed out by Schein (1978), the mentor can fulfill a number of distinguishable and important functions for the new employee including behaving as a teacher or coach, role model, developer of talent, protector, sponsor, and as an opener of doors.

6. In his work on police socialization, for example, Van Maanen (1975) describes how police recruits, highly motivated and "gung-ho" at the start, are completely metamorphosed within a few short months by their veteran partners into identifying and accepting a complete opposite and contrary set of behavioral and attitudinal assumptions and expectations, mostly around a cultural ethos that emphasizes "keeping a low profile" and "not making waves."

7. Unfortunately, the attitudinal data reported in this experiment are more difficult to interpret because attitude survey and job design instruments were distributed and completed both before and after the experimental manipulations. Under such circumstances, Campbell & Stanley (1963) strongly suggest that one worry about questionnaire contamination, that is, the possibility that in answering the questions prior to the treatment, the subjects have essentially been cued as to how they are expected to respond to the posttreatment version.

8. Much of experimental social psychology, for example, is replete with laboratory experiments demonstrating how presumably voluntary subjects redefine and reorient the meaning of the situation or themselves when subjected to limited and often well-controlled kinds of information that are, or at least seem to be, contrary to the original set of beliefs or expectations held by the subjects. In their recent discourse on the limitations of individual dispositional explanations of behavior, Salancik & Pfeffer (1978) rely heavily on this social information processing perspective to assert that individuals, in general, use the various salient and relevant cues and information within the social environment to construct and interpret events. Based on the temporal perspective of this essay, it will also be argued later on that the way in which social information is processed and utilized by actual workers will vary according to their particular stage of job longevity!

9. In their careful search for correlates of scientific success, Pelz & Andrews (1966) also note that scientists are likely to be more successful if they establish more job security through solid feelings of technical expertise and psychological success by spending their initial years of employment on one main project.

10. Perhaps one of the best examples of what can happen when socialization is inhibited by preventing social relationships and interactions is Schein's (1956) fascinating comparisons of the structural differences between the German and Chinese prisoner-of-war camps.

11. If, on the other hand, the individual is able to maintain or even increase his or her own sense of task challenge and stimulation on a given job for a long period of time, then instead of moving toward adaptation, the process might be the reverse—continued growth and innovation; the scientist or engineer, for example, who never seems to tire of working at the "bench" or the academic professor who shifts his or her major areas of research every 5 or so years.

12. Two other recent longitudinal studies also checked for systematic changes in the broad construct of growth need strength but without success (i.e., Hall, Goodale, Rabinowitz, & Morgan, 1978; and Hackman, Pearce, & Wolfe, 1978). However, both studies were of rather short duration, less than 1 year. This may not have been sufficiently long to capture shifts in needs and values that gradually emerge from incompatible changes in circumstances; or perhaps, the likelihood of discovering shifts in such a broad and abstract construct may not be sufficiently realistic.

13. One notable exception is Hackman et al.'s (1978) recent longitudinal findings. The study, however, could only examine very short-term results, less than 6 months. The time lag, however, is most likely sufficient to negate the problem of possible questionnaire contamination between the pre- and post-administrations of the data collection instruments.

14. In fact, veterans who are not able to determine or ascertain how they are actually performing vis-à-vis their new job duties will often turn outside the organization for such feedback, seeking advice, help, and support from external consultants or purported experts. While one's ability to communicate effectively with external consultants is questionable in the first place, Allen (1977) found in a real-time study of matched R&D groups, working on parallel technical problems for the U.S. Government, that the use of external consultants, especially in the initial months of the project, was strongly and negatively related to overall group performance.

15. The present chapter has focused on the importance of only two kinds of breakpoints: socialization after entering the organization or after shifting jobs in the same organization. Clearly, other kinds of breakpoints also occur over time. For example, a change in one's supervisor, new replacements in the composition of one's work group, or one's temporary assignment to a new group or task force represent additional breakpoints that can create uncertainty for the individual. Furthermore, changes in one's technical environment, such as new automated equipment or the implementation of computers, can be a rather uprooting experience. Even the addition of a major client, customer, or product line can cause considerable uncertainty in one's working life. A constant complaint by U.S. corporations, for example, is that the technical monitors, assigned to help and watch over their corporation's government contracts, change too frequently often resulting in considerable instability and confusion—"by the time you get to know them, they've moved on."

16. In many corporations, such as in the automotive industry, the idea of periodic job changes for all employees seems a bit prohibitive in which case the organization must face the likelihood of meeting worker demands via increases in pay, vacations, job security, bonuses, and other benefits or through other contextual improvements.

17. In addition to these two studies, Smith (1970) has also published results to support the purported decline in group performance with extended group longevity. However, it is possible that the 49 projects analyzed by Smith are the same 49 projects comprising the industrial part of Wells's (1962) original sample. The main findings from Wells's study can be found in Chapter 13 of Pelz & Andrews (1966).

18. Whether group performance tends to decline with group longevity will of course depend on the kinds of trends that seem to materialize as a group functions for an extensive period of time as well as on the degree to which such trends are dysfunctional. In the case of R&D groups, it is likely that shifts toward more homogeneity and more NIH will tend to hinder the group's creativity and resultant technical performance (Katz & Allen, 1977). Generally speaking, performance may be perfectly acceptable with long-term job and group longevities *provided* the nature of the problems and the nature of potential solutions and technologies are both fairly stable.

19. Based on our previous discussions, it is likely that pressures toward conformity may also be strong during the adaptation stage for both individuals and groups with long-term job and group longevities, respectively.

20. Although employees are usually told in the group meetings that their participation is strictly voluntary, it is likely that very few would feel sufficiently comfortable to leave the room given what we know about situational control and group pressures on individual behavior. On the other hand, if employees are really given the opportunity to decide whether to participate or not, it is likely that participants will become significantly more committed to the "success" of the survey. Of course, if individuals are allowed to decide for themselves, one runs the risk of a lower response rate, or in the case of a group decision, the whole sample may be lost.

21. Monetary rewards may also be especially important during anticipatory and early socialization (Merton, 1957) in the sense that it can provide a comparative means of locating oneself within a new situational context.

## REFERENCES

Allen, T. J. (1977) *Managing the flow of technology*. Cambridge, Mass.: M.I.T. Press.

Argyris, C. (1957) *Personality and organization*. New York: Harper Torch Books.

———. (1975) "Dangers in applying results from experimental social psychology." *American Psychologist*, April, 469–485.

Asch, S. E. (1955) "Opinions and social pressure." *Scientific American* November, 31–34.

Bailyn, L. (1977) "Involvement and accommodation in technical careers: An inquiry into the relation to work at mid-career." In J. Van Maanen (ed.), *Organizational careers: Some new perspectives*. New York: Wiley.

Berlew, D. E., & Hall, D. T. (1966) "The socialization of managers: Effects of expectations on performance." *Administrative Science Quarterly*, *11*, 207–223.

Blake, R. R., & Mouton, J. S. (1969) *Building a dynamic corporation through grid organization development*. Reading, Mass.: Addison-Wesley.

Blumer, H. (1969) *Symbolic interactionism: Perspective and method*. Englewood Cliffs, N.J.: Prentice-Hall.

Bray, D. W., Campbell, R. J., & Grant, D. L. (1974) *Formative years in business: A long-term study of managerial lives*. New York: Wiley.

Brim, O. G. (1966) "Socialization through the life cycle." In O. G. Brim & S. Wheeler (eds.), *Socialization after childhood: Two essays*. New York: Wiley.

Brown, J. A. C. (1963) *Techniques of persuasion*. Baltimore: Penguin Books.

Buchanan, B. (1974) "Building organizational commitment: The socialization of managers in work organizations." *Administrative Science Quarterly*, *19*, 533–546.

Campbell, D. T. & Stanley, J. G. (1963) *Experimental and quasi-experimental designs for research*. Chicago: Rand McNally.

Campbell, J. P., Dunnette, M. D., Lawler, E. E., & Weick, K. E. (1970), *Managerial behavior, performance, and effectiveness*. New York: McGraw-Hill.

Comer, R. & Laird, J. D. (1975), "Choosing to suffer as a consequence of expecting to suffer: Why do people do it?" *Journal of Personality and Social Psychology*, *32*, 92–101.

Cooley, C. H. (1922) *Human nature and the social order*. New York: Scribner and Sons.

Davis, L. E. (1966) "The design of jobs." *Industrial Relations*, *6*, 21–45.

———. (1976) "Developments in job design." In P. B. Warr (Ed.), *Personal goals and work design*. London: Wiley.

Deci, E. L. (1975) *Intrinsic motivation*. New York: Plenum.

Denisi, A. S. & Pritchard, R. D. (1978) "Implicit theories of performance as artifacts in survey research: A replication and extension." *Organizational Behavior and Human Performance*, *21*, 358–366.

Dubin, S. S. (1972) *Professional obsolescence*. Lexington, Mass.: Lexington Books, D. C. Heath.

Durkheim, E. (1964) *Rules of the sociological method* (Translated by S. Solovay & J. Mueller), New York: Free Press.

Erikson, E. H. (1963) *Childhood and society*. New York: Norton.

Evan, W. M. (1963) "Peer group interaction and organizational socialization: A study of employee turnover." *American Sociological Review*, *28*, 436–440.

Farris, G. F. (1969) "Organizational factors and individual performance: A longitudinal study." *Journal of Applied Psychology*, *53*, 87–92.

Faucheux, C. & Mackenzie, K. D. (1966) "Task dependency of organizational centrality: Its behavioral consequences." *Journal of Experimental Social Psychology, 2,* 361–375.
Garfinkel, H. (1967) *Studies in ethnomethodology,* New York: Prentice-Hall.
Glaser, B. G., and Strauss, A. (1964) "Awareness contexts and social interaction." *American Sociological Review, 29,* 669–679.
Goffman, E. (1967) *Interaction ritual.* Chicago: Aldine.
Gomberg, W. (1973) "Job satisfaction: Sorting out the nonsense." *AFL-CIO American Federationist,* 14–19.
Graen, G. (1976) "Role-making processes within complex organizations." In M. D. Dunnette (ed.), *Handbook of industrial and organizational psychology.* Chicago: Rand McNally.
Grusky, O. (1966) "Career mobility and organizational commitment." *Administrative Science Quarterly, 10,* 488–503.
Hackman, J. R. (1978), "The design of work in the 1980s." *Organizational Dynamics,* Summer, 3–17.
———. & Lawler, E. E. (1971) "Employee reactions to job characteristics." *Journal of Applied Psychology Monograph, 55,* 259–286.
———, and Oldham, G. R. (1975) "Development of the job diagnostic survey." *Journal of Applied Psychology, 60,* 159–170.
———, Pearce, J. L., & Wolfe, J. C. (1978) "Effects of changes in job characteristics on work attitudes and behaviors: A naturally occurring quasi-experiment." *Organizational Behavior and Human Performances, 21,* 289–304.
Hall, D. T. (1976) *Careers in organizations.* Pacific Palisades, California: Goodyear.
———, Goodale, J. G., Rabinowitz, S., & Morgan, M. (1978), "Effects of topdown departmental and job change upon perceived employee behaviors and attitudes: A natural field experiment." *Journal of Applied Psychology, 63,* 62–72.
———, & Nougaim, K. E. (1968) "An examination of Maslow's need hierarchy in an organizational setting." *Organizational Behavior and Human Performance, 3,* 12–35.
——— & Schneider, B. (1973) *Organizational climates and careers.* New York: Seminar Press.
Herzberg, F. (1966) *Work and the nature of man.* Cleveland: World.
Hillery, J. M. & Wexley, K. N. (1974) "Participation effects in appraised interviews conducted in a training situation." *Journal of Applied Psychology, 59,* 168–171.
Homans, G. C. (1961) *Social behavior: Its elementary forms.* New York: Harcourt, Brace, and World.
House, R. J. & Wigdor, L. A. (1967) "Herzberg's dual-factor theory of job satisfaction and motivation: A review of the evidence and a criticism." *Personnel Psychology, 20,* 369–389.
———. (1971) "A path-goal theory of leader effectiveness." *Administrative Science Quarterly, 16,* 321–338.
Hughes, E. C. (1958) *Men and their work.* Glencoe, Ill.: Free Press.
Kahn, R. L., Wolfe, D. M., Quinn, R. P., Snoek, J. D., & Rosenthal, R. A. (1964) *Organizational stress: Studies in role conflict and ambiguity.* New York: Wiley.
Kanter, R. M. (1977) *Work and family in the United States.* New York: Russell Sage.
Katz, R. (1977) "The influence of group conflict on leadership effectiveness." *Organizational Behavior and Human Performance, 20,* 265–286.
———. (1978a) "Job longevity as a situational factor in job satisfaction." *Administrative Science Quarterly, 23,* 204–223.
———. (1978b) "The influence of job longevity on employee reactions to task characteristics." *Human Relations, 31,* 703–725.
——— & Allen, T. (1977) "The technical performance of long duration R&D project

groups." Technical report to the chief of studies management office, Department of the Army.

——— & Tushman, M. (1979) "Communication patterns, project performance, and task characteristics: An empirical evaluation and integration in an R&D setting." *Organizational Behavior and Human Performance, 23,* 139–162.

——— & Van Maanen, J. (1977) "The loci of work satisfaction: Job, interaction, and policy." *Human Relations, 30,* 469–486.

Kelly, H. H. (1971) *Attribution of social interaction.* Morristown, New Jersey: General Learning Press.

King, A. S. (1974) "Expectation effects in organizational change." *Administrative Science Quarterly, 19,* 221–230.

Kopelman, R. E. (1977) "Psychological stages of careers in engineering: An expectancy theory taxonomy." *Journal of Vocational Behavior, 10,* 270–286.

Kotter, J. (1973) "The psychological contract: Managing the joining-up process." *California Management Review, 15,* 91–99.

Latham, G. P. & Yukl, G. A. (1975) "A review of research on the application of goal setting in organizations." *Academy of Management Journal, 18,* 824–845.

Lawrence, P. R. & Lorsch, J. W. (1967) *Organization and environment.* Boston: Harvard Business School, Division of Research.

Levinson, D. J. (1978) *The seasons of a man's life.* New York: Alfred A. Knopf.

Maier, N. R. F. (1973) *Psychology in industrial organizations.* Boston: Houghton Mifflin.

Manning, P. K. (1977) "Rules, colleagues, and situationally justified actions." In R. Blankenship (ed.), *Colleagues in organizations: The social construction of professional work.* New York: Wiley.

Maslow, A. H. (1954) *Motivation and personality.* New York: Harper.

McClelland, D. C. (1961) *The achieving society.* Princeton: Van Nostrand.

McGregor, D. (1960) *The human side of enterprise.* New York: McGraw-Hill.

McHugh, P. (1968) *Defining the situation: The organization of meaning in social interaction.* Indianapolis: Bobbs-Merrill.

Mead, G. H. (1932) *The philosophy of the present.* Chicago: Open Court.

Merton, R. K. (1957) *Social theory and social structure.* New York: Free Press.

Messinger S. L., Sampson, H., & Towne, R. D. (1962) "Life as theatre: Some notes on the dramaturgic approach to social reality." *Sociometry, 25,* 98–110.

Meyer, H. H. (1966) "The communication of evaluations." *Proceedings of the Executive Study Conference.* Princeton: Educational Testing Service.

Milgram, S. (1963) "Behavioral Study of Obedience." *Journal of Abnormal and Social Psychology, 67,* 371–378.

Odiorne, G. S. (1965) *Management decision by objectives.* Englewood Cliffs, N.J.: Prentice-Hall.

Parsons, T. (1951) *The social system.* New York: Free Press.

Pelz, D. C. (1967) "Creative tensions in the research and development climate." *Science, 157,* 160–165.

——— & Andrews, F. M. (1966) *Scientists in organizations.* New York: Wiley.

Perrow, C. (1972) *Complex organizations: A critical essay.* New York: Scott Foresman.

Porter, L. W. & Steers, R. M. (1973) "Organizational, work, and personal factors in employee turnover and absenteeism." *Psychological Bulletin, 80,* 151–176.

Quinn, R. P., Staines, G. L., & McCullough, M. R. (1974) "Job satisfaction: Is there a trend?" Washington, D.C.: U.S. Government Printing Office, Document 2900-00195.

Rubenstein, J. (1973) *City police.* New York: Farrar, Strauss, and Giroux.

Ronen, S. (1978) "Job satisfaction and the neglected variable of job seniority." *Human Relations, 31,* 297–308.

Salancik, G. R. & Pfeffer, J. (1977) "An examination of need satisfaction models of job attitudes." *Administrative Science Quarterly, 22,* 427–456.

——— & Pfeffer, J. (1978) "A social information processing approach to job attitudes and task design." *Administrative Science Quarterly, 23,* 224–253.

Schacter, S. (1959) *The psychology of affiliation.* Palo Alto, California: Stanford University Press.

Schein, E. H. (1956) "The Chinese indoctrination program for prisoners of war." *Psychiatry, 19,* 149–172.

———. (1964) "How to break in the college graduate." *Harvard Business Review, 42,* 68–76.

———. (1968) "Organizational socialization and the profession of management." *Industrial Management Review, 9,* 1–15.

———. (1969) *Process consultation: Its role in organization development.* Reading, Mass.: Addison-Wesley.

———. (1971) "The individual, the organization, and the career: A conceptual scheme." *Journal of Applied Behavioral Science, 7,* 401–426.

———. (1973) "Personal change through interpersonal relationships." In W. G. Bennis, D. E. Berlew, E. H. Schein, & F. I. Steele (eds.), *Interpersonal dynamics: Essays and readings on human interaction.* Homewood, Illinois: Dorsey Press.

———. (1978) *Career dynamics: Matching individual and organizational needs.* Reading, Mass.: Addison-Wesley.

———. (1979) *Organizational psychology* (3rd edition). Englewood, Cliffs, N.J.: Prentice-Hall.

Schutz, A. (1967) *The phenomenology of the social world.* Evanston, Illinois: Northwestern University Press.

Scott, W. E. (1966), "Activation theory and task design." *Organizational Behavior and Human Performance, 1,* 3–30.

Shaw, M. E. (1971) *Group dynamics: The psychology of small group behavior.* New York: McGraw-Hill.

Shepard, H. A. (1956) "Creativity in R&D teams." *Research and Engineering,* October, 10–13.

Sims, H. P. & Szilagyi, A. D. (1976), "Job characteristic relationships: Individual and structural moderators." *Organizational Behavior and Human Performance, 17,* 211–230.

Smith, C. G. (1970), "Age of R and D groups: A reconsideration." *Human Relations, 23,* 81–96.

Smith, E. E. (1961), "The power of dissonance techniques to change attitudes." *Public Opinion Quarterly, 25,* 626–639.

Sofer, C. (1970), *Men in mid-career.* Cambridge, England: Cambridge University Press.

Staw, B. M. (1975), "Attribution of the "causes" of performance: A general alternative interpretation of cross-sectional research on organizations." *Organizational Behavior and Human Performance, 13,* 414–432.

———. (1977) "Motivation in organizations: Toward synthesis and redirection." In B. M. Staw & G. R. Salancik (eds.), *New directions in organizational behavior.* Chicago: St. Clair Press.

Stogdill, R. (1974 *Handbook of leadership: A survey of theory and research.* New York: Free Press.

Strauss, G. (1977) "Managerial practices." In J. R. Hackman & J. L. Suttle (eds.), *Improving life at work: Behavioral science approaches to organizational change.* Santa Monica, California: Goodyear.

Thompson, P. H. & Dalton, G. W. (1976) "Are R&D organizations obsolete?" *Harvard Business Review, 54,* 105–116.

Thompson, J. D. (1967) *Organizations in action.* New York: McGraw-Hill.

Umstot, D. D., Bell, C. H., & Mitchell, T. R. (1976) "Effects of job enrichment and task goals on satisfaction and productivity: Implications for job design." *Journal of Applied Psychology, 61,* 379–394.

Van Maanen, J. (1975) "Police socialization." *Administrative Science Quarterly, 20,* 207–228.

———. (1977) "Experiencing organizations: Notes on the meaning of careers and socialization." In J. Van Maanen (ed.), *Organizational careers: Some new perspectives.* London: Wiley.

——— & Katz, R. (1976), "Individuals and their careers: Some temporal considerations for work satisfaction." *Personnel Psychology, 29,* 601–616.

——— & Katz, R. (1979), "The cognitive organization of police perceptions of their work environment: An exploratory study into organization space and time." *Sociology of Work and Occupations, 6,* 31–58.

——— & Schein, E. H. (1977), "Career development." In J. R. Hackman & J. L. Suttle (eds.), *Improving life at work: Behavioral science approaches to organizational change.* Santa Monica, California: Goodyear.

——— & Schein, E. H. (1978), "Toward a theory of organizational socialization." In B. Staw (ed.), *Research in Organizational Behavior.* Greenwich, Connecticut: JAI Press.

Varela, J. A. (1971), *Psychological solutions to social problems.* New York: Academic Press.

Vicino, F. L. & Bass, B. M. (1978), "Lifespace variables and managerial success." *Journal of Applied Psychology, 63,* 81–88.

Walker, C. R. & Guest, R. H. (1952), *The man on the assembly line.* Cambridge: Harvard University Press.

Wanous, J. (1976), "Organizational entry: From naive expectations to realistic beliefs." *Journal of Applied Psychology, 61,* 22–29.

———. "Organizational entry: Newcomers moving from outside to inside." *Psychological Bulletin, 84,* 601–618.

Warr, P. (1976) "Theories of motivation." In P. Warr (Ed.), *Personal goals and work design.* London: Wiley.

Weick, K. E. (1969) *The social psychology of organizing.* Reading, Mass.: Addison-Wesley.

Wells, W. P. (1962) "Group age and scientific performance." Unpublished University of Michigan Dissertation.

Wheeler, S. (1966) "The structure of formally organized socialization settings." In O. G. Brim & S. Wheeler (eds.), *Socialization after childhood: Two essays.* New York: Wiley.

Young, P. T. (1961) *Motivation and emotion.* New York: Wiley.

Zimbardo, P. (1972) "On the ethics of intervention in human psychological research: With special reference to the Stanford prison experiment." *Cognition, 2,* 243–256.

——— & Ebbesen, E. B. (1969) *Influencing attitudes and changing behavior.* Reading, Mass.: Addison-Wesley.

# COLLECTIVE BARGAINING AND ORGANIZATIONAL BEHAVIOR RESEARCH*

Thomas A. Kochan

CORNELL UNIVERSITY

## ABSTRACT

The purpose of this paper is to lay a foundation for behavioral research on topics of interest within the field of collective bargaining. The first part of the paper reviews the differences between the normative premises, theoretical frameworks, and research methodologies commonly used by collective bargaining and organizational behavior researchers and briefly explores the reasons for the historical separation of research in these two fields.

The second part of the paper presents a theoretical framework for applying organizational behavior concepts, theories, and methods to collective bargaining topics. The issues discussed include the (1) behavioral dynamics of the negotiations process, (2) models of strikes and impasses, (3) alternative systems of conflict resolution in union-management settings, (4) union

**Research in Organizational Behavior, Volume 2, pages 129–176**

**ISBN: 0-89232-099-0**

impact on managerial decision-making and the personnel function, (5) organizational change in a union-management setting, and (6) unions as complex bargaining organizations. The benefits of more integrative research for both the fields of organizational behavior and collective bargaining are discussed.

## I. INTRODUCTION

Despite the centrality of trade unions and the collective bargaining process to organizational life, relatively little attention has been paid to this area by American organizational behavior researchers in the last two decades. Likewise, many traditional collective bargaining researchers tend to discount the relevance of organizational behavior research to collective bargaining issues and problems. This, however, was not always the case. At various points in the development of the study of collective bargaining and organizational behavior the interface between these areas received more attention than it had during the last decade or so. Fortunately, there are encouraging signs that the situation is changing once again. Thus, the interface between these two areas of research appears ripe for discussion and analysis.

The purpose of this paper is to lay a foundation for the orderly development of behavioral research on topics of interest within the field of collective bargaining. In doing so an effort will be made to identify research opportunities for organizational researchers interested in using their theories, concepts, and methodologies to conceptualize and explain collective bargaining phenomena. It will be argued that the collective bargaining area offers an ideal laboratory for the study of organizational behavior concepts such as conflict and conflict resolution, bargaining theory, power, organizational change, environment-structure relations, boundary spanning roles, and the evaluation of the effectiveness of social systems. Collective bargaining also offers one of the best opportunities for examining the interactions of economic, political, and interpersonal forces in decision making and, therefore, illustrates the need for an integration across the individual, organizational, and interorganizational perspectives that currently divide organizational behavioral researchers. Finally, those interested in conceptualizing organizations as political systems composed of shifting coalitions and interest groups can benefit immensely from the rich descriptive research that already exists on collective bargaining.

A major premise underlying the approach taken in this paper is that before organizational behavior researchers can significantly contribute to research on collective bargaining problems, they must first understand the substantive questions and problems that lie at the heart of this field. In addition, researchers must understand the normative values that underlie the study and practice of industrial relations and collective bargaining in the United States. Finally, they must overcome the obstacles that have historically kept organizational behavior researchers outside the mainstream of collective bargaining research. The first part of the paper, therefore, reviews differences in the normative premises, theoretical frameworks, and research methodologies commonly used by collective bargaining and organizational behavior researchers. Several reasons for the historical separation of research in these fields will be suggested. The second part of the paper presents a theoretical framework for the study of collective bargaining and illustrates where behavioral scientists can make important contributions.

Collective bargaining issues have historically been studied from one of three alternative perspectives: (1) legal; (2) institutional/historical; or (3) neoclassical economics. Behavioral scientists can complement and expand the potential of each of these approaches by building a framework and providing a methodology for addressing the critical questions in this field at the microlevels of analysis.[1] Specifically, behavioral science concepts and research techniques can extend the quantitative methodologies of neoclassical economics to more microlevels by measuring what economists tend to define as "unobservable" variables and can thereby help bring quantitative research more directly to bear on the central issues in the field. For example, the addition of behavioral variables to economic models of strikes or impasses (see Section VI, B below) has allowed researchers to use these models to explain variations across specific bargaining relationships whereas previously the economic models could only accurately explain variance in aggregate (economy wide) strike activity over time. Behavioral science can advance the institutional approach by organizing the vast array of factual and descriptive material into a manageable conceptual framework and a set of testable propositions. Finally, behavioral science can advance traditional legal analysis by helping to empirically test the validity of the behavioral assumptions and predictions made by legislatures, courts, and regulatory agencies on the behavior of individuals or organizations.

Very broad definition of the domains of collective bargaining and organizational behavior are used in this paper. The domain of collective bargaining research will be outlined later in Figure 1 and will be discussed in detail throughout the paper. The terms *organizational behavior research* or *behavioral science research* will be used interchangeably and are assumed to encompass the study of individual or group behavior within organizations, the behavior of total organizations, and/or relations between or among organizations.

## II. HISTORICAL DEVELOPMENT OF COLLECTIVE BARGAINING RESEARCH

Collective bargaining research owes its heritage to the development of institutional labor economics in the United States that emerged around the turn of the century. Commons (1924) described institutional economics as a shift away from the study of *individual* labor inputs as commodities similar to other factors of production, to the study of transactions between organized *groups* engaged in collective action. Institutional economics developed partly as a reaction to the overly deductive methodology of classical economics. In contrast to the classical economists' search for stable mathematical laws, institutionalists adopted the German tradition of historical, legal, and empirical analysis. This inductive approach emphasized description and explanation of the actual impact of market forces on groups and organizations in society.

In addition to the differences in theory and methodology, the institutionalists adopted more of a social reformist normative perspective than did the classical economists. Three important normative premises were central to the institutional tradition. These premises still underlie most collective bargaining research today. First, institutionalists rejected the view that labor was simply a commodity that should be subject to the same economic laws as other factors of production. Because of the centrality of labor to the life interests and welfare of workers, the American institutionalists, along with their British counterparts (Webb, 1897), were not only concerned with the worker as an instrumental factor of production, but also with workers' needs, goals, and welfare. Second, an inherent conflict of interests was assumed to exist between the needs of workers for job security and economic advancement, and employer and society needs for efficiency (Perlman, 1928; Barbash, 1964). Unlike Marx or the British Fabian Socialists (Webb, 1897), however, the American tradition stressed the *structural* rather than the class sources of this conflict of interests. They argued that (similar to Dahrendorf's, 1959, argument) conflict of interests is a structural characteristic of industrial society and is not limited to capitalist economic systems. The conflict was not organized around class lines as Marx argued but over the differences in the specific job-related goals of workers and employers. The difference in interests was assumed to be a permanent fixture of an employment relationship and would not be eliminated either by replacing the capitalist economic system or by destroying trade unions. Third, the institutionalists believed that workers and employers should both have the right to assert their interests individually or collectively and should seek to resolve their conflicts through bargaining and compromise without destroying either the capitalist economic system or the trade unions. In short, the institutionalists believed it was necessary to strike a balance among the needs and goals of labor, management, and the public. The early labor scholars first studied the history of labor organizations (Commons et al., 1951) and then argued for policies to protect rights of workers to join unions and bargain collectively. They also were early and vocal advocates for much of the protective labor legislation that became law in the 1930s.

The public policy, descriptive, and normative orientations of this research produced a close interchange between the researchers and practitioners of industrial relations. Therefore, researchers of this tradition have always been intensely interested in doing research that has short run relevance for public policy and/or practical problem solving. This orientation continues to dominate collective bargaining research. While this applied orientation provides a clear sense of relevance to the field, it may have inadvertently hindered the development of theoretical models of the industrial relations and collective bargaining system.

The concern for short-run relevance may have been part of the reason why most collective bargaining researchers ignored the advances in theory building and quantitative analysis techniques that spread throughout other areas of the social sciences in the 1950s and 1960s. Collective bargaining scholars also distrusted quantitative research for fear of missing the nuances, complexities, dynamics, and adversarial nature so important to describing and understanding developments in their field. Instead, these researchers have traditionally favored case studies and "thick description."

Research on labor problems and collective bargaining reached national prominance in the 1940s. The growth of the labor movement following the passage of the National Labor Relations Act in 1935 made collective bargaining and union-management relations one of the most visible and important social and economic developments of this time period. Furthermore, the high rate of strike activity that followed World War II encouraged more social scientists to examine labor problems. By the 1950s and 1960s, however, labor as a social problem declined in importance and the study of labor problems and industrial conflict lost status with both behavioral scientists and economists (Strauss, 1977). Consequently, social psychologists, sociologists, and psychologists all moved back toward issues more closely identified with their mother disciplines. The lack of theoretical focus that characterized most collective bargaining research further discouraged most scholars of the 1960s and 1970s from taking up these issues. Instead, they turned toward the more established research paradigms and issues within their disciplines.

## III. DEVELOPMENT OF ORGANIZATIONAL BEHAVIOR RESEARCH

A comprehensive history of the development of thought and research in organizational behavior need not be presented here (cf Blau, 1965; Massie, 1965). Instead, the normative, theoretical, and methodological perspectives of collective bargaining researchers will be contrasted with the perspectives of those who influenced the development of research in organizational behavior.

Shortly after the turn of the century American psychologists began to examine the behavior of workers in industry. Vitales (1932) described industrial psychology as an effort to help individuals adjust to their industrial situation and thereby advance the efficiency of industry. Consequently, the primary goal of these early efforts was to improve the efficiency of industry. Later, the development of scientific management represented an effort to blend economic incentives and industrial engineering techniques to produce the "one best way" for organizing work. By tying wages of workers to their output it was assumed that the interest of workers (for economic rewards) and the interest of employers (for productivity) could be made compatible. Management's function, therefore, was to design jobs and supervise and compensate the work force so as to eliminate any potential conflict of interests between workers and their employers. These early industrial psychologists and scientific management advocates, therefore, saw no need nor useful role for trade unions (Nadworny, 1955).

The emergence of industrial sociology in the 1930s and 1940s rejected the psychologists' focus on the isolated individual worker, yet retained the underlying normative assumption that the appropriate management of social relations at the workplace could make worker-employer interests compatible. The human relations movement represented a shift from the individual level of analysis and from the emphasis on structuring the appropriate economic incentives to achieve acceptance of organizational policies to an emphasis on group-related strategies for the same purpose (Kerr & Fisher, 1964). Unlike the scientific management advocates, the human relations scholars did see a potential role for trade unions. Kerr & Fisher (1964) summarized the posture of these human relations theorists: unions were believed to be useful when they cooperated with management's efforts and dysfunctional when they opposed management. Of course, as Kerr & Fisher note, unions would be unnecessary if their sole purpose is to cooperate with management.

While scientific management focused on the individual and human relations focused on the work group, another group of management theorists attempted to identify principles for structuring, directing, and controlling the overall organization. Fayol (1949), for example, developed fourteen "principles" of management. Like the scientific management and human relation advocates, these management theorists placed a high value on inducing cooperation toward the goals of the organization and stressed the supremacy of managerial objectives over worker interests. For example, one of Fayol's principles stated that individual interests must be subordinated to the general interest of the organization; another stated that conflict is an undesirable feature of organizations.

The antipathy of the forerunners of modern organizational behavior toward conflict within organizations, their overriding concern for the goals of the organization, and their belief in the fundamental compatibility of interests between workers and employers differentiated their approach from that of the early collective bargaining researchers. This outlook also meant that unions became skeptical of the motives of most early management researchers (Nadworny, 1955: pp. 48–67). This skepticism has carried over to the present. Some leaders of the American labor movement tend to equate most behavioral scientists with "union busters" (Payne, 1977). Not surprisingly, some behavioral scientists have had difficulty cultivating access in trade unions (Lewicki & Alderfer, 1973).

## IV. PRESSURES FOR INTEGRATIVE RESEARCH

The growing national importance of trade unions and collective bargaining in the 1940s and the visibility of the social and economic problems caused by the strike wave that followed the end of World War II motivated efforts to bridge the gap between the trade union-collective bargaining and the management researchers. The development of a number of industrial relations programs at major United States universities in the 1940s and the founding of the Industrial Relations Research Association in 1947 were both designed to foster integrative research by economists, management specialists, and behavioral scientists that focused on all aspects of labor problems and considered the needs and interests of labor, management, and the public (Derber, 1949). Although the infrastructure to support these integrative research efforts produced a number of efforts and an increase in the number of behavioral scientists studying union and collective bargaining problems (Kornhauser, Dubin, & Ross, 1954; Derber, 1948; Sayles & Strauss, 1953), the differences in methodological approaches and the failure to develop a coherent body of theory for linking organizational behavior and collective bargaining led behavioral scientists to move away from these problems in the 1960s.

Fortunately, a number of developments in the last decade suggest that the potential for building the bridge between these two fields is once again emerging. For example, Walton & McKersie (1965) used behavioral theories to conceptualize the labor negotiations process; Stagner & Rosen (1965) analyzed the psychological aspects of union-management relations; Fox (1971) presented a comprehensive theoretical integration of organizational theory and industrial relations; and, Child, Loveridge, & Warner (1973), Edelstein & Warner (1977), and Anderson (1977) have examined the organizational aspects of trade unions. In addition, the effects of unions on traditional dependent variables in organizational behavior such as organizational change (Kochan & Dyer, 1976) and job satisfaction and performance (Hammer, 1978) are beginning to be examined.

Equally important, organizational behavior researchers are becoming increasingly willing to view conflict (Schmidt & Kochan, 1972; Kochan, Huber, & Cummings, 1975), power, (Zald, 1970; Hinnings, Hickson, Pennings & Schneck, 1974), and organizational politics (Cyert & March, 1963; Bacharach, 1978) as concepts that are critical to an understanding of behavior within organizations. Thus, it appears that organizational theorists are now willing to accept the institutionalists' assumption regarding the inherent nature of and legitimacy of conflicts of interests within organizations. In several respects, therefore, the stage appears to be set for bridging the gap between these two fields of research.

One of the reasons the integrative effects of the immediate post World War II period were not sustained was the failure of a theoretical framework to emerge that was capable of organizing the disparate strands of research that evolved. Therefore, to avoid making our current integrative efforts another short-lived affair, a theoretical framework is needed for organizing research. The approach taken in this paper is to describe a framework for organizing the critical research questions within the field of collective bargaining and to then suggest where organizational behavior research has contributed or could contribute to an understanding of these issues. This approach was chosen in order to illustrate the range of opportunities that collective bargaining offers behavioral scientists for testing and enriching their theories and ideas. Since the application of behavioral science to collective bargaining problems is still in its infancy, the emphasis will focus more on future research needs and opportunities than on past contributions.

## V. A GENERAL MODEL OF COLLECTIVE BARGAINING

A general conceptual framework for the study of collective bargaining is presented in Figure 1. Most traditional textbooks in collective bargaining (Chamberlain & Kuhn, 1965; Sloane & Whitney, 1972; Davey, 1972; Beal,

*Figure 1.* Conceptual Framework for the Study of Collective Bargaining

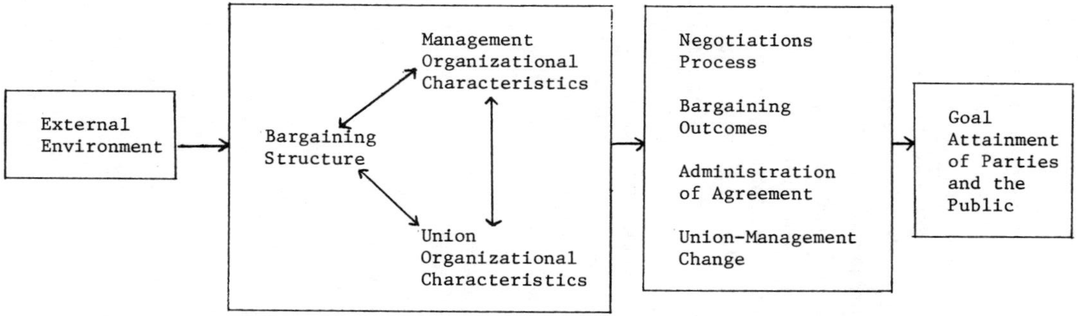

External Environment

Bargaining Structure

Management Organizational Characteristics

Union Organizational Characteristics

Negotiations Process

Bargaining Outcomes

Administration of Agreement

Union-Management Change

Goal Attainment of Parties and the Public

Wickersham & Kienast, 1976) focus on four major sets of issues, namely (1) characteristics of the negotiation process, (2) the substantive outcomes or terms included in collective bargaining agreements, (3) the administration of the bargaining relationship, and (4) the union-management change process. These issues, therefore, still represent the central sets of dependent variables in the study of collective bargaining. Since the early institutionalists and their followers *assumed* that collective bargaining is the most preferred mechanism for determining wages, hours, and conditions of employment, little interest was shown in examining the impacts of this process on the goals of the various parties affected by it.

The framework presented here goes one step farther than this traditional approach and suggests that collective bargaining researchers should examine the impact of the bargaining relationship on the goal attainment of individual workers, unions (as organizations), employers, and the public. Thus, the traditional "dependent variables" identified above should not be viewed solely as the ultimate criterion variables, but as the major means by which collective bargaining relationships seek to balance the needs and goals of the major actors in the system. In essence, the entire collective bargaining system can be viewed as an aggregate independent variable to be related to the goals of the actors. Because of the inherent conflict between the goals of the parties, no single overriding effectiveness criterion can be identified. Instead, a subsystems approach to conceptualizing and measuring the effectiveness of a collective bargaining system is required. The effectiveness of the overall system can then be evaluated by determining the extent to which it responds to the goals and needs of each of these various interest groups.

The major explanatory variables in the framework are grouped as characteristics of (1) the environmental context of bargaining, (2) the structure of the bargaining relationship, (3) the union as a bargaining organization, and (4) management as a bargaining organization. The environment of bargaining has been conceptualized in a variety of ways. Dunlop (1958) focused on economic, technological, and locus of power or the public policy contexts. Others have examined the political, social, and demographic contexts of bargaining relationships. In general, most efforts to explain or predict events in a collective bargaining relationship start by conceptualizing the environmental conditions that appear to be important. Thus, collective bargaining research has historically recognized the importance of an "open systems" approach.

The structure of bargaining is defined as the scope of the units (employers and employees) covered under or affected by the collective bargaining agreement. Thus "bargaining structure" as used here refers to the interorganizational structure that links the union and employer organizations. In contrast to the normal use of this term in organizational behavior (i.e., to describe the structure of a single organization) it is used in collective bargaining to describe the structure of the overall bargaining relationship. Bargaining structures can vary from highly decentralized (e.g., a contract covering only a single plant) to highly centralized (e.g., a contract covering all unionized employees of several employers within an industry). The structure of bargaining along with the union and management organizational characteristics, play important roles as intervening variables in most collective bargaining models. These aspects of the model are all affected by the external environment and in turn affect the process, outcomes, administration, and change processes in a bargaining relationship.

The relationships among the major concepts in this framework are not viewed as static or as subject only to a simple one-way causal flow. Instead, changes in one component of the bargaining system may set off changes in others and over time may produce feedback effects on the initiating part of the system. Thus the direction of the arrows depicted in the model should only be interpreted as representing what is proposed as the dominant direction of the causal chain within the system. The interdependence and dynamic nature of the model will be illustrated as the research issues central to different components of the model are discussed. The major tasks for researchers lies in identifying the links among these explanatory factors, the traditional dependent variables, and the impacts of the system on the goals of parties and the public.

In the sections to follow, research on the central questions involving the traditional dependent variables in the model will be reviewed. Major emphasis will be given to a number of critical questions pertaining to the negotiations process. The outcomes, administration, and change processes will then be discussed in a somewhat briefer fashion. The final component of the model that will be discussed directly will be research on unions as bargaining organizations. Although specific sections will not be devoted to research on the environment, structure, or management as a bargaining organization, the linkages between each of these components of the model and the traditional dependent variables will be noted wherever relevant.

## VI. RESEARCH ON THE NEGOTIATIONS PROCESS

The negotiations process is perhaps the central focus of research in collective bargaining. This is partly a function of the value that United States scholars and policy makers have placed on *free* collective bargaining (free from government intervention). In addition it reflects society's concern for the effects of work stoppages. The major questions that a theory of the negotiations process must address are: (1) How can the dynamics of the negotiations process or the bargaining behavior of the parties in negotiations be described, explained, and predicted? (2) What are the causes of strikes or impasses? (That is, what explains variations in the frequency, scope and impact of strikes and/or impasses across industries, union-employer relationships, and over time?) (3) What alternative intervention strategies are available for resolving impasses and under what conditions are they effective? Here the processes of mediation, factfinding, arbitration, and other more continuous bargaining or problem solving techniques have been examined. Although no single overriding or comprehensive theory addresses all of these questions, a number of models have drawn on behavioral research to deal with each of them separately. These will be briefly reviewed below.

*A. Dynamics of Negotiations*

Perhaps no other area of collective bargaining research has been more influenced by behavioral theories and concepts than the study of the dynamics of the negotiations process. Figure 2 outlines a standard set of concepts that have been built into theories of collective bargaining that use a "bargaining or contract zone" approach (Stevens, 1963; Walton & McKersie, 1965; Reynolds, 1970). These contract zone models make the following assumptions. First, it is assumed that each negotiator establishes a *resistance point,* "a level of achievement below which the negotiator would choose to sustain a strike or an impasse over the issues that are still unresolved" (Walton & McKersie, 1965, p. 41). The Walton & McKersie model also suggests that negotiators establish *target points,* a more optimistic estimate of the point at which the negotiator would prefer to settle; i.e., it reflects what the negotiator would like to achieve more than what the negotiator realistically expects to achieve. Second, negotiations are assumed to begin with a gap between the stated positions of the parties; i.e., the employer's initially stated offer is assumed to be less than the initially stated demand of the union. The bargaining process therefore involves movement or compromising from parties' initial or opening offers until the gap is closed and a settlement is reached. For strategic reasons, however, the parties are likely to hold back concessions that they are willing to make until they can determine how far the other party is willing to concede; i.e., what the other party's resistance point is, whether a settlement is possible within the limits set by the party's resistance point, or whether an impasse or a strike is imminent and therefore some concessions need to be saved for postimpasse negotiations and/or dispute resolution.

The Walton & McKersie (1965) and Stevens (1963) models draw heavily on behavioral decision theory by using the concepts of Subjective Expected Utility (SEU) to determine the conditions under which each party

*Figure 2.* Contract Zone Model of the Bargaining Process

$U_I$ — $U_T$ — $U_R$ — $M_R$ — $M_T$ — $M_I$

$U_I$ = Union Initial Offer

$U_R$ = Union Target Point

$U_R$ = Union Resistance Point

$M_I$ = Management Initial Offer

$M_T$ = Management Target Point

$M_R$ = Management Resistance Point

If $U_R > M_R$ a negative contract zone exists

If $M_R > U_R$ a positive contract zone exists

will make concessions in negotiations. Both models present a series of tactical propositions and introduce a range of behavioral constraints on the ability and willingness to make concessions in negotiations. Here the behavioral or institutional elements of the bargaining process that arise out of intraorganizational or constituency pressures or conflicts, attitudinal or interpersonal hostilities, personality characteristics, structural characteristics of bargaining, communication breakdowns, premature commitment, bluffing, etc., are introduced into these models. Thus, the careful blend of concepts from rational decision making models (SEU) and the more behavioral or institutional components of a bargaining relationship make these useful models for conceptualizing the dynamics of a bargaining process.

Although these models have been widely used to conceptualize the strategic aspects of negotiations, very little empirical research has attempted to apply or test them in actual collective bargaining settings. Several field studies have attempted to look at concession rates in the bargaining process (Hammermesh, 1973; Bowlby & Shriver, 1978) and at least two field studies (Wheeler, 1978; Kochan & Jick, 1978) have attempted to look at the impact of alternative impasse procedures on the incentive to make concessions in negotiations. In addition, Balke, Hammond & Meyer (1973) simulated an actual negotiations process using negotiators who were involved in an actual case and analyzed their bargaining behavior and use of information. None of these, however, have made much progress in developing the independent variable side of a model of bargaining behavior.

On the other hand, an enormous amount of laboratory research has been conducted on the bargaining process by a wide variety of behavioral scientists (cf. Rubin & Brown, 1975; Hamner, 1978; Magenau & Pruitt, 1978). Most of this research has focused on identifying the structural, strategic, constituency, and personality variables associated with the behaviors of negotiators and the settlement process. While this research is useful for suggesting a set of general concepts that can be hypothesized to affect the behavior of negotiators in collective bargaining, and while some of the laboratory experiments have used collective bargaining games, there has been little transfer of the laboratory findings to field research settings. Consequently, the next step in integrating these findings into collective bargaining research would be to use the empirical results of laboratory studies to develop propositions for predicting movement in bargaining and other bargaining behavior that can be empirically tested outside the laboratory. The development of an empirically based theory using the bargaining zone framework would also provide the necessary theoretical background for assessing the impacts of alternative impasse resolution techniques that will be discussed in a later section.

*B. Models of Impasses*

A good deal of theoretical and empirical research has been done by collective bargaining researchers on the determinants of strikes or impasses. Instead of working directly from the bargaining process models, however, these studies have been moving toward a systems framework or what Thomas (1976) refers to as a structural determinants approach. Only recently, has progress been made in integrating some of the ideas from the bargaining process models into models of the determinants of strikes and impasses.

The majority of research on the determinants of strikes has been carried out at a very aggregate level of analysis, even though the theories have been based on hypotheses concerning individual and/or union organizational behavior; Ashenfelter & Johnson (1969), and Skeels (1971) used a set of economic variables to examine the aggregate rate of strike activity in the United States economy. These economic models have shown that at the aggregate level, strike frequency is directly related to declining real wages, lower levels of unemployment, and the passage of the Landrum-Griffin Law. Another stream of research approaches the analysis of strikes from a sociological perspective (Britt & Galle, 1972, 1974; Shorter & Tilly, 1974; Snyder, 1975; Stern, 1976). The sociological studies have shown that strikes are related to the degree of unionization, the level of industrialization, the nature of the industry, the structure of collective bargaining, and the size of the community. More recently, Roomkin (1976) extended both the economic and sociological approaches by incorporating hypotheses concerning the effects of union structure.

Neither the economic nor the sociological variables can accurately predict strike rates (Burton & Krider, 1975) or impasses (Huettner & Watkins, 1974); Anderson & Kochan, 1977; Kochan & Baderschneider, 1978) at more disaggregated levels of analysis. Thus, in order to predict the probability of an impasse or a strike occurring in a given bargaining relationship, one needs to supplement these economic and other environmental variables with other characteristics of the bargaining relationship (Stern, 1978). Here is where behavioral science research on bargaining becomes most useful. A recent study of impasses in the public sector has attempted to build this more comprehensive type of model by integrating characteristics of the structure of bargaining, the organizational characteristics of unions and employers, and the personal and interpersonal aspects of the bargaining process (Kochan & Baderschneider, 1978). It was found that the probability of impasse in a bargaining relationship could be accurately predicted when the traditional economic variables were supplemented with organizational, interpersonal, and personal characteristics of the bargaining relationship. For example, the probability of impasse was found to be positively related to the interpersonal hostility between the parties, the degree of constituency pressure on the union leaders, the use of union pressure tactics, and the use of professional negotiators; and negatively related to the degree of power delegated to the management negotiator (*boundary spanner*).

The importance of these variables further illustrate both the need for better conceptualization and measurement of these behavioral characteristics, and the fertile ground that this area offers for behavioral researchers interested in studying variables of interest to them in the context of collective bargaining. Research on strikes and impasses also illustrates the need to cross multiple disciplines and levels of analysis before a comprehensive model of the phenomena can be developed. It would be just as dysfunctional for behavioral researchers to attempt to study the determinants of strikes and impasses by ignoring the effects of the environmental context as it has been for economists and sociologists to ignore the organizational and interpersonal dimensions of bargaining relationships. In fact, many of the experimental social-psychological studies suffer from exactly this problem.

*C. Conflict Resolution Research*

Although the study of procedural alternatives for resolving collective bargaining disputes has always been of interest to collective bargaining researchers (Cullen, 1968; Northrup, 1966), the growth of collective bargaining in the public sector has added a host of new opportunities to examine the effects of a wide range of different types of procedures. Consequently, the comparative analysis of the effectiveness of these procedures is now at the forefront of collective bargaining research (Stern, Rehmus, Loewenberg, Kasper & Dennis, 1975; Wheeler, 1978; Kochan & Jick, 1978).

The three most popular types of procedures built into public sector laws are mediation, factfinding, and arbitration. *Mediation*—the most informal of the three alternatives—is a procedure in which a neutral third party assists the union and management negotiators in reaching voluntary agreement. A mediator has no power to impose a settlement but rather acts as a facilitator for the parties. *Factfinding* is a somewhat more formal process in which the neutral holds a hearing and issues a set of written recommendations for resolving the dispute. Again, however, the neutrals have no power to impose their recommendations on the parties. The parties are free to accept, reject, or use the recommendations as a basis for further negotiations. *Arbitration* is analogous to factfinding except that the neutral's award is binding on the parties. There are a number of alternative types of arbitration. *Voluntary arbitration* is a procedure in which the parties voluntarily choose to be bound by the neutral's decision. *Compulsory arbitration* is a procedure in which a state or federal statute requires that a dispute be resolved by arbitration. Recently, a procedure known as *final offer arbitration* (Stevens, 1966) has been used in several states for resolving public disputes. Under final offer arbitration the arbitrator is constrained to choose either the employer's or the union's final offers. In contrast to conventional arbitration, the arbitrator may not formulate a compromise solution of his or her own choosing.

There is a great need for better theories of each of these dispute resolution procedures. In addition, there is a tremendous demand for policy-relevant research regarding the relative effectiveness of these alternatives. More than two-thirds of the states now provide some combination of these procedures for resolving disputes between public employees and local and state governments. The national emergency disputes procedures embedded in our two major private sector laws, the Taft-Hartley Act and the Railway Labor Act (Cullen, 1968), also rely on various combinations of these procedures. It is important to improve our understanding of the advantages and limitations of each of these procedures; thus opportunities for naturally occurring field experiments exist for those interested in doing evaluation research.

Research on these conflict resolution techniques (Thomas, 1976; Robbins, 1974; Filley, 1976; Walton, 1969) should be of particular interest to the growing number of organizational behavior researchers interested in conflict resolution or conflict management. Mediation, for example, is similar to the role that organizational development interventionists often play. Factfinding and arbitration are quasi-judicial procedures for resolving protracted conflicts and therefore offer a number of alternatives for those interested in designing more formal conflict resolution systems (Thibaut & Walker, 1975).

*Mediation.* The mediation process is an extremely fruitful ground for integrating collective bargaining and organizational behavior research. The works of Kerr (1954), Douglas (1962), Landsberger (1955), Stevens (1963; 1967), Rehmus (1965), Simkin (1971), Kressel (1972), and Robbins & Dennenberg (1976) provide a number of implicit and explicit hypotheses about the nature of the labor mediation process and a description of how mediation works in the context of collective bargaining. The studies by Walton (1969) and research summaries by Rubin & Brown (1975), Deutsch (1973), Thibault & Walker (1975), and Hamner (1978) identify a number of general propositions about mediation that can be derived from the behavioral literature. Taken together, these works suggest at least the following basic propositions about mediation:

(1) Mediators must be acceptable to the parties and perceived to be trustworthy. This proposition is noted in every discussion of mediation by collective bargaining researchers and is verified by the results of laboratory studies (Rubin and Brown, 1975).

(2) Mediation is at least partially an "art" that must be learned through experience, therefore, experienced mediators should be more effective than inexperienced. Again this point is stressed repeatedly in the collective bargaining literature.

(3) Mediation works best when the parties are motivated or under strong pressure to resolve their dispute; it is likely to be less successful if the parties are not "ready" to settle. Thus the timing of mediation efforts in the larger cycle of negotiations is crucial.

(4) The strategies mediators employ vary over the course of the intervention process and over the different stages of the negotiation cycle. In earlier stages of negotiations, the mediator is likely to adopt a more passive (listening, gaining trust, facilitating negotiations, etc.) strategy while as the negotiations move closer toward the settlement the mediator may adopt a more active, assertive, and aggressive strategy aimed at changing the expectations of the parties and suggesting substantive compromises for a settlement (Kressel, 1972).

(5) Mediator strategies vary widely across mediators (Simkin, 1971).

(6) Mediators adopt different strategies contingent upon the "type" or sources of conflict that are impeding a settlement (Stevens, 1963).

(7) Mediation is more likely to be successful in resolving some types of conflict than others. For example, Thibaut and Walker (1975) found mediation was more successful in resolving minor, less intense conflicts than major disputes. Stevens (1963) argued mediation should work best in "positive contract" type disputes, i.e., where the parties' resistance points (See Figure 2) overlap but because of some breakdown in the bargaining process, the parties are holding back concessions that they are willing to make.

These and several other propositions were recently organized into a model of the mediation process and tested using data from 130 public sector labor mediation cases (Kochan & Jick, 1978); this model is illustrated in Figure 3. The structure of the model and the classification of variables included in it was derived from the literature summarized above and is similar to McGrath's (1966) model of the negotiation process. The effectiveness of the mediation process serves as the dependent variable in the model. Four operational measures of effectiveness are considered: (1) probability of settlement in mediation; (2) the percentage of issues resolved through the mediation effort; (3) the amount of movement or concessions made; and (4) the extent to which the parties hold back concessions. The explanatory variables included in the model were classified as sources of impasse, mediator strategies, mediator characteristics, and situational variables. The test of this model confirmed several of the propositions summarized above. For example, it was found that mediation effectiveness was negatively related to the intensity of the dispute and that when an inability to pay existed it was an important source of impasse. Mediation effectiveness was positively related to the aggressiveness, perceived quality, and experience of the mediator, the motivation of the parties to settle, and where the sources of impasse reflected a breakdown in the process of negotiations. After controlling for these factors, it was also found that mediation was marginally more effective under a compulsory arbitration statute than under a factfinding statute.

The empirical findings of this study have only begun to scratch the surface of our understanding of the dynamics of this conflict resolution technique. The careful application of experimental designs in the laboratory along with additional field studies that capture the more dynamic aspects of the mediation process represent the next steps in the development of our understanding of this process. For example, laboratory studies would be the best way to test more hypotheses about the impact of alternative sources or types of impasses on the settlement of disputes

*Figure 3.* Model of the Mediation Process

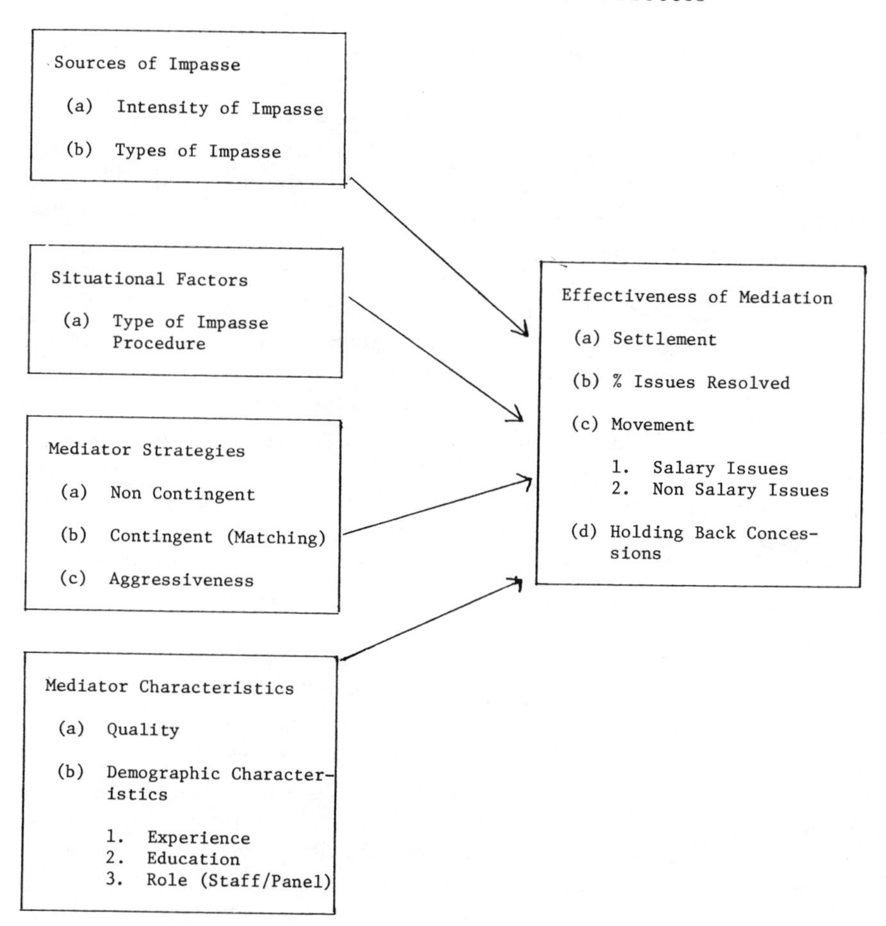


through mediation. Different types or sources of impasses (e.g., economic, structural, intraorganizational, interpersonal, etc.) could be manipulated in the laboratory relatively easily. The effects of mediation in positive and negative contract zones could also be tested.

Laboratory studies provide an opportunity to test the effectiveness of mediation when followed by a number of alternative types of formal impasse procedures (Johnson & Pruitt, 1972; Bigoness, 1976) or when followed by the right to strike. This is one of the central public policy questions surrounding the mediation process. Policymakers would like to know whether mediation would be more or less successful if followed in the sequence of dispute resolution techniques by factfinding, alternative forms of arbitration, or by the right to strike. Laboratory experiments may be the most efficient way to provide the necessary control over other causal variables to identify the net differences in the effectiveness of mediation that can be expected under these different situational conditions.

Case studies, on the other hand, would be useful for understanding the dynamic aspects of the mediation process that cannot be captured through cross-sectional comparative research nor through controlled laboratory studies. Better description and conceptualization of variations of mediation strategies and their effects should be particularly appropriate for case study research. Although participant observation is an expensive and time-consuming methodology, it offers the greatest potential for understanding how the mediation process changes from the early stages of intervention to the initialing of the settlement. Clearly, it is again time to (cf. Douglas, 1962; Landsberger, 1955) apply this methodology in order to improve our understanding of these dynamic components of mediation.

*Factfinding and Arbitration.* Factfinding and arbitration are somewhat analogous procedures except that a factfinder's recommendations are not binding on the parties. Thus, the theoretical and policy issues of interest regarding these conflict resolution procedures are somewhat similar. These issues include: (1) the impact of the presence of an impasse procedure on the parties' incentives to bargain and reach agreement without use of the procedure; (2) the effectiveness of a procedure as a strike deterrent, and (3) effects of various structural characteristics of the procedure on the decision making process and outcomes.

Concern for the impact of factfinding or arbitration on the incentive to settle arises out of the value that collective bargaining practitioners, researchers, and policymakers have historically placed on the ability of unions and employers to resolve their differences without third-party intervention. Consequently, this question has often been addressed within the context of a larger model of the determinants of impasses in collective bargaining. That is, the nature of the impasse procedure has been treated as one independent variable where the dependent variable is the probability of settling without reliance on the procedure.

The theoretical proposition that is most commonly tested in these studies is that the greater the expected economic or political risks, or costs associated with the procedures, the stronger the parties' incentive to settle without using the procedure. The argument is that procedures do not pose as severe a "cost of disagreement" (Chamberlain & Kuhn, 1965) on the parties as does the strike and, therefore, the parties are likely to use them more frequently than they would engage in a strike. Thus, the more procedures appear to be "strikelike" (impose high risks and costs on the parties) the more the parties should be disinclined to use them. It has also been hypothesized that over time clear patterns of repeated usage of the procedures are likely to develop in the absence of strong disincentives. This "narcotic hypothesis" suggests that use of the procedure leads negotiators to adapt their expectations in future rounds of bargaining in ways that further "chills" the bargaining process. Over time the expectation that little progress will be made without outside intervention becomes a self-fulfilling prophecy and a pattern of repeated usage develops.

The most intriguing theoretical and political development in this area that has occurred in the last decade is the experimentation with final offer arbitration. Stevens (1966) introduced the theory of final offer arbitration. He presented it as a system designed to reduce the chilling effect of conventional arbitration by raising the risks or potential losses associated with invoking the procedure. Under a final offer system the arbitrator must choose either the final offer of the union or the employer—the arbitrator may not split the difference or fashion any other compromise award. This system both raises the risks of going to arbitration (the other party's position may be imposed rather than some compromise solution), and reduces the disincentive to hold back concessions by putting a premium on the "reasonableness" of each party's final position. Thus, if final offer systems work according to the theory, we should observe (1) less of a chilling effect (more movement in bargaining by both parties), (2) a higher settlement rate short of arbitration, and (3) fewer unresolved issues going to arbitration.

The empirical evidence from both laboratory (Notz & Starke, 1978) and field research (Kochan, Mironi, Ehrenberg, Baderschneider, & Jick, 1979; Feuille, 1975) generally supports the proposition that a higher settlement rate prior to invoking arbitration occurs under final offer arbitration than under conventional arbitration or factfinding. As yet, however, few controls for confounding environmental or other causes of dependence on impasse procedures have been built into these studies. Thus, the future research that examines the performance of these alternatives while controlling for other causes of impasses and use of procedures is needed. Surprisingly, no research has yet compared final offer arbitration with the right to strike. Since these two alternatives exist in one jurisdiction (the State of Wisconsin) an opportunity is available to make this important comparison.

While the short-run performance of final offer arbitration has been favorable, there are some who fear that this procedure may have an extremely short half-life. Since it is easier to identify winners and losers under this system, and since the neutral arbitrator is constrained from fashioning an award that balances the political and organizational needs of each party, it may be that final offer arbitration will be less innovative and acceptable to the parties over the longer run than its more flexible alternatives. A combination of longitudinal field studies and creative laboratory designs that simulate repeated trials of alternative arbitration and/or factfinding systems and that broaden the subjects to include actual practitioners could shed meaningful light on these questions.

An equally important question for evaluating the effectiveness of factfinding and arbitration procedures is their effectiveness as strike deterrents. An important reason for establishing these procedures is that the public (through its elected officials) and/or the concerned parties (through their own agreement) have decided that the strike is not a preferred option. Thus, in addition to examining the impact of these procedures on the incentive to reach an agreement, their performance as strike deterrents must be compared. This requires cross-sectional research designs that compare strike rates under different procedures or longitudinal designs that track strike activity before and after a policy change. Both designs require controls for other than policy causes of strikes. Where longitudinal data are available for periods before and after a policy change, each bargaining unit could serve as its own control. Although the lack of adequate national data on bargaining and strike activity under different dispute resolution systems has hindered the use of sophisticated research designs, there is enough evidence from a variety of different studies to at least tentatively conclude that arbitration has been a more effective strike deterrent in the public sector than has factfinding (Kochan, 1979). Studies that control for other nonpolicy related causes of strikes, that draw on larger samples, and that track the performance of dispute resolution systems in the same bargaining relationships over a longer period of time are needed before this tentative conclusion can be stated with any degree of confidence.

The effectiveness of dispute resolution procedures in deterring strikes is perhaps the one issue of greatest importance to public policymakers and the public. While industrial relations professionals may value dispute resolution systems for their ability to avoid overuse and maximize acceptability, the larger society is primarily interested in whether they can avoid costly strikes. Thus it is critical that researchers improve the state of research in this area so the public's concern for avoiding disruptive strikes is addressed.

*D. Structural Design of Dispute Resolution Systems*

Since not all disputes will be resolved without using factfinding or arbitration procedures, it is important to examine the operational features of alternative dispute resolution systems. As in other conflict resolution research there is an implicit theme running through the factfinding and arbitration literature that suggests a set of systematic relationships between the structural design of the system and the decision making process and outcomes occurring in the system. A framework that relates the structural characteristics of alternative systems to their expected impacts on the decision-making process and outcomes is beginning to emerge from the results of empirical studies of these systems (Kuhn, 1952; France & Lester, 1951; Bowers, 1973; Stern, et al., 1975; Kochan et al., 1979; Grahem, 1978). These findings will be summarized below in a way that will allow them to be subjected to more refined empirical tests. The term "arbitration" will be used throughout this discussion although the same system options could be applied to factfinding procedures. The only difference would be that the decisions of the factfinders would be advisory while the decisions of arbitrators would be binding.

The major structural, process and outcome characteristics included in the framework are shown in Figure 4. The major structural options that vary across arbitration systems include: (1) final offer versus conventional arbitration procedure; (2) tripartite panel versus neutral-only arbitrators; (3) permanent versus ad hoc arbitrators, and (4) arbitrators selected by the parties or appointed by some public body.

Two polar types of decision-making processes have been debated as being appropriate for arbitration. The first approach sees the decision-making process within arbitration as an extension of collective bargaining and mediation. Emphasis is placed on achieving a decision that is acceptable to the parties and sensitive to their economic and political needs and reflects the balance of power between the parties. This has been popularly labeled the "med-arb" approach. The countervailing approach sees arbitration as a quasi-judicial procedure in which the arbitrator searches for a "rational" decision and applies specific standards or decision criteria in predictable and systematic ways.

The outcomes of arbitration processes are most frequently judged by: (1) their acceptability to the parties; (2) the satisfaction of the parties with the outcome; (3) the ability of the parties to innovate (break new ground through arbitration); and (4) the extent to which the procedures are not consistently biased in favor of one of the disputants.

The structural characteristics of the procedure affect the process and outcomes in at least two important ways. First, the structure determines how much control the disputants have over the discretion of the arbitrator. Thibaut & Walker (1975) see the amount of control by the disputants as the key to procedural justice in any conflict-resolution system. In collective bargaining this is often felt to be important so that the arbitrator is constrained from imposing a decision that is unworkable or unresponsive to the economic and political problems of the parties. Second, the structural variations are expected to influence the extent to which the decision-making process takes on the med-arb characteristics versus the judicial approach to decision-making. In turn, the degree to which the

*Figure 4.* The Relationship Between the Structure, Process, and Outcome Components of an Arbitration System

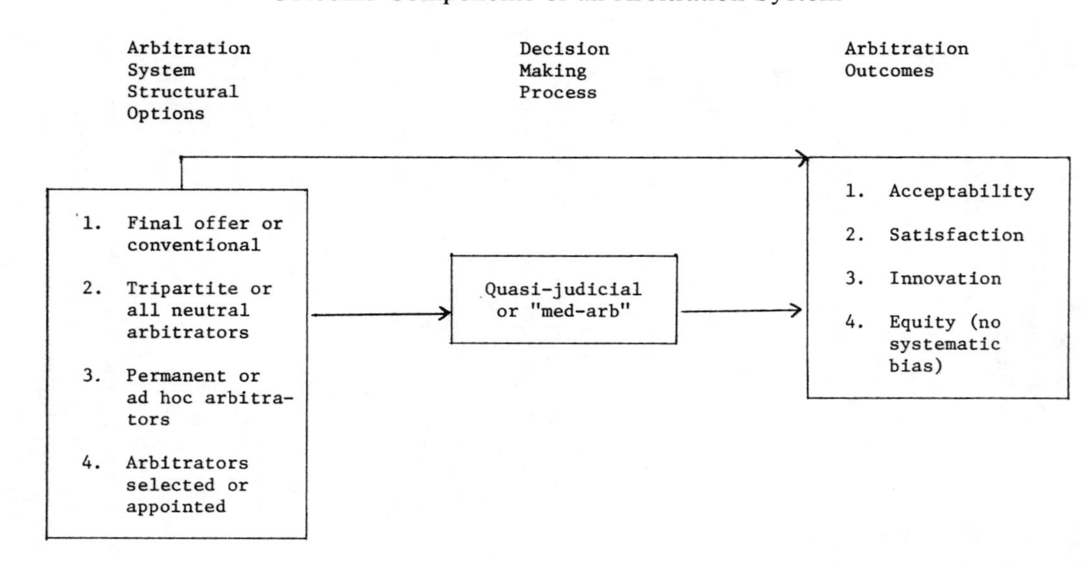

parties control the process and the extent to which the process reflects the med-arb approach is expected to be directly related to the acceptability of the procedure to the parties and inversely related to the ability of the parties to achieve major innovations in arbitration.

The major hypotheses that can be derived from the empirical studies of arbitration systems can be summarized as follows:

(1) Final offer systems provide greater control to the disputants than conventional arbitration systems since the arbitrator must choose between the offers of the parties rather than fashion an award of his or her own choosing (Thibaut & Walker, 1975).

(2) Conventional arbitration offers greater opportunities to employ the med-arb decision-making process and is therefore more likely to produce results that are more acceptable to the parties.

(3) Tripartite arbitration provides more control to the disputants and increases the use of a med-arb process and therefore produces results that are more acceptable to the parties and reflect the balance of power between them (Kochan, Mironi, Ehrenberg, Baderschneider & Jick, 1978).

(4) Selection of the arbitrators by the parties provides greater control to the disputants (Simkin, 1952). Selection by the parties also increases the acceptability of the outcomes to the parties since the arbitrator depends on continued acceptability to get chosen by the parties in the future (Horton, 1975). The arbitrator is less likely to impose an outcome that does not reflect the needs of the parties. Furthermore, selection by the parties should keep the outcomes of a procedure from being systematically biased in favor of one of the disputants, since an arbitrator who is perceived as biased will not be chosen by that party again in the future.

(5) Permanent arbitration panels are less likely to engage in the med-arb type of decision-making process and provide less control to the parties than do systems that use ad hoc arbitrators (Graham, 1978). Permanent arbitration panels are also likely to be less concerned with the acceptability of their awards (Graham, 1978). The evidence is ambiguous concerning whether permanent panels are likely to be more innovative than ad hoc arbitrators. On the one hand, the security of the arbitrators and the disputants' lack of control give the permanent panels more freedom to be innovative (Graham, 1978). On the other hand, permanent panels are also likely to depend heavily on precedent and the development of systematic and consistent standards for decision making (Barnes & Kelly, 1975) and look to comparable settlement in framing awards (Anderson, 1977).

As the above summary suggests, more evidence exists on the impacts of alternative structures on decision-making process than is known about their impacts on the outcomes. Even less is known about the long-run stability of these systems. Since it has been hypothesized that these arbitration systems may suffer from a natural half-life effect (Anderson & Kochan, 1977), it is important that future research track the behavior and survivability of these systems over time.

A wide variety of possibilities exist for behavioral science research on these conflict resolution systems. Many of these structure-process-outcome propositions could be effectively studied in the laboratory. The results of laboratory studies that can more carefully isolate cause-effect relations would provide useful supplements to the case study and survey findings, and general theoretical analysis that policymakers have relied on in the past to choose among these alternative systems. Furthermore, the natural variations that exist across states that provide arbitration for public employees provides a multitude of field sites for comparative analysis and evaluation. Mironi (1977) has developed and tested an instrument for measuring the satisfaction of disputants with an arbitration system. This instrument could be utilized to assess the effects of alternative systems on the process and outcomes of interest. Thus, the theoretical framework, empirical sites, and measurement tools are available to those interested in addressing these topics.

Before leaving this section, it should be noted that a vast majority of the research on negotiations, strikes and impasses, and conflict resolution procedures has treated these issues as dependent variables. Remarkably little attention has been paid to the effects of these aspects of the bargaining system on the other components of the model outlined in Figure 1. Clearly, however, a number of interesting propositions concerning the effects of the negotiations and/or conflict resolution experiences of interest to organizational behavior and to collective bargaining researchers could be generated. For example, what is the effect of a consistent reliance on strikes or on impasse procedures on the satisfaction of employees with their jobs and with their union? Does the absence of the right to strike and the reliance on dispute resolution systems lead to greater bureaucratization (e.g., formalization, centralization, and professionalization) within union and management organizations? If so, what effect do these changes in union and management structure have on individual member militancy, and participation in and commitment to the union? Over time, what effects will these conflict resolution systems have on the effectiveness of the union as a bargaining organization? Does reliance on outsiders to settle impasses lead to a reduction in the problem solving potential during the administration of the bargaining agreement and greater difficulty in adapting to change?

Finally, it is clearly time to evaluate more systematically the extent to which the negotiations process is responding to the central needs, interests, and goals of union members (Lawler & Levin, 1968; Kochan, Lipsky & Dyer, 1974). Behavioral scientists should not be satisfied with the traditional institutionalist presumption that the bargaining process is the optimal mechanism for dealing with all worker interests. Instead, research that identifies the conditions under which bargaining is most and least responsive to worker concerns deserves high priority by organizational behavior researchers.

## VII. BARGAINING OUTCOMES, AGREEMENT ADMINISTRATION, AND UNION/MANAGEMENT CHANGE

The process of negotiations and dispute resolution has received more attention from collective bargaining researchers than any other subject. Equally important, however, are the outcomes of these processes, the administration of the agreement, and the adaptability of the bargaining relationship. These three aspects of the model exert a strong impact on the responsiveness of a collective bargaining system to the goals and needs of the parties. Most traditional collective bargaining researchers, however, have been content to simply study the determinants of wages and other bargaining outcomes under collective bargaining. Researchers have only recently begun to conceptualize the impacts of bargaining outcomes on the goals of the parties. Attempts to identify impacts of the outcomes, administration, and change process on the parties will be analyzed in this section.

### *A. Impact of Collective Bargaining on Management*

There are two competing sets of propositions regarding the impact of negotiated increases in wages and other contract provisions that affect labor costs or reduce the discretion of management. Standard neoclassical economic theory suggests that increases in labor costs under conditions of equilibrium will result in a reduction in the scale of output and employment, an increase in the price of the product, and/or a substitution of capital for labor. All of these are mechanisms for bringing the system back into equilibrium. A countervailing proposition argues that collective bargaining has a "shock-effect" on management (Slichter, 1941) i.e., the presence of a union and/or of a union-negotiated wage increase induces management to search for more efficient means of managing the firm. Thus, the effects of negotiated increases in wages and other contract terms on productivity depends on the response of management to the increase (Freeman, 1977).

Clearly both forces outlined in the competing propositions are at work in any organization. Although these competing propositions have been discussed by economists and the effects of collective bargaining on productivity and other organizational outputs is currently being studied empirically (Freeman, 1977), behavioral scientists have yet to systematically address these issues. While economists can examine these issues at an aggregate level and provide estimates of the differences in organizational outputs in union and nonunion settings, behavioral research techniques will be more useful in identifying how management adjusts to the presence of a union and reorganizes itself to bargain effectively. Consequently, one fruitful area of research would be to address either the impact of a union or the impact of negotiated settlements on the economic performance, organizational structure, and decision making of employers. To do this, one needs to examine how the employer adapts to union pressure in general and to new contract provisions in particular.

*Impact on Management Structure and Process.* Slichter et al. (1960) argued that collective bargaining leads to greater: (1) formalization of managerial policies, (2) specialization of management decision making in labor relations, (3) redistribution of power within the management structure, and (4) need for internal coordination of decision making. In a more general sense, the presence of a union can be conceptualized as a specific form of environmental uncertainty or pressure that induces differentiation within the organizational structure (Lawrence & Lorsch, 1967) and encourages the establishment of specialized boundary spanning units (Thompson, 1967). Two empirical studies have shown that the power of these units is a direct function of the intensity or strength of union pressure (Goldner, 1970; Kochan, 1975). As the external pressure from the union induces management to relocate power, however, internal political opposition is likely to develop (Child, 1972), and internal conflicts or opposition to the organizational changes occur (Kochan et al., 1974). Thus, for management to successfully adapt its structure to cope with unionization, these internal conflicts must be resolved and the labor relations policies must be successfully coordinated with other managerial actions.

While there is enough empirical evidence from both the general organizational research on boundary units (Aldrich & Herker, 1977) and the research examining boundary unit power in labor negotiations (Goldner, 1970; Kochan, 1975) to conclude that strong unions will exert pressure on organizations to make these structural adaptations, there is no empirical evidence on whether these adaptations in fact are instrumental to labor relations performance and/or general organizational performance. In short, the structural adaptations hypothesized by Slichter et al. (1960) do have empirical support, however, the effects of these structural changes on employer goal attainments or on the negotiations process have yet to be systematically examined.

*Impact on the Content of Management Policy.* The growth of unionization and the labor movement has been cited as one of the major factors underlying the growth of personnel departments in American organizations (Ling, 1965). Typical union contracts cover a wide range of issues that affect the personnel function such as promotion, layoffs, seniority, leave policies, discipline and discharge training, wage payment systems and structures, job evaluation procedures, safety and health conditions, grievance procedures, etc. All of these policies are jointly negotiated and formalized into legally enforceable contract terms. In addition, the existence of a grievance procedure ending in arbitration induces further formalization and standardization of personnel policies since it increases the importance of precedents and past practices.

There is no a priori reason to predict that these provisions should either increase or decrease the productivity or organizational performance of an organization. For example, these contract provisions may reduce organizational performance to the extent that they impose unproductive work rules or constrain management's flexibility to assign, transfer, promote, or reward employees based on merit or performance. On the other hand, there is some evidence that unions reduce turnover (Block, 1978) and increase the probability that employees on layoff who are recalled will return to the organization (Medoff, 1976). Such contract provisions may, therefore, have positive effects on productivity, to the extent that these effects help organizations maintain productive workers and reduce organizational recruitment and training costs. Furthermore, union wage increases may also help attract higher quality workers and thus again increase worker productivity and organizational performance. Despite the importance of collective bargaining provisions to the personnel and human resource functions of the firm, there has been surprisingly little examination of their effects on organizational performance indicators such as productivity, turnover, absenteeism, accidents and illnesses, etc. Thus, the direct impact of alternative contract provisions on these indicators of organizational performance appear to be an important and relatively untapped area for future research.

The empirical evidence available to date is not sufficient to determine whether the neoclassical or the shock effect propositions more accurately captures the average effects of unions and collective bargaining on management goals. The task of behavioral scientists could be not only to attempt to answer this question but also to examine the conditions under which management is successful in adjusting to and working with the presence of unions to make organizations more productive and effective.

*B. Impact on Goals of Workers, Unions, and the Public*

Although the above sections focused on the effects of contract provisions on employer goals, analogous questions can be raised concerning the impacts of the results of collective bargaining on the goals of the union as an organization, the goals of the public (as expressed in the policies of federal, state, and local governments), and the goals of individual workers. For example, the federal government is playing an increasingly active role in monitoring the performance of firms and unions in the areas of equal employment opportunity, safety and health, pension management, and other areas. Clearly, the results of collective bargaining affect these aspects of the employment relationship. Thus, the effects of collective bargaining need to be examined for their impact on these increasingly important public policy goals. Similarly, the content of the bargaining agreement influences the security of the union as an organization. Finally, the terms of the agreement are likely to exert a very strong and direct effect on workers' job and income security, control over their work environment, and other goals. The study of these issues can be nicely integrated into the mainstream of microorganizational behavior research. Hammer (1978), for example has used an expectancy theory framework to examine the impact of collective bargaining on worker and employer goals at the individual level of analysis. Application of this framework in other unionized settings would be a logical next step in addressing these issues.

*C. The Administration of the Bargaining Relationship*

Although many of the economic terms of the collective bargaining contract are self-enforcing and have a direct effect on individual and employer goals, most of the provisions have their impact through the administration of the bargaining relationship. The centerpiece of the contract administration process is the formal grievance procedure. Thus, the effectiveness of the grievance procedure may be another critical determinant of the extent to which collective bargaining has a positive or negative impact on the goals of employers, unions, and individual workers.

Some industrial relations scholars view the grievance procedure as the most significant innovation of the United States collective bargaining system (Thomas & Murray, 1976). It is the major mechanism for providing industrial justice at the workplace in unionized settings (Selznick, 1969). The major functions of the grievance procedure are: to serve as an orderly strike substitute for resolving day-to-day problems arising out of interpretations and application of the collective bargaining agreement; and to provide an avenue of appeal for individuals who feel their rights have been violated. The procedure should serve the needs of the union and the employer by developing a "common law" of the shop by providing a consistent and uniform application of the agreement, while protecting the rights of individual employees.

The concept of the effectiveness of a grievance procedure is not yet fully developed. A good deal of research has examined the determinants of grievance rates and/or evaluated the internal mechanics of the grievance procedure (e.g., the time delays, costs, grievance rates, and settlement rates short of arbitration) (Anderson, 1977; Thomson & Murray,

1976; Fleming, 1964). These internal mechanics thus provide operational measures of the performance of a procedure once a problem gets defined as a grievance and introduced into the procedure. Most studies to date have conceptualized and measured the effectiveness of a grievance procedure against these operational indicators.

This approach does not fully test the effectiveness of the grievance procedure as a system of industrial justice since it ignores problems that never get introduced into the procedure. Surprisingly little attention has been given to whether major problems of workers, supervisors and higher level union and management officials are dealt with or excluded from the grievance procedure. To what extent, for example, do the central problems workers face on their jobs get introduced into the grievance procedure? Is the procedure dealing with the full range of worker concerns and problems or only the subset clearly covered by the terms of the collective bargaining agreement? What other informal (Thomson & Murray, 1976) or formal avenues for coping with problems are needed to provide an effective system of industrial justice? Analyses of these questions would improve our theoretical understanding of the role of grievance procedures in organizations and the extent to which individuals, unions, employers, and/or the government must supplement the formal procedure with other means of responding to individual and organizational problems at the workplace. This question has become increasingly important in recent years since various court decisions and other governmental actions have provided workers with alternative channels for appealing employer actions affecting safety and health, equal employment opportunity, and a host of other employment problems. Although the standard evaluation frameworks and criteria noted above are important and certainly deserve additional research by collective bargaining and organizational behavior researchers, it is also time to examine how the formal grievance procedure relates to other mechanisms for problem solving at the workplace.

*D. Union-Management Change Efforts*

A growing number of issues of concern to union members and employers require more complex and *continuous* problem solving and are not subject to the legalistic structure of the formal grievance procedure. Thus, there has been an increase of interest in recent years in examining the conditions under which unions and employers can effectively engage in joint organizational change efforts (Kochan & Dyer, 1976; Schlesinger & Walton, 1977; Goodman & Lawler, in press; Drexler & Lawler, 1977). Most of the empirical studies to date have been reports of efforts to introduce Quality Of Work (QOW) experiments into unionized settings. Although some of the early reports of these experiments have been favorable, most programs have experienced a number of strategic difficulties in getting started (Schlesinger & Walton, 1977; Drexler & Lawler, 1977), and in maintaining union and employer commitment to the program over time (Goodman & Lawler, in press).

A model of the union-management change process was recently presented (Kochan & Dyer, 1976) and will be briefly reviewed here in order to illustrate why these efforts often encounter difficulty. The model suggested that unions and employers will be reluctant to embark on any ambitious program of organizational change unless they are experiencing significant external and internal pressures to do so. This proposition is consistent with most models of organizational change and also reflects the traditional American adversarial pattern of industrial relations—unions seek to promote formal collective bargaining and grievance handling and employers seek to limit the scope of union involvement and influence in decision making. Despite the call for "new" forms of labor-management cooperation contained in highly publicized documents such as the *Work in America* (1972) and the aggressive efforts of the National Commission on Productivity and Quality of Working Life, Goodman & Lawler (in press) estimate that less than 20 out of the approximately 70 known QOW experiments initiated between 1970 and 1976 involved unions. Thus, efforts to reorganize work are still largely taking place in nonunionized organizations.

The calls by government officials and academics for reform of labor-management relationships through joint QOW experiments fell largely on deaf ears partly because of poor timing. The *Work in America* report and the expansion of the National Commission on Productivity into the quality of working-life area came largely in response to the tight labor markets of the late 1960s and resultant expression of worker power through several well-publicized wildcat strikes. By the time the federal government and its agents were actively soliciting unions and employers to embark on these programs (1973/76), however, unemployment and inflation had increased. Therefore, the priorities of union leaders and members centered on the traditional issues of job and income security and the visible pressure for improving the quality of working conditions and satisfaction with the content of jobs subsided. Both the external pressure on employers (tight labor markets) and the internal pressures on union leaders (constituency pressure) needed to stimulate these changes were no longer present.

A number of unionized organizations have, however, embarked on ambitious efforts to supplement the formal bargaining system with joint change efforts; others may do so in the future, especially if the labor market tightens and the bargaining power of workers at the local level once again intensifies. Over the next several decades, as the educational level of the work force increases one can expect unions to be under greater pressure to increase the participation and control of workers over their jobs. For those unions and employers that do initiate these efforts, the model of the change process (Kochan & Dyer, 1976) suggests that these joint efforts will have to overcome the following obstacles to be successful and to survive over time:

(1) both parties must be able to see the change programs as being instrumental to the attainment of goals valued by their respective organizations and/or constituents;

(2) the internal political risks to union leaders and management officials must be overcome;

(3) the programs must produce tangible, positive results in the short-run and must demonstrate a high probability of being able to continue to achieve valued goals in the future;

(4) the initial stimulus or pressure to embark on the program must continue to be important and the initial goals of the programs must continue to be of high priority to the parties;

(5) the gains or benefits from the initial program must be equitably distributed among the workers and the employer;

(6) the union must be perceived as being an instrumental force in achieving the goals or benefits of the program and the union leaders must be protected from getting overidentified as part of management;

(7) the change process must be successfully integrated with the formal structure and procedures of contract negotiations and administration.

These are difficult organizational and political hurdles to overcome. Although the literature is ripe with case study reports of "success" stories, interpretation of these reports are complicated by the problem that different reporters of the same case have made different assessments (Goodman & Lawler, in press). The most carefully evaluated union-management quality of work experiment completed to date, a study of a QOW effort to improve productivity, safety, and skill levels in a unionized mine, provides a mixed picture of success and failure. Goodman & Lawler (in press), reported that the experiment resulted in: (1) only a slight marginal impact on productivity, (2) no reduction in accidents but improvements in other safety indicators, (3) improved skill levels of workers, (4) improved communications between labor and management, (5) no significant change in the performance of the formal labor-management relationship, and (6) increased internal conflict within the union. The most serious conflicts developed between those individuals who participated in and benefitted financially from the experiment and those who served as members of a control group. Ultimately, the internal union conflicts led the local union to vote against continuation of the program.

The purpose of the above discussion is neither to advocate nor disparage these innovative efforts to modify the traditional pattern of union-management relations. Clearly, additional experimentation and systematic evaluation is needed before any firm conclusions can be drawn about the long-run potential of such efforts. The quality of work area is probably responsible for more attention being given to unions by organizational behavior researchers and consultants in the 1970s than any other issue discussed in this paper. Clearly, this will continue to be an arena for integrating organizational behavior and collective bargaining research and practice in the future. The challenge will be to improve the objectivity and quality of evaluations of these efforts.

Additional case study research as well as comparative case study research on these types of efforts are badly needed. These joint change efforts provide an excellent opportunity for examining the potential for collaborative problem solving in a setting where clear structural sources of power and conflict are an inherent feature of the relationship. Thus, this type of research not only can contribute to our understanding of organizational change in a unionized setting, but also to the effectiveness of alternative types of change strategies in situations characterized by conflicts of interest, shared power, and high potential for conflict. This research could also help to assess the potential for supplementing the formal collective bargaining relationship with more continuous forms of problem solving and conflict resolution that depart from the traditional patterns of labor-management relations. In the long run, research in this area should assess the potential for using the union-management relationship as a way of adapting to the changing needs of individual workers and employers. In general, it should help determine the extent to which unions and employers can effectively adapt to a changing environment and changing characteristics and expectations of the labor force.

While the quality of work movement provides the most visible and in many ways the most ambitious form of labor-management change programs, a number of other longstanding and some newly emerging forms of problem solving could also be examined. The passage of the Occupational Safety and Health Act of 1970, for example, has helped revive union and management interest in joint safety and health committees at the plant level (Kochan, Dyer & Lipsky, 1977). Experiments with community level labor-management committees are active in at least seventeen cities around the country (Blondman, 1977). Industry level labor-management committees can also be found (Rees, 1975). The bitter 112-day coal strike in 1977/78 has led President Carter to establish a national commission to attempt to improve labor relations, productivity, and safety in this important sector of the economy. Thus, the next decade should provide a host of field sites for those interested in exploring the potential for introducing systematic change into formal collective bargaining relationships.

## VIII. UNIONS AS COMPLEX BARGAINING ORGANIZATIONS

The characteristics and behavior of unions as bargaining organizations play an important intervening role in the theoretical framework presented in this paper. Since the pioneering study by the Webbs (1897) which defined *trade unions* as "continuous associations of wage earners for the purpose of maintaining or improving the conditions of their working lives," there has been a tremendous volume and variety of research examining the goals, growth, internal government and structure, and strategies of trade unions. The literature can be classified into three separate streams of research. The first is the traditional institutional/historical research that documented the development of the American labor movement and its component unions over the last century. This research was at the center of attention of the early institutional schools in the field. In the past decade this tradition has faded considerably. Few intensive studies of the structures, goals, and internal government of trade unions have been conducted since the 1950s (Strauss, 1977). The second stage of union research developed as economists began (1) using economic theory to build models of union bargaining and strike behavior, and (2) applying econometric techniques to the analysis of union growth (Ashenfelter & Pencavel, 1969). The third stream of research is still emerging as behavioral scientists apply organizational theories to trade unions. Although the majority of this research has been done in Britain (Child, Loveridge & Warner, 1973; Donaldson & Warner, 1974), organizational behavior research on trade unions is now gaining increased attention in North America as well (Anderson, 1977).

None of these three approaches to research have developed a comprehensive model of unions as bargaining organizations. The institutional research provides a rich body of descriptive and factual material on unions which has not been woven into a systematic theoretical framework. The economic and behavioral models are more systematic but have only dealt with a limited array of issues from limited perspectives. Consequently, this section will attempt to outline the essential questions that need to be included in a comprehensive model of unions as organizations.

The study of unions as organizations provides a tremendous opportunity to integrate the traditional approach in industrial relations, economic theory, and the comparative analysis frameworks of organization theorists. Unions, like employers, must adapt to their external environment and engage in strategies that promote their effectiveness as bargaining and service organizations. In addition, unions are representative-administrative bureaucracies that expected—by society as well as by union members—to be governed in a democratic fashion. Thus, traditional variables important in the studies of environment, organizational structure, internal organizational processes, and organizational effectiveness, are also central to the investigation of union organizations.

As Strauss & Warner (1977, p. 116) caution, the drive for more systematic analysis of trade union organizational issues should be careful to avoid losing the major strength of the traditional institutional/historical approach, namely, an appreciation of the larger environmental context. However, while industrial relations research has included the environment, the attention has mainly been limited to the impact of economic and political aspects of the environment on union goals and the outcomes of the bargaining process (Dunlop, 1944; Ross, 1948), the growth in union membership (Ashenfelter & Pencavel, 1969), and industrial conflict and dispute resolution (Ashenfelter & Johnson, 1969; Anderson & Kochan, 1977). There has been much less attention to such issues as how the environment impacts union structure or how unions interact with other organizations in attempting to achieve their goals. This is surprising since the outcomes of negotiations may depend on coordination with or the support of other unions; legislative change may depend on interaction and cooperation among central labor bodies, unions, political bodies, and community groups; and the overall impact of the labor movement is in part a function of cooperation among union, political, legal, and community interests in society. Thus, research on interorganizational relations may be useful in building models to explain union behaviors.

Unions, as strategic organizations, also need to develop a repertoire of tactics to use in adapting to or controlling their environments. For example, it has been discovered that union mergers may be explained in a similar fashion to business mergers (Freeman & Britton, 1977) as a strategy for handling interorganizational dependence. It would be interesting to examine the extent to which other strategies such as boundary spanning, cooptation, coalition bargaining, representation, forecasting, joint ventures, exchange of personnel, information or resources (Salancik & Pfeffer, 1977) are also used by unions to reduce environmental uncertainty and dependence. Moreover the conditions under which these strategies are used and their impact on union effectiveness should also be examined.

Unions traditionally have been classified as voluntary associations

(Tannenbaum, 1965; Blau & Scott, 1962) because of their internal representational system. Recently, several authors (Barbash, 1969; Bok & Dunlop, 1970; Shirom, 1975), have emphasized that unions need to be run by experts to become more effective as administrative agencies. It appears that more and more unions are beginning to be viewed as administrative organizations (Child, Loveridge & Warner, 1973). In fact, several studies have found that both national (Donaldson & Warner, 1974) and local unions (Anderson, 1977) may develop administrative bureaucracies similar to those found in manufacturing organizations (Pugh, Hickson, Hinings & Turner, 1968). Additional theoretical and empirical work is needed which examines unions as bureaucracies, the conditions under which bureaucratic structure develops, and the impact of bureaucracy on union effectiveness.

Edelstein & Warner (1977) suggest that as unions become larger and more complex organizations requiring greater technical expertise to effectively perform their collective bargaining and other representational functions, union democracy may be difficult to maintain. Others (Lester, 1958; Lipset, Trow & Coleman, 1956) also argue that the "iron law of oligarchy" (Michels, 1949) tends to develop within trade unions over time. Despite the centrality of this thesis to theories of union democracy, most of the research has focused on individual rather than organizational correlates of internal democracy (Perline & Lorenz, 1970; Spinrad, 1960). One study (Anderson, 1977) found that measures of organizational complexity (i.e., specialization, vertical and horizontal differentiation) and bureaucratic control mechanisms (i.e., centralization, standardization and formalization of activities) significantly reduce the degree of union democracy as measured by participation, closeness of elections, and membership influence over the way the union is run. Thus, one of the critical problems facing a modern trade union is the need to design a structure for maximizing effectiveness in bargaining, while at the same time, maintaining internal democracy.

Ultimately, a comprehensive model of the organizational behavior of unions must incorporate a concept of union effectiveness. Unfortunately, union effectiveness has only recently been introduced as a dependent variable in empirical research in this field. Anderson (1977, p. 282) defines *union effectiveness* as "the attainment of valued goals and fulfillment of members' needs on a day to day basis." He examined union performance in (1) collective bargaining, (2) grievance processing, (3) political action, and (4) union-management committees and other informal interactions between the parties to a bargaining relationship. More research is needed on the various dimensions of union effectiveness and the way in which a union can strategically use them to achieve its members' goals. Finally, additional theoretical and empirical research is required on the environmental and organizational structure, and process determinants of union effectiveness.

A comprehensive model of the behavior of unions as organizations must therefore address at least the following questions or issues: (1) Why do individuals join unions? (Discussed in Jeanne Brett's chapter in this book.) (2) What factors influence overall trade union growth, certifications and decertifications? (3) What influences the goals of trade unions? (4) What influences the internal structure of unions? (5) What influences the internal democracy of unions? (6) What influences the organizational effectiveness of unions? A wide variety of variables have been used to describe and explain variations in these dimensions of unions. Most studies incorporate some combination of independent variables that measure characteristics of the economic, political, and legal environment; the management or employer the union(s) interact with; and the characteristics of the rank-and-file workers or members. A combination of the industrial relations, economic, and organizational theories may help us to develop a more comprehensive model of union behavior on the critical dependent variables outlined above. The next step in union organizational research, therefore, should be to conduct comparative organizational studies of trade unions and examine variations in these critical union dimensions with models which incorporate the key environmental, organizational, and individual causal forces.

## IX. THE ROLE OF STRATEGY

Although, for simplicity, various components of the bargaining system were presented in this paper as discrete topics, in reality they are likely to be closely interrelated. It must be recalled that collective bargaining is an adversarial process involving the interplay of organizations (and subgroups) with partially conflicting goals. Changes in the system often come about, therefore, through the exercise of power. This implies that strategic considerations play an extremely important role in the behavior of the parties in all aspects of a bargaining system. The need for strategic consistency is likely to bind the different components of the framework presented in this paper together more closely than one might otherwise expect. At this point an analogy can be made to what Boulding (1962) described as a "conflict trap"—when two adversaries conflict on one issue the probability increases that they will also experience conflict on other seemingly unrelated issues. Thus, bargaining relationships with high strike or impasse rates in negotiations can be expected to also experience high grievance rates, take legalistic rather than problem solving approaches to grievance and arbitration processes, rely heavily on management rights and/or strict contract enforcement doctrines, and have little potential for success in joint change programs. Efforts to either build models or change the outcomes of a single component of a bargaining system thus need to recognize the importance of strategic consistency in collective bargaining.

## X. A CURRENT ILLUSTRATION

Prior to summarizing the major arguments presented in this paper, it may be helpful to provide a concrete example of one current societal problem that could benefit from research that integrates collective bargaining and organizational behavior concepts and methodologies. From December, 1977 to April 1978 the United States economy endured a long and bitter nationwide strike in the coal industry. The causes of this extended strike were complex, involving a combination of external economic pressures, the history of union-management relations in the industry, leadership and internal conflict problems within both the union and management bargaining organizations, and, some might add, lack of skill in handling the strike once it began. There is widespread agreement that the industrial relations problems of the industry are long-run in nature and did not disappear with the end of the strike. As noted earlier in the paper, the president has therefore established a National Commission to study these industrial relations problems. Suppose behavioral scientists were turned loose on this project. Where would they start in their search for an understanding of the underlying problems and for strategies for improving the performance of the collective bargaining and industrial relations system in this industry? What kind of research, and ultimately, what organizational change strategies are needed in this industry? What kinds of research designs and data collection strategies are necessary to make meaningful, empirically and theoretically grounded prescriptions for change?

Development of a research and action program for the coal industry is obviously beyond the scope of this paper. However, these are the types of questions and problems that behaviorally trained and institutionally sensitive researchers ought to be able to effectively and routinely tackle. Clearly, a working knowledge of the history of union-management relations in this industry, an understanding of the impact of this history on the current attitudes and beliefs of rank-and-file miners, union leaders, and management officials would be essential to this task. Ability to design surveys across firms and union districts to assess the existing variations in performance of the collective bargaining system would be essential. Finally, the political skills and willingness to work with union, management, and government representatives who hold the keys to change in this setting would be needed if the research was to have an impact on practice.

Thus, a combination of the training, skills, and interests of traditional students of collective bargaining and organizational behavior is clearly required to address these type of problems.

## XI. SUMMARY AND CONCLUSIONS

### *A. Benefits for Collective Bargaining*

Collective bargaining research has been dominated by economists, lawyers, and historians ever since the early institutionalists began documenting and commenting on the development of the American labor movement. As the discipline of economics turned more toward mathematical model building and quantitative analysis, a new generation of analytical labor economists developed who attempted to model labor collective bargaining and union-management issues. Unfortunately, the quantitative economists incorporated in their models only those variables which were measurable with existing data. Consequently, they tended to test their models at aggregate levels of analysis (models of strikes used annual economy-wide strike rates, models of union growth used economy-wide rate of change in union membership, etc.). The institutionalists criticized this approach by arguing that it ignored the most important political, structural, organizational, and power-related variables that affect events in labor-management relations at the microlevels. Yet the institutionalists failed to develop alternative theory building and quantitative research techniques to systematically capture the effects of these additional variables. An assumption underlying this paper is that theories and methodologies of the behavioral sciences can play an important role in overcoming the limitations of both the economic and the institutional approaches to research in this area. If the behavioralists use their theory building, measurement, and analysis techniques to address the critical substantive issues that have been mapped out for the field, they may help to end a debate that has outlived its usefulness.

The issues discussed in this paper clearly do not exhaust the range of possibilities for integrative research among students of collective bargaining and organizational behavior. Instead, the discussion illustrates the broad range of research needs and opportunities open to those willing to cross this boundary. The material was presented in a way that was designed to encourage organizational behavior researchers to address more of the traditional questions of interest to those in the area of collective bargaining. Organizational behavior theories, concepts, and methodological techniques can introduce more systematic thinking to the diverse issues normally studied by collective bargaining researchers. Ultimately, the application of behavioral research techniques and concepts to this area should result in better predictions and explanations of behavior in union-management settings. This should help collective bargaining researchers have a greater impact on public policy and private practice.

*B. Benefits for Organizational Behavior*

There are also a number of ways in which research on collective bargaining topics can enrich the thinking and understanding of students of organizational behavior. First, the theoretical framework outlined in this paper clearly illustrates the issue-oriented nature of collective bargaining research. Collective bargaining research has always been problem oriented. Thus it has always been difficult to abstract the research from its real-world context. Attention to collective bargaining issues could therefore reverse an undesirable trend that seems to have emerged in organizational behavior research in the last decade, i.e., a movement toward more abstract thinking on a narrower range of topics that grow increasingly far removed from actual problems and critical concerns to workers, managers, and society. In short, attention to collective bargaining problems may increase the relevance of organizational behavior research.

It is impossible to study collective bargaining problems without directly facing very difficult questions concerning values and one's own normative framework. Since conflicting interests are involved in collective bargaining problems, researchers must identify their own normative framework in sorting out these interests. This again should help to reawaken many organizational behavior researchers who have unconsciously taken the interest of the employers as their predominant framework for structuring their research.

Because of the political nature of the most controversial collective bargaining problems and because of the impact of environmental changes on these problems, research on collective bargaining helps one to recognize the limits of cross-sectional designs and quantitative methodologies. Collective bargaining researchers have always recognized the need to adapt an "open system" approach by incorporating the impact of the economic, legal, and political context into their analyses. Consequently, the more behavioral researchers get into analysis of these problems, the more they will be encouraged to broaden their models and thinking to incorporate these external aspects of organizations. While a good deal of progress has been made on adapting the open systems notion to organizational research in the last two decades, much remains to be done to add a strong historical and dynamic component to the open systems analyses.

Collective bargaining research clearly is an avenue for improving the understanding of the political aspects of organizational life. Concepts such as conflicts of interest and goals, power, overt conflict behavior, and conflict resolution are central to a political analysis of organizations and have been central concepts in collective bargaining theories and research for decades. Thus, studying these concepts in a collective bargaining context should enrich the emerging political models of organizations.

Finally, there are two overriding contributions—one, pragmatic, the other, philosophical—that collective bargaining can make to the field of organizational behavior. To the extent that organizational researchers are willing to accept trade unions as permanent and legitimate institutions and are willing to devote their efforts to improving the quality of union-management relationships, a whole new array of research sites and issues will be open to them. As the supply of competent and creative researchers addressing these problems grows, they may create their own demand in universities, consulting firms, government agencies, and union and management organizations. Thus, to the extent that effective research is carried out on these issues, a whole new set of job opportunities may develop. On the more philosophical side, most of what has been said in this concluding section may be summarized very simply: stepping into these new areas of research should increase the range of contributions that behavioral scientists can make to a better society, the ultimate purpose of any social scientist.

## FOOTNOTES

* I wish to thank John Anderson, Samuel Bacharach, Jeanne Brett, Lee Dyer, Ray Scannell, Robert Stern, and Lawrence Williams for their comments and suggestions on an earlier draft of this paper. Partial support for this research was provided by the National Science Foundation (Grant No. APR 77-17120). The content of the paper is the sole responsibility of the author and does not reflect the official policies of the National Science Foundation.

1. *Micro* is defined here as encompassing the bargaining relationship down to the individual levels of analyses. As such the micro level of analysis for collective bargaining research encompasses the interorganizational, organizational, group and individual levels of analysis—essentially all of organizational behavior. *Macro* is defined as research that takes the national economy or some other aggregation of bargaining relationships as its unit of analysis. Thus, our distinction parallels the distinction between micro and macro economics and not micro and macro organizational behavior.

## REFERENCES

Aldrich, H. & Herker D. (1977) "Boundary spanning roles and organizational structure." *Academy of Management Review*, 2, 217–230.

Anderson, J. N. (1977) *Union effectiveness: An industrial relations systems approach.* Unpublished Ph.D. Dissertation. Ithaca, N.Y.: Cornell University.

Anderson, J. C. & Kochan, T. A. (1977) "Impasse procedures in the Canadian federal service: Effects on the bargaining process." *Industrial and Labor Relations Review*, 30, 283–301.

Ashenfelter, O. & Johnson, G. (1969) "Bargaining theory, trade unions, and industrial strike activity." *American Economic Review*, 59, 35–49.

——— & Pencavel, J. H. (1969) "American trade union growth: 1900–1960." *Quarterly Journal of Economics*, 83, 434–448.

Bacharach, S. (1978) Power, consensus, and conflict: The dimensions of a political analysis of intraorganizational structure. Unpublished manuscript, Cornell University.

Balke, W. M., Hammond, K. R., & Meyer, G. D. (1973) "An alternative approach to labor-management negotiations." *Administrative Science Quarterly*, 18, 311–327.

Barbash, J. (1964) "The elements of industrial relations." *British Journal of Industrial Relations*, 2, 66–78.

——— (1969) "Rationalization in the American union." In G. G. Somers (ed.), *Essays in industrial relations theory*. Ames, Iowa: The Iowa State University Press, 147–162.

Barnes, L. W. C. S., & Kelly, L. A. (1975) *Interest arbitration in the federal public service of Canada*. Kingston, Ontario: Queen's University, Industrial Relations Center.

Beal, E., Wickersham, E., & Kienast, P. (1976) *The practice of collective bargaining*. Homewood, Ill.: Irwin.

Bigoness, W. J. (1976) "The impact of initial bargaining position and alternative modes of third party intervention in resolving impasses." *Organizational Behavior and Human Performance*, 17, 185–198.

Blau, P. (1965) "The comparative study of organizations." *Industrial and Labor Relations Review*, 18, 323–338.

——— & Scott, R. (1962) *Formal organizations*. San Francisco: Chandler.

Block, R. N. (1978) "The impact of seniority provisions on the manufacturing quit rate." *Industrial and Labor Relations Review*, 31, 474–488.

Blondman, M. (1977) "The development of community labor-management committees." Unpublished Masters Thesis. Ithaca, N.Y.: Cornell University.

Bok, D. & Dunlop, J. T. (1970) *Labor and the American community*. New York: Simon and Schuster.

Boulding, K. B. (1962) *Conflict and defense*. New York: Harper and Row.

Bowers, M. H. (1973) "A study of legislated arbitration in the public safety services in Michigan and Pennsylvania." Unpublished Ph.D. Dissertation. Ithaca, N.Y.: Cornell University.

Bowlby, R. L. & Schriver, W. R. (1978) "Bluffing and the split-the-difference theory of wage bargaining." *Industrial and Labor Relations Review*, 31, 161–171.

Brett, J. M. & Goldberg, S. B. (1979) "Wildcat strikes in bituminous coal mining." *Industrial and Labor Relations Review*, 32, 465–483.

Britt, D. & Galle, O. (1972) "Industrial conflict and unionization." *American Sociological Review*, 37, 46–57.

——— & Galle, O. (1974) "Structural antecedents of the shape of strikes: A comparative analysis." *American Sociological Review*, 39, 642–651.

Burton, J. & Krider, C. (1975) "The incidence of strikes in public employment." In D. Hamermesh (ed.), *Labor in the public and non-profit sectors*. Princeton, N.J.: Princeton University Press.

Chamberlain, N. W., & Kuhn, J. W. (1965) *Collective bargaining*. New York: McGraw-Hill.

Child, J. (1972) "Organizational structure, environment and performance: The role of strategic choice." *Sociology*, 3, 1–22.

———, Loveridge, R., & Warner, M. (1973) "Toward an organizational study of trade unions." *Sociology*, 7, 71–91.

Commons, J. R. & others. (1951) *History of Labor in the United States*. New York: Macmillan.

———. (1924) *Legal foundations of capitalism*. New York: Macmillan.

Cullen, D. (1968) *National emergency disputes*. Ithaca, N.Y.: New York State School of Industrial and Labor Relations.

Cyert, R. M., & March, J. G. (1963) *A behavioral theory of the firm*. Englewood Cliffs, N.J.: Prentice Hall.

Dahrendorf, R. (1959) *Class and class conflict in industrial society*. London: Routledge.

Davey, H. W. (1972) *Contemporary collective bargaining*. Englewood Cliffs, N.J.: Prentice Hall.

Derber, M. Preface (1949) *Proceedings of the First Annual Meeting of the Industrial Relations Research Association*. Urbana, Ill.: Industrial Relations Research Association.

———, Chalmers, W. E., & Edelman, M. T. (1965) *Plant union-management relations: From theory to practice*. Urbana, Ill.: University of Illinois Press.

Deutsch, M. (1973) *The resolution of conflict*. New Haven, Conn.: Yale University Press.

Donaldson, L. & Warner, M. (1974) "The structure of occupational interest organizations." *Human Relations*, 27, 721–38.

Douglas, A. (1962) *Industrial peacemaking*. New York: Columbia University Press.

Drexler, J. A. & Lawler, E. E. (1977) "A union-management cooperative project to improve the quality of work life." *Journal of Applied Behavior Science*, 13, 373–386.

Dunlop, J. T. (1958) *Industrial relations systems*. New York: Holt.

———. (1944) *Wage determination under trade unions*. New York: A. M. Kelley.

Edelstein, J. D. & Warner, M. (1977) "Research areas in national union democracy." *Industrial Relations*, 16, 186–198.

Fayol, H. (1949) *General and industrial management*. London: Pitman and Sons.

Feuille, P. (1975) "Final offer arbitration and the chilling effect." *Industrial Relations*, 14, 311–317.

Filley, A. (1976) *Interpersonal conflict resolution*. (Glenview, Illinois: Scott Foresman.

Fleming, R. W. (1964) *The arbitration process*. Urbana, Ill.: University of Illinois Press.

Fox, A. (1971) *A sociology of work in industry*. London: Collier MacMillan.

Freeman, R. B. (1977) Productivity and collective bargaining: the impact and mediating factors. Research proposal to the National Science Foundation.

Freeman, J. & Britton, J. (1977) "Union merger process and industrial environment." *Industrial Relations*, 16, 173–185.

France, R. R. & Lester, R. A. (1951) *Compulsory arbitration of utility disputes in New Jersey and Pennsylvania*. Princeton, N.J.: Industrial Relations Section, Princeton University.

Goodman, P. S. & Lawler, E. E. (in press) *New forms of work organization in the U.S.* Geneva: International Labor Organization.

Goldner, R. (1970) "The division of labor: Process and power." In M. Zald (ed.), *Power in organizations*. Nashville, Tenn.: Vanderbilt University Press, pp. 97–143.

Graham, J. C. (1978) Decision making in dispute resolution: A study of the federal service impasses panel. Unpublished Masters Thesis. Ithaca, N.Y.: Cornell University.

Hammer, T. (1978) "The role of the union in organizational behavior: A study of relationships between local union characteristics and worker behavior and attitudes." *Academy of Management Journal*.

Hammermesh, D. S. (1973) "Who wins in wage bargaining." *Industrial and Labor Relations Review*, 26, 1146–1149.

Hamner, W. C. (1978) "The role of the union in organizational behavior: A study of relationships between local union characteristics and worker behavior and attitudes." *Academy of Management Journal*.

———. (1978) "The influence of structural, individual, and strategic differences on bargaining outcomes: A review." In D. L. Harnett & L. L. Cummings (eds.), *Bargaining and personality: An international study*.

Heuttner, D. A. & Watkins, T. L. (1974) "Public sector bargaining: An investigation of possible environmental influences." *Proceedings of the 26th Annual Winter Meeting of the Industrial Relations Research Association,* Madison, Wisconsin, IRRA, 178–187.

Hinnings, C. R., Hickson, D. J., Pennings, J. M., & Schneck, R. G. (1974) "Structural conditions of intraorganizational power." *Administrative Science Quarterly,* 19, 22–44.

Horton, R. (1975) "Arbitration, arbitrators, and the public interest." *Industrial and Labor Relations Review,* 28, 497–507.

Johnson, D. F. & Pruitt, D. G. (1972) "Pre-intervention effects of mediation vs arbitration." *Journal of Applied Psychology,* 56, 1–10.

Kerr, C. (1954) "Industrial conflict and its mediation." *American Journal of Sociology,* 60, 230–245.

——— & Fisher, L. H. (1964) "Plant sociology: The elite and the aborigines." In C. Kerr (ed.), *Labor-management and industrial society.* Garden City, N.Y.: Doubleday.

Kochan, T. A. (1979) "The dynamics of dispute resolution." In B. Aaron, J. Grodin, & J. L. Stern (eds.), *Public sector bargaining.* Madison, Wis.: Industrial Relations Research Association.

——— & Baderschneider, J. (1978) "Determinants of the reliance on impasse procedures: The case of New York police and firefighters." *Industrial and Labor Relations Review,* in press.

——— & Dyer, L. (1976) "A model of organizational change in the context of union-management relations." *Journal of Applied Behavioral Science,* 12, 57–78.

———, Dyer, L., & Lipsky, D. B. (1977) *The effectiveness of union-management safety and health committees.* Kalamazoo, Mich.: W. E. Upjohn Institute for Employment Research.

———, Huber, G. P., & Cummings, L. L. (1975) Determinants of intraorganizational conflict in collective bargaining in the public sector. *Administrative Science Quarterly,* 20, 10–23.

——— & Jick, T. (1978) "A theory of the public sector mediation process." *Journal of Conflict Resolution.*

———, Lipsky, D. B., & Dyer, L. D. (1975) "Collective bargaining and the quality of work: The views of local union activists." *Proceedings of the 27th Annual Winter Meeting of the Industrial Relations Research Association,* Madison, Wisconsin, IRRA, 150–162.

———, Mironi, M., Ehrenberg, R. G., Baderschneider, J., & Jick, T. (1979) *Dispute resolution under factfinding and arbitration: An empirical analysis.* New York: American Arbitration Association.

Kornhauser, A. R., Dubin, R. & Ross, A. (1954) (eds.) *Industrial conflict.* New York: McGraw-Hill.

Kressel, K. (1972) *Labor mediation: An exploratory survey.* Albany, N.Y.: Association of Labor Mediation Agencies.

Kuhn, A. (1952) *Arbitration in transit: An evaluation of wage criteria.* Philadelphia: Pennsylvania University Press.

Landsberger, H. A. (1955) "Interaction process analysis of the mediation of labor-management disputes." *Journal of Abnormal Social-Psychology,* 51, 522–558.

Lawler, E. E. & Levin, E. (1968) "Union officers perceptions of members' pay preferences." *Industrial and Labor Relations Review,* 21, 509–517.

Lawrence, P. R. & Lorsch, J. W. (1967) *Organizations and environments: Managing differentiation and integration.* Boston: Harvard Business School.

Lester, R. A. (1958) *As unions mature.* Princeton, N.J.: Princeton University Press.

Lewicki, R. J. & Alderfer, C. P. (1973) "The tensions between research and intervention in intergroup conflict." *Journal of Applied Behavioral Science,* 9, 424–449.

Ling, C. (1965) *The management of personnel relations.* Homewood, Ill.: Irwin.

Lipset, S. Trow, M., & Coleman, J. (1956) *Union democracy.* Garden City, N.Y.: Doubleday.

Magenau, J. M. & Pruitt, D. C. (1978) "The social psychology of bargaining: A theoretical synthesis." In G. M. Stephenson & C. J. Brotherton (eds.) *Industrial Relations: A Social Psychological Approach*. London: Wiley.

Massie, J. L. (1965) "Management theory." In J. G. March (ed.), *Handbook of organizations*. Chicago: Rand-McNally, 387–422.

McGrath, J. E. (1966) "A social psychology approach to the study of negotiations." In R. V. Bowers (ed.), *Studies on behavior in organizations*. Athens, Georgia: University of Georgia Press.

Medoff, J. L. (1976) "Layoffs and alternatives under trade unions in United States manufacturing." Discussion Paper No. 525, Cambridge, Mass.: Harvard Institute of Economic Research.

Michels, R. (1949) *Political parties*. Glencoe, Ill.: The Free Press.

Mironi, M. (1977) "Compulsory arbitration of public safety interest disputes in New York: An analysis and performance evaluation." Unpublished Ph.D. dissertation. Ithaca, N.Y.: Cornell University.

Nadworny, M. J. (1955) *Scientific management and the unions*. Cambridge, Mass.: Harvard University Press.

Northrup, H. (1966) *Compulsory arbitration and government intervention in labor disputes*. New York: Labor Policy Association.

Notz, W. W. & Starke, F. A. (1978) "Final-offer versus conventional arbitration as means of conflict management." *Administrative Science Quarterly*.

Payne, P. (1977) "The consultants who coach the violators." *The Federationist,* 84, 22–29.

Perline, M. & Lorenz, V. (1970) "Factors influencing participation in trade union activities." *American Journal of Economics and Sociology*, 29, 425–437.

Perlman, S. (1928) *A theory of the labor movement*. New York: A. M. Kelley.

Pugh, D. S. Hickson, D. J., Hinings, C. R., & Turner, C. (1968) "The context of organizational structure." *Administrative Science Quarterly,* 14, 91–114.

Rees, A. (1975) "Tripartite wage stabilizing in the food industry." *Industrial Relations,* 14, 250–258.

Rehmus, C. (1965) "The mediation of industrial conflict: A note on the literature." *Journal of Conflict Resolution,* 9, 118–125.

Reynolds, L. (1970) *Labor economics and labor relations*. Englewood Cliffs, N.J.: Prentice Hall.

Robbins, S. P. (1974) *Managing organizational conflict*. Englewood Cliffs, N.J.: Prentice Hall.

Robbins, E., with Dennenberg, T. S. (1976) *A guide for labor mediators*. Hawaii: University of Hawaii.

Roomkin, M. (1976) "Union structure, internal control, and strike activity." *Industrial and Labor Relations Review,* 29, 198–217.

Ross, A. M. (1948) *Trade union wage policy*. Berkeley: University of California Press.

Rubin, J. Z. & Brown, B. R. (1975) *The social psychology of negotiations*. New York: Academic Press.

Salancik, C. R. & Pfeffer, J. (1977) An examination of need-satisfaction models of job attitudes. *Administrative Science Quarterly,* 22, 427–456.

Sayles, L. & Strauss, G. (1953) *The local union*. New York: Harper and Brothers.

Schlesinger, L. A. & Walton, R. E. (1977) "Work restructuring in unionized organizations: Risks, opportunities, and impact on collective bargaining." *Proceedings of the 29th Annual Winter Meeting of the Industrial Relations Research Association,* Madison, Wisconsin, IRRA, 329–337.

Schmidt, S. M. & Kochan, T. A. (1972) "Conflict: Toward conceptual clarity." *Administrative Science Quarterly,* 17, 359–370.

Selznick, P. (1969) *Law society and industrial justice*. New York: Russel Sage.

Shirom, A. (1975) "Union use of staff experts: The case of the Histadrut. *Industrial and Labor Relations Review*, 29, 107–20.

Shorter, E. & Tilly, C. (1974) *Strikes in France 1830 to 1968*. Cambridge: Cambridge University Press.

Simkin, W. E. (1952) *Acceptability as a factor in arbitration under an existing agreement*. Philadelphia: Pennsylvania University Press.

———. (1971) *Mediation and the dynamics of collective bargaining*. Washington: Bureau of National Affairs.

Skeels, J. W. (1971) "Measures of U.S. strike activity." *Industrial and Labor Relations Review*, 24, 515–525.

Slichter, S. (1941) *Union policies and industrial management*. Washington, D.C.: Brookings Institution.

———, Healy, J. J., & Livernash, E. R. (1960) *The impact of collective bargaining on management*. Washington, D.C.: Brookings Institution.

Sloane, A. A. & Witney, F. (1972) *Labor relations*. Englewood Cliffs, N.J.: Prentice Hall.

Snyder, D. (1975) "Institutional setting and industrial conflict: Comparative analysis of France, Italy, and the United States." *American Sociological Review*, 40, 259–278.

Spinrad, W. (1960) "Correlates of trade union participation: A summary of the literature." *American Sociological Review*, 25, 237–244.

Stagner, R. & Rosen, H. (1965) *Psychology of union-management relations*. Belmont, Cal.: Wadworth.

Stern, J. L., Rehmus, C., Loewenberg, J., Kasper, H., & Dennis, B. (1975) *Final offer arbitration*. Lexington, Mass.: Lexington Books.

Stern, R. N. (1976) "Intermetropolitan patterns of strike frequency." *Industrial and Labor Relations Review*, 29, 218–235.

———. (1978) "Methodological issues in quantitative strike analysis." *Industrial Relations*, 17, 32–42.

Stevens, C. M. (1966) "Is compulsory arbitration compatible with collective bargaining?" *Industrial Relations*, 5, 38–52.

———. (1967) "Mediation and the role of the neutral." In J. T. Dunlop & N. Chamberlain (eds.), *Frontiers of collective bargaining*. New York: Harper and Row, 271–290.

———. (1963) *Strategy and collective bargaining negotiations*. New York: McGraw-Hill.

Strauss, G. (1977) "The study of conflict: Hope for a new synthesis between industrial relations and organizational behavior." *Proceedings of the 29th Annual Winter Meeting of the Industrial Relations Research Association*, Madison, Wisconsin, IRRA, 329–337.

———. (1977) "Union government in the U.S.: Research past and future." *Industrial Relations*, 16, 215–242.

Strauss, G., & Warner, M. (1977) "Symposium introduction." *Industrial Relations*, 16, 115–125.

Tannenbaum, A. S. (1965) "Unions." In J. March (ed.), *Handbook of organizations*. Chicago: Rand-McNally.

Tannenbaum, & Kahn (1958).

Thibaut, J. & Walker, L. (1975) *Procedural justice: A psychological analysis*. New Jersey: Lawrence Erlbaum Associates.

Thomas, K. (1976) "Conflict and conflict management." In M. Dunnette (ed.), *Handbook of industrial and organizational psychology*. New York: Rand McNally, 889–936.

Thompson, J. D. (1967) *Organization in action*. New York: McGraw-Hill.

Thomson, A. W. & Murray, V. V. (1976) *Grievance procedures*. Lexington, Mass.: Lexington Books.

U.S. Department of Health, Education, and Welfare. (1972) *Work in America: Report of a special task force*. Cambridge, Mass.: M.I.T. Press.

Vitales, M. S. (1932) *Industrial psychology.* New York: Norton.

Walton, R. E. (1969) "Interpersonal peacemaking: *Confrontation and third party consultation.* Reading, Mass.: Addison-Wesley.

——— & McKersie, R. B. (1965) *A behavioral theory of labor negotiations.* New York: McGraw-Hill.

Webb, S. & Webb, B. (1897) *Industrial democracy.* London: Longmans.

Wheeler, H. N. (1978) How compulsory arbitration affects compromise activity. *Industrial Relations,* 17, 80–84.

*Work in America: Task Force Report to the Secretary of Health, Education, and Welfare* (1972) Cambridge, Mass.: MIT Press.

Zald, M. (1970) *Power in organizations.* Nashville, Tenn.: Vanderbilt University Press.

# BEHAVIORAL RESEARCH ON UNIONS AND UNION MANAGEMENT SYSTEMS*

Jeanne M. Brett

NORTHWESTERN UNIVERSITY

## ABSTRACT

Many of the chapters in this volume discuss issues familiar to organizational behavior researchers. This chapter is intended to introduce two underutilized research settings—unions and union-management systems. The chapter describes the organizational characteristics of unions and both traditional, adversarial union-management relationships and nontraditional, cooperative relationships. Propositions are presented for research on union organizing, and the effectiveness of adversarial and cooperative union-management relationships in order to illustrate the interface between organizational behavior theories and industrial relations issues.

**Research in Organizational Behavior, Volume 2, pages 177–213**

**ISBN: 0-89232-099-0**

# INTRODUCTION

This chapter describes the organizational characteristics of unions and the interorganizational characteristics of union-management relationships. It also develops behavioral research propositions on three major industrial relations issues: union organizing, the effectiveness of traditional union-management relationships, and the effectiveness of nontraditional, cooperative union-management relationships.

The chapter is intended for a variety of different uses. For industrial relations researchers, the research propositions illustrate a behavioral research approach which delves into the reasons underlying hypothesized relationships. For organizational behavior researchers, who are not interested in industrial relations research per se, the sections describing unions and union-management relationships will introduce these research sites, and should stimulate ideas about the kinds of organizational behavior research issues for which these sites are particularly appropriate. Finally, the chapter is intended to encourage organizational behavior researchers to use their behavioral theories and research methods to study the industrial relations issues which psychologists and sociologists studied in the 1950s (Tannenbaum & Kahn, 1958; Kornhauser, Dubin & Ross, 1954), and which organizational behavior researchers are again studying in the 1970s (Edelstein & Warner, 1975; Getman, Goldberg & Herman, 1976).

The chapter is in three major segments. The first segment is on unions. It begins with a description of the organizational characteristics of unions and then develops behavioral research propositions for research on union organizing. The second segment focuses on traditional, adversarial union-management relationships. The interorganizational characteristics of these relationships are described and alternative approaches to evaluating their effectiveness are discussed. This second segment concludes with a series of behavioral research propositions identifying factors influencing the effectiveness of union-management relationships. The third segment focuses on cooperative union-management relationships. The alternative structures that these cooperative relationships may take are described and a set of behavioral research propositions linking structure, process, and effectiveness are developed.

# UNIONS

### *The Nature of Unions*

Unions have both voluntary and involuntary, democratic and nondemocratic characteristics. The process of union organizing is voluntary and democratic. It typically begins when a group of employees conclude that they would be better off if they could deal collectively, rather than individually, with their employer. Frequently, these employees will contact an international union (usually one which represents other employees in their industry) to help them organize. If they can then persuade a majority of their fellow workers to designate the union as their representative in bargaining with the employer about wages, hours, and conditions of employment, that union will become the employees' exclusive agent in bargaining with the employer.[1]

Joining the union which already represents the employees at one's place of work may be involuntary, if the collective bargaining agreement contains a union security clause requiring union membership as a condition of employment.[2] Of course, the contract is unlikely to have a security clause unless union members want such a clause, since the terms of the contract are negotiated by union representatives and typically are subject to member ratification.

When union membership is not required as a condition of employment, membership is voluntary but there are strong psychological reasons to join the union. The union is the exclusive bargaining agent for all employees and no employee can negotiate individual terms and conditions of employment. The only way an employee can influence his terms and conditions of employment is by joining the union and participating in the process by which the union decides what conditions it will seek in bargaining with the employer.

While there are opportunities for participation in union affairs, since unions transact their affairs by majority vote at open membership meetings, participation in union affairs is typically low. Barbash (1967) reported that rarely more than five percent of the membership attends a routine meeting. Tannenbaum & Kahn (1958) put the figure at ten percent, and Anderson (1977) at eighteen percent. Even at a crisis meeting, for example, when a strike vote is to be taken, attendance is unlikely to be more than 60 percent of the membership (Barbash, 1967; Tannenbaum & Kahn, 1958).

To some extent, the low level of employee participation in union affairs can be explained by the involuntary aspects of union membership. Employees who join a union because they must do so to keep their jobs may not be highly committed to the union. Tannenbaum (1969) argued that most members view the union in instrumental terms (so long as the union is able to obtain satisfactory working conditions, they see no need to participate in union affairs). Anderson's (1977) data which show a negative correlation between participation (attendance at local meetings) and effectiveness in collective bargaining, support Tannenbaum's explanation.

The level of participation in local union affairs may also be due to a perception by individual members that their views can have little impact on the terms of the collective bargaining contract (Olsen, 1965). The development of centralized bargaining, in which a single contract is negotiated to cover many operations of a single employer (as in the automobile industry) or the operations of many employers (as in the trucking industry) has probably contributed to this perception even though many unions have elaborate procedures for local input into national bargaining demands and provisions for contract ratification at the local level. In industries in which bargaining takes place on the local level local control may be limited if, as is frequently the case, the actual negotiations are directed by a representative of the international union. Additionally, the international frequently insists that the terms of the local contract conform in many respects to those contracts negotiated by other locals of that international. To the extent that effective power to determine the union's bargaining position lies with the international union, union members are not likely to be motivated to participate in local union affairs.

While unions are governed by democratic principals, they have a tendency to become oligarchies. The Labor Management Reporting and Disclosure Act of 1959 requires unions to provide members certain democratic rights, including the right to run for office, to nominate candidates for office, and to speak freely in support of candidates. But, union constitutions typically fail to provide executive boards and conventions with enough power to balance that of the union president (Barbash, 1967). Thus, while unions are, in principal, democratic, in practice they tend to be oligarchical.

In sum, unions have an unusual set of organizational characteristics. While they are aptly characterized at the time of formation as voluntary, participatory, and democratic, they subsequently tend to become involuntary, oligarchic, and marked by a low level of membership participation. The relationship of a union member to the union is similar to the relationship of a citizen to the state. Just as the citizen is required to pay taxes to support the state and is bound by the laws passed by the legislature, so is the union member required to pay union dues and bound by the contract negotiated by the union. Just as the affairs of the state are handled by full time, elected leaders and appointed administrators, so are those of the union. And, just as the level of citizen participation in the affairs of state is low—except in time of crisis—so is the level of member participation in the affairs of the union low—except in times of crisis.

*Behavioral Research on Unions*

There has been substantial research on union democracy (see Strauss, 1977 for a recent review), but there are other aspects of union activities on which there has been much less research; among these are union organizing and the relationship between union democracy and collective bargaining effectiveness. The following section develops propositions for research on union organizing.

### *Union Organizing*

*The Psychological Contract in a Nonunion Firm.* In nonunion firms the relationship between employer and employee can be conceptualized in terms of a psychological contract (Schein, 1965; Stagner & Rosen, 1965). While some conditions of the employment relationship (like wages and hours) are specified, most conditions are not. The employment relationship is based primarily on unstated expectations of employer and employee about the other's behavior. The employee expects the employer to provide reasonable working conditions, and the employer expects the employee to produce a reasonable amount of work. In addition, when the employee joins the organization he implicitly accepts the employer's authority to direct his work (Schein, 1965; Weick, 1969).

While each party to the psychological contract recognizes that the other party expects him to behave in a certain way, as long as those behaviors are not specified, their interpretation is ambiguous. When the employer's expectations about employee behavior are violated, he can enforce his interpretation of the psychological contract through his authority system (Schein, 1965). When the employee's expectations about employer behavior are violated, he has little power to enforce his view of the psychological contract. He can threaten to quit but such a threat will not be particularly powerful if he is easily replaceable (Pfeffer, 1978). Depending on his job, the employee may be able to commit sabotage, slow down production, or lower production quality. None of these behaviors is legitimate, however, because each violates the employee's implicit agreement to accept the employer's authority to direct his work (such illegitimate behaviors may not be particularly powerful when committed by individuals). While the organization cannot function unless employees consent to its authority system, it can function very well if only isolated individual employees withhold their consent. Thus, the employer has greater influence over the terms of the psychological contract than do individual employees. Individual employees, however, can increase their control over the psychological contract by pooling their influence. One way for them to accomplish this goal is to organize a union.

> *Proposition 1*: An employee's motivation to organize a union is based on dissatisfaction with the employer's failure to fulfill his part of the psychological contract.

There is substantial evidence for Proposition 1 in four recent empirical studies. Employees who are dissatisfied with economic and working conditions say they would vote for union representation, if given the chance. Satisfied employees say they would not (Kochan, 1978). Organizational units in which employees are dissatisfied with economic and working conditions are more likely to experience union organizing activity (card signing drives, elections, etc.) than units in which employees are satisfied (Hamner and Smith, 1978). Employees who are dissatisfied with pay, supervision, and promotional opportunities actually vote for union representation in significantly greater number than employees who are satisfied with these characteristics of work (Getman, Goldberg & Herman, 1976). Dissatisfaction with job content is a less-important factor in union organizing than dissatisfaction with working conditions (Getman, Goldberg & Herman, 1976; Kochan, 1978; Schriesheim, 1978).

> *Proposition 2*: An employee's motivation to organize a union is based on dissatisfaction with his individual ability to change his situation in the organization.

Schein (1965) argued that an employee's acceptance of the organization's authority system rests on his "sense of being able to affect the authority directly and to change his situation in the organization" (p. 11). When an employee's expectations are violated, he is likely to try to enforce his view of the psychological contract by asking the employer to change the unsatisfactory condition. If the employer refuses to make the change, the employee may conclude that he has no influence over the terms and conditions of his employment. In this case reactance theory (Wortman & Brehm, 1975) implies that the employee is likely to respond by becoming increasingly motivated to influence working conditions, and by becoming increasingly hostile or aggressive toward the employer (the agent responsible for the loss of influence). Kochan (1979) reported significant correlations between employees' desire for on-the-job influence and perceived difficulty in exerting influence and willingness to vote for union representation if an election were held on their job.

> *Proposition 3*: Efforts to organize a union will be preceded by the formation of a coalition of employees who share the opinion that the employer has violated their individual psychological contracts.

Initially, a dissatisfied employee will probably attempt to add meaning to his opinion that the employer has violated the psychological contract by seeking to confirm it among other employees (Weick, 1969). If the employee finds no support among other employees, he may revise his opinion. Other employees, however, may have had their own expectations violated and support the first employee's opinion. It does not matter if the opinions of different employees were drawn from different incidents. The opinions of each employee reinforce those of the other employees (Newcomb, 1956). Perceiving that another person has similar attitudes toward an object (in this case the employer) not only validates one's own interpretation of that object (Weick, 1969), but also increases the attraction among the persons sharing the attitudes (Newcomb, 1956). Shared opinions provide a basis for new social structures (Allport, 1962). Individuals coalesce around primitive emotional issues because the simplest cognitive structure for interpreting events is affect, the primary determinant of interpersonal attraction is affective similarity, and the best way to get a stable interpretation of events is to find others with the same opinion (Weick, 1969). Dissatisfied employees form anti-employer coalitions because their shared attitudes toward their employer enhance their interpersonal attraction. The coalition, once formed, should be stable, because the shared attitudes are mutually reinforcing and provide a basis for interpreting subsequent employer actions.

*The Determinants of Collective Action.* Dissatisfied employees, even after they have coalesced, do not necessarily organize unions. If the employer acts promptly to deal with the issues that caused the dissatisfaction, the shared dissatisfaction may dissipate and the coalition may dissolve; if he does not, the situation is ripe for union activity. The coalition, formed on the basis of shared opinions, will itself provide a basis for collective action.

> *Proposition 4*: The likelihood that a coalition of employees will try to organize a union depends on (1) their understanding of collective action; (2) their belief about the relative value and likelihood of positive versus negative outcomes of unionization.

The concept of collective action is simple. An individual employee who withholds his labor in an effort to influence employer action is likely to be unsuccessful, either because the loss of his labor has little impact on the employer, or because one person can easily be replaced. Whereas a group of employees withholding their labor has a much greater impact, both because of the group's ability to affect productivity, and because of the difficulty of replacing the entire group. Thus, collective action provides a means for employees to exert control over their working conditions. Such collective action is the very essence of unionization.

Employees will not unionize unless they believe that to do so is likely to improve the unsatisfactory aspects of their working conditions. There is no certainty that collective bargaining will improve wages or working conditions. Some employers cannot increase their labor costs and remain competitive. Other employers will not meet union demands unless the unions can compel them to do so by a strike. Some employees may fear economic losses or loss of their jobs if strike activity is necessary. Other employees, who recognize the power of collective action may, nevertheless, find the concept distasteful. Collective action implies a loss of individuality. The employee's individual relationship with the employer is replaced by a collective relationship. Skilled employees may be less well off under a collective system in which their individual advantages will go to serve the collective good. Finally, some employees may fear employer reprisals for union activity. Certainly, employees know that employers do, on occasion, discharge employees for union activity, even though such discharges are illegal (NLRA §8 (a)(3); see also Greer & Martin, 1978, for a discussion of the reasons why an employer might find it economically advantageous to engage in such illegal activity).

Whether or not a coalition of dissatisfied employees will try to organize a union depends upon whether, on balance, they believe that the likelihood of the positive outcomes of unionization outweigh the likelihood of negative outcomes. This belief may not be related to previous experience with unions. Forty-three percent of the employees in the Getman, Goldberg & Herman study had been union members on other jobs, and 75 percent reported that a member of their immediate family had been a union member. Yet, these employees were not more favorable toward unions than employees with no direct contact with unionization. Employees who are favorable toward unions in general, and who believe that a union can be instrumental in changing unsatisfactory working conditions, indicate they are willing to vote for union representation if an election were held on their job (Kochan, 1979); and actually do vote for union representation (Getman, Goldberg & Herman, 1976; see also Schriesheim, 1978 for post election data).

*Moving from a Coalition to a Majority.* In order for the coalition to organize a union, it must have the support of a majority of employees. If the majority is dissatisfied with working conditions and in favor of unionization, the coalition must protect the majority's solidarity in the face of a possible anti-union campaign. If a majority is dissatisfied but uncertain of their willingness to engage in collective bargaining, the coalition must convince them that the benefits of collective action are more valuable and more likely than the adverse consequences. If a majority of employees is satisfied, the coalition may be able to convince them to change their opinions, but the task would be formidable.

> *Proposition 5*: The coalition's success at organizing a union depends on (1) the coalition's ability to gain the support of a majority of the employees; (2) the coalition's ability to protect the solidarity of the majority in the face of the employer's anti-union campaign.

*Gaining the Support of the Majority.* The coalition's most effective organizing tactic is probably an emotional appeal. Simmel (1950) argued that because group members are initially linked by shared affect, they will be persuaded more by emotional than intellectual appeals. Theoretically, the union campaign is supposed to provide employees who know little about unionization, with information about unions. In fact, the coalition probably gains most of its supporters through emotional appeals, emphasizing that many employees are dissatisfied with working conditions, that all have a common fate working for the same employer, and that the way to change the situation is through collective action.

The union meeting is the best forum for making this emotional appeal because it provides a graphic demonstration of the appeal. It shows employees the degree of support for the union among their fellow employees and it provides an opportunity to vent anger toward the employer. The reciprocal interactions among employees at union meetings reinforce anti-employer and pro-union opinions. Getman, Goldberg & Herman (1976) found that employees who were not initially in favor of union representation were more likely to vote for union representation if they attended a union meeting. Their opinions may have changed before the meeting, but in that case the meeting could have served to reinforce those opinions. It is likely that union meetings had such a reinforcement function for employees who were initially in favor of union representation since those who attended union meetings were more likely to vote for union representation than those who did not attend.

The union meeting may also be an effective forum for influencing employees' beliefs about the likelihood of positive versus negative consequences of unionization. The meeting may provide information about the kinds of benefits that the union has obtained for employees elsewhere—information intended to increase employees' beliefs in the likelihood that the union can get them benefits. The meeting may also allay fears of the likelihood of negative consequences of collective action. The fact that strikes are only called after a majority vote of union members, and that the employer cannot by law take reprisals against employees for union activity are frequent themes at union meetings (Getman et al., 1976).

While the information about the likelihood of benefits from collective action may influence some employees, Getman, Goldberg & Herman (1976) believe that those employees who fear the negative consequences of collective action will never get to the union meetings in the first place.

*Protecting the majority from the employer's anti-union campaign.* Employers use three basic arguments to dissuade employees from collective action: threats and acts of reprisal (e.g., threats to close the plant if the union wins, firing union supporters); promises and grants of benefits (e.g., promises to change or the actual change of unsatisfactory conditions contingent on the employees voting the union down); and appeals to the uncertainty of union representation (e.g., "you don't know what it is going to be like with the union around").

Persuasive communications are notoriously ineffective in changing attitudes and behavior (McGuire, 1968a; Sears & Freedman, 1967). Early political researchers (Campbell, Converse, Miller, & Stokes, 1964; Lazarsfeld, Berelson & Goudet, 1968) found evidence of selective exposure to political campaigns, and inferred that persuasive communications fail to change attitudes and behavior because people do not expose themselves to communications that are anticipated to be contrary to their opinions. Getman, Goldberg & Herman (1976) found no evidence that any employer campaigning was effective in influencing union supporters to vote against union representation. Herman (1976) discussed the utility of three theories of attitude change—consistency, complexity, and satiation—to explain the failure of employer attempts to dissuade employees from voting for union representation. *Consistency theory* assumes that there is a motivated tendency to seek out supportive and avoid nonsupportive information (McGuire, 1968b). When nonsupportive information is encountered, it is rejected as unreliable, or distorted to fit into the current belief system (Pepitone, 1968). *Complexity theory* assumes that some people prefer social stimuli that are multidimensional and even nonsupportive with their own belief system (Bieri, 1968). These people can integrate reliable but nonsupportive information by differentiating or adding dimensions to their current belief system. *Satiation theory* predicts that people will avoid exposure to familiar information regardless of whether it is expected to be supportive or nonsupportive of the current belief system (McGuire, 1968b). If an issue has few viable alternatives (e.g., vote for or against union representation) people may not expect to learn any information from a particular communication that would stimulate them to change their attitudes or behavior. Thus, they will exhibit a pattern of underexposure to persuasive communications and little attitude or behavior change. Herman (1976) concluded that consistency, complexity, and satiation theories may each account for the failure of persuasive communications to influence some people under some conditions, but none provides a general explanation for the frequent failure of persuasive communications to change attitudes and behaviors in union organizing campaigns.

*a) Threats and acts of reprisal.* Getman, Goldberg & Herman (1976) found no evidence that threats and acts of reprisal cause employees who have initially decided to support union representation to switch and vote against union representation. Union supporters in the Getman, Goldberg & Herman (1976) study were, nevertheless, particularly sensitive to the employer's statements about his opposition to unionization and his economic power over them. Many union supporters interpreted these statements as threats yet they voted for union representation. These union supporters may have rejected the employer's campaign as unreliable. Alternatively the employer's campaign may have reinforced their prior beliefs that only unionization would provide them with sufficient power to balance the employer's power.

*b) Promises and grants of benefits.* Getman, Goldberg & Herman (1976) found no evidence that promises and grants of benefits cause initial supporters to switch and vote against union representation. Union supporters frequently reported that the employer stated he was willing to improve unsatisfactory conditions without a union, yet, they voted for union representation. If union supporters were distorting the employer's campaign statements they would be expected to report promises and grants of benefits less frequently than company supporters, yet they did not. Apparently, some union supporters rejected the promised changes as insufficient. Others recognized that they had no assurance that the employer would continue to provide good working conditions unilaterally and would have no influence over him to do so once the union organization drive was defeated.

This discussion of the ineffectiveness of threats and acts of reprisal and promises and grants of benefit in dissuading employees from voting for union representation relies on both consistency and complexity theories. *Consistency theory* predicts that nonsupportive campaign information will be either rejected as unreliable or distorted to fit into the prior cognitive structure. *Complexity theory* predicts that nonsupportive information will be assimilated by differentiating the cognitive structure. Employees apparently rejected some of the employer's campaign statements as unreliable; and accepted other statements, but interpreted them as evidence in support of their pro-union position, not as reasons to vote against the union.

*c) Appeals to uncertainty.* The union cannot assure employees that conditions which they find unsatisfactory will be alleviated by collective bargaining. The employer may be able to take advantage of this uncertain situation by emphasizing that employees do not know what benefits or consequences will result from union representation. Employees who are dissatisfied with particular working conditions may be unwilling to trade the known status quo for the uncertain future with the union since uncertainty is, in theory, an aversive state (Weick, 1969).

*Summary.* The process of union organizing has its beginning in employee dissatisfaction with the way the employer is fulfilling his part of the psychological contract. A dissatisfied employee realizes that he has no power over the employer to enforce his view of the psychological contract when acting alone. He also learns that other employees share his opinions of the employer. Shared opinions provide a stable basis for an anti-employer coalition. The anti-employer coalition, in turn, provides a basis for union organizing. The decision to organize is based on the coalition's understanding of collective action, and beliefs about the relative value and likelihood that a union will provide benefits versus negative consequences. The coalition's success at organizing a union depends on its ability to gain and protect the support of a majority of employees.

## TRADITIONAL ADVERSARIAL UNION-MANAGEMENT RELATIONS

### *The Nature of Union-Management Relationships*

*Interorganizational Characteristics.* Union-management relationships are a particular type of interorganizational relationship. They have some features of exchange relationships (Levine & White, 1961; Van de Ven, 1975; Turk, 1973), some features of power-dependency relationships (Benson, 1975; Schmidt & Kochan, 1977), and some features that neither *exchange relationships* or *power-dependency relationships* have.

In a pure *exchange relationship,* two or more organizations recognize that by sharing information or resources, each can facilitate the attainment of a common goal. To some extent union and management have a common goal—to maximize the organization's profits. When profits are high, it is easier for management to satisfy the union's financial demands. Of course, the more management does satisfy those demands, the less will be available for management salaries, dividends, and investment. At this point the goals of management and the union diverge and the relationship becomes more similar to a *power-dependency relationship* in which two or more organizations recognize that they are mutually interdependent because each can facilitate the other's goal attainment. In a union-management relationship, management needs the labor of the union-represented employees to function, and the union needs management to provide jobs so that it will have employees to represent.

In general, the exchange and power-dependency relationships are voluntary associations among organizations. Each organization enters into such a relationship because it wants the support or service that the other can provide. Union-management relationships are not voluntary. If a union represents a majority of the employees, management is required by law to bargain with that union. Indeed, the government regulates all aspects of the union-management relationship to a far greater extent than any other interorganizational relationship. In addition to requiring the union and management to bargain, the law specifies the subjects about which they must bargain and their bargaining practices. The outcome of their bargaining is also subject to regulation through legislation dealing with minimum wages, pensions, health and safety, and fair employment practices.

*A Structural Model.* Figure 1 represents a structural model of a union-management relationship. (See Benson, 1975 for a discussion of structural models of interorganizational relationships, and Dunlop 1958 for a discussion of another structural model of an industrial relations system.) Figure 1 shows two overlapping shapes: the shape on the right represents the organization; the shape on the left represents the union. The overlap between the two shapes shows that unionized employees have organizational ties to both the employing organization and the union. Local union leaders may also be employees and are therefore located within the boundaries of the organization. Figure 1 shows government regulation of three relationships: that between the union and the firm (regulated by the National Labor Relations Act); that between union leaders and union members (regulated by the National Labor Relations Act, the Labor Management Reporting and Disclosure Act, and Title VII of the Civil Rights Act); and that between the firm and its employees (regulated by the National Labor Relations Act and Title VII of the Civil Rights Act).

*The Effectiveness of a Union-Management Relationship.* Industrial relations scholars typically have used the parties' goals as criteria for evaluating the effectiveness of a union-management relationship. (See Derber, Chalmers & Edelman, 1961 for a review of this literature.) A relationship in which the parties are achieving their goals is assumed to be *effective*.

There are several problems with the goal criterion. While all the parties share the goal of the continuance of the relationship many of their other goals are in conflict. Table 1 presents goals typically attributed to each of the parties in an industrial relations system. Union and management are assumed to be adversaries in this system. Management's goal of profitability, for example, may be in direct conflict with the union's goal of increasing wages or improving working conditions. The government's social policy goals may also conflict with union or management goals. The

*Figure 1.* An industrial relations system

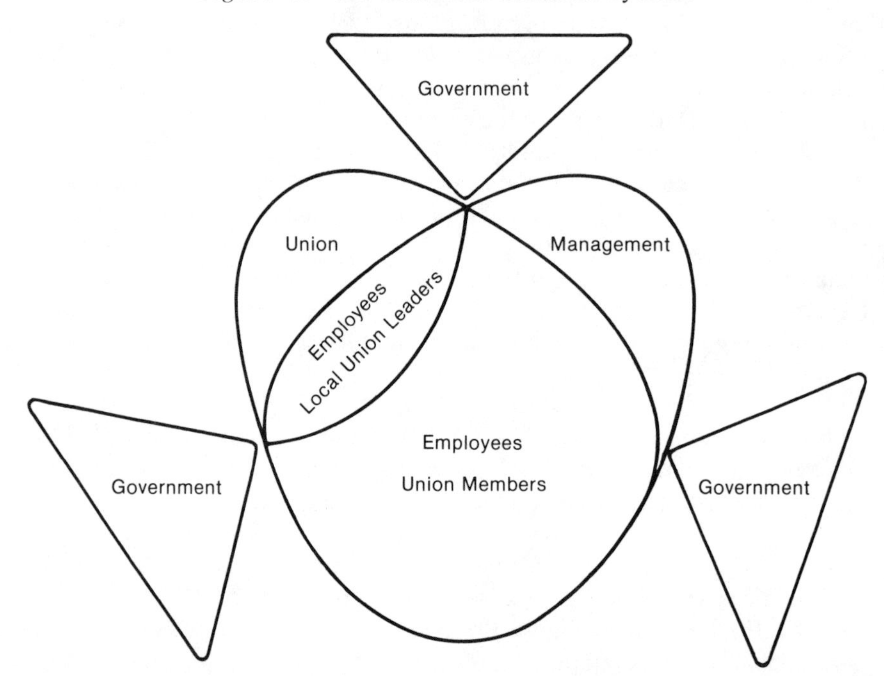


government's desire for union democracy, for example, may conflict with the union leaders' desire for political security; and management's desire for the unhampered right to direct the work force may be limited by the government's definition of fair employment practices. Thus, one problem with the goal criterion is that it is impossible to determine whether the system is effective when the parties' goals conflict—where one party is satisfied with its goal attainment and another party is dissatisfied.

There is another perhaps more basic problem with the goal criterion of organizational effectiveness. Hall (1972) points out that it is almost impossible to define what an organization's goals are. Different people in the organization (e.g., top management, middle management, stockholders, etc.) view the organization's goals differently because they have different perspectives and values. Hall argues that even if a statement of official goals exists, what the organization actually does is determined by operative goals, which, because they must be relevant to the organization's environment, may be unrelated to official goals.

The major alternative to the goal approach is the systems theory perspective (Katz & Kahn, 1978; Yuchtman & Seashore, 1967; Evan, 1976). From this perspective, an organization is a social system which, in interaction with its environment, activates input, transformation, output,

*Table 1.* Traditional Goals of Workers, Union, Management, and Government

| |
|---|
| WORKERS |
| Good extrinsic working conditions—wages, benefits, fair supervision, safe and pleasant working conditions, job security |
| Good intrinsic working conditions—interesting work that provides a sense of accomplishment |
| Participation in decisions which affect work and working conditions |
| System for redressing grievances |
| UNIONS |
| Achievement of members' goals for working conditions |
| Survival and growth |
| Political security for union leaders |
| MANAGEMENT |
| Profitability |
| Preservation of management's prerogative to direct the work force |
| Job security/advancement |
| GOVERNMENT (public policy goals) |
| Democratic unionism |
| Healthy economy |
| Safe working conditions |
| Profitable firms |
| Fair wages |
| Fair and nondiscriminatory employment |
| Noninflationary wage agreements and pricing policies |

and feedback processes. An organization's effectiveness is determined by its performance on all four processes and the interrelationships (Evan, 1976). One solution to the dilemma of measuring the effectiveness of a union-management relationship is to adopt the system's perspective and focus on intermediate outcomes of the parties' interactions as criteria of effectiveness. Table 2 presents two sets of such outcomes: those associated with collective bargaining and those associated with contract administration.[4] Some of these interactions are between management and union leaders (contract negotiations, grievance administration); between employees and management (strikes, turnover, absenteeism); between the government and management (NLRB orders to bargain in good faith); between the government and the union (Taft-Hartley antistrike injunctions); or between nongovernment third parties, such as arbitrators, and both the union and management.

*Table 2.* Criteria for Evaluating Union-Management Relationships: Intermediate Outcomes of the Parties' Interactions

| *Collective Bargaining* | *Contract Administration* |
|---|---|
| Duration of negotiations | Frequency of grievances |
| Contract ratification vote outcome | Level in grievance procedure at which grievances are settled |
| Frequency and duration of strikes at the termination of the contract | Frequency of strikes during the term of the contract (wildcats) |
| Use of mediation, arbitration or both | Absenteeism |
| Government intervention, e.g., court orders to bargain; Taft-Hartley | Sabotage |
| | Turnover |
| | Slowdowns |
| | Government intervention, e.g., to enforce an arbitrator's award, or to enjoin a wildcat strike |

Intermediate outcomes are appropriate, perhaps essential, for comparative research. Goals, if they can be specified, will have different meanings in different union-management relationships. The meaning of intermediate outcomes should remain constant across different union-management relationships. Intermediate outcomes are also useful for evaluating the effectiveness of a particular union-management relationship. The general character of intermediate outcomes allows the researcher to place the particular relationship in context. The intermediate outcome is, however, unlikely to have the same impact on each of the parties (Kochan, personal communication). For example, a strike, which as an intermediate outcome indicates a failure of union-management bargaining, may, nevertheless, result in the achievement of the union's goals. Thus, in evaluating the effectiveness of a particular union-management relationship, research must determine how intermediate outcomes impact on each party's goals.

An effective industrial relations system is not necessarily a low conflict system. Brickman (1974) characterizes conflict situations as: unstructured, in which there are no social restraints on either party; partially structured, in which a definite area of conflict exists within an agreed set of social restraints; and fully structured, in which each party's behavior is prescribed by a social norm. A traditional industrial relations system is a partially structured conflict situation. An effective partially structured conflict relationship is one in which the conflict can be dealt with through normative channels. Hence, an effective relationship might include: occasional brief strikes at the termination of the collective bargaining contract; a grievance system in which most grievances are settled promptly. Such a system does not need government intervention nor is it without conflict; but the conflict is contained by the norms of the collective bargaining system. During negotiations each party has a normative right to exert power over the other. But once a contract is signed, management knows it must adhere to the contract and the union recognizes that it has no rights other than those provided by the contract. There are bound to be disagreements over contract interpretation, but in an effective relationship the parties will resolve these disagreements in the early stages of the grievance procedure.

*Behavioral Research on Union-Management Relationships*

Why are some union-management relationships more effective than others? Much of the existing research on this question has been done by labor economists and industrial relations scholars, and does not take a behavioral approach. (See preceding chapter in which Thomas Kochan reviews this research.) The following section presents a rather eclectic set of propositions for behavioral research on the effectiveness of industrial relations systems. The propositions attempt to delineate how the social structures of the communities in which workers live and the characteristics of their jobs, working conditions, and union organization are related to strike behavior.

These propositions are not meant to be a complete enumeration of the factors contributing to union-management effectiveness or even a complete enumeration of behavioral as opposed to economic factors contributing to union-management effectiveness. Rather, the propositions were chosen to illustrate a behavioral research approach. For the sake of simplicity, the dependent variable used throughout the section is strikes. Other intermediate outcomes of the parties' interactions may, of course, also be used in behavioral research on the effectiveness of union-management relationships.

*Social Structure.* The preeminent systemic theory of industrial unrest is Kerr & Siegal's (1954) theory of social conditions. They concluded that when social conditions cause workers to cluster into a homogeneous mass isolated from the rest of society, the workers will be particularly strike-prone. Kerr & Siegal (1954) treat strikes as spontaneous collective action born of shared frustration with common working or living conditions. Such a "riot" model of strikes fails to take account of the fact that shared conditions and attitudes are stimuli to coalition formation (see the discussion of Proposition 3). Hence, workers clustered into an isolated, homogeneous mass may be strike-prone because their social conditions support the formation of a social organization which can, in turn, support strike action.

*Proposition 6*: Workers who form an occupationally homogeneous mass isolated from the rest of society will be more strike-prone than workers who are integrated into a heterogeneous social milieu.

Kerr & Siegal's theory has been tested infrequently. Shorter & Tilly (1974) grouped French *arrondissements* (like U.S. counties) into mono-industrial, poly-industrial and metropolitan, and found strike rates to be highest in the metropolitan group. Accordingly, they rejected the Kerr & Siegal theory. Lincoln (1978) also rejected the theory. Although he found strike rates in 78 SMSAs to be related to occupational homogeneity, and occupational homogeneity related to the degree of unionization. The validity and reliability of the operationalizations of Kerr & Siegal's major constructs, "social isolation" and "occupational homogeneity" are questionable in these two studies. Actually because each study used community as the unit of analysis, social isolation was never measured directly.

A proper test of Kerr & Siegal's theory must use occupational groups in a number of different communities as the unit of analysis. A behavioral test of the theory would require the collection of data on group members' attitudes and frame of reference as well, since the psychological theory of relative deprivation (Crosby, 1976) predicts the opposite of proposition 6. Crosby (1976 p. 85) argues that "deprivation is relative not absolute . . . it is often true that those who are the most deprived in an objective sense are not the ones most likely to experience relative deprivation." Marx agreed. "A house may be large or small; as long as the surrounding houses are equally small it satisfies all social demands for a dwelling. But, let a palace arise beside the little house, and it shrinks from a little house to a hut" (quoted in Crosby, 1976, p. 85). The theory of relative deprivation predicts greater strike-proneness among workers who live in occupationally heterogeneous communities and who are not socially isolated. The social conditions of such communities make the contrasts greater. Some groups of workers will feel deprived and since deprivation is assumed to be an adversive state, relative deprivation should lead to strikes.[5]

*Job Power*. The major industrial relations theory used to explain strike frequency is power. Employees derive more power from their jobs in some industries than in others (Dunlop, 1958). Among the most powerful job holders are those whose jobs are crucial to continued operation in an integrated production process. Hence, if employees who have power use it, an organization's technology will be related to strike frequency.

*Proposition 7*: Assuming that workers with power will use it, workers will strike more frequently when their jobs are so strategically located in the production process that they can easily halt production.

Butler's (1976) time series study of strike rates in the American and British coal mining industries provides strong support for the assumption underlying Proposition 7 (employees who have power will use it). Butler (1976) reported that the number of strikes over a twenty-six-year period reflected the miners' power (measured by the proportion of total annual energy produced from coal and annual coal consumption in each country). The switched replication design, used in this study, is very powerful in ruling out alternative third variable explanations. While the time series analysis used by Butler does not make a strong statement about causation, the alternative explanation for his data (that strikes cause miners' power as measured by coal consumption) is illogical. A Fourier analysis of Butler's data, such as the one Losada (1978) used to show the lead and lag relationships in consumer attitudes and behavior would complete Butler's study.

There is some systematic empirical research relevant to Proposition 7. Woodward (1965) reported that organizations with large batch and mass technologies have poorer labor relations than organizations with unit and small batch or process technologies (though her dependent variable is not specified). Eisele (1974) found that large organizations with batch or mass technologies had more strikes per year than large organizations with other technologies (though this conclusion is not supported by some of the data in his study). Finally, Kuhn (1961), studying strikes during the term of the contract, concluded that strike frequency by a work group was a function of the technological conditions of their work. The more vulnerable production was to distruption by the group, the more frequently the group struck.

*Dissatisfaction and Alienation*. Two different psychological theories may account for strike action. According to frustration-aggression theory, workers strike to vent their feelings of frustration with conditions over which they have no control and to experience brief moments of power. In this case a strike is a spontaneous revolt by an alienated group. (See Shorter & Tilly, 1974, pp. 336–349; Dollard, Doob, Miller, Mower & Sears, 1939; Berkowitz, 1961, 1962, 1972.) According to instrumentality theory, a strike is a calculated action by a group of workers who believe that through their collective power they can change an unsatisfactory working condition (Vroom, 1964).

*Proposition 8a*: A strike is a spontaneous revolt by an alienated group.
*Proposition 8b*: A strike is a calculated action by a dissatisfied group.

The key point distinguishing these propositions is that the dissatisfied group in 8b believes the strike will help resolve the problem while the frustrated group in 8a has no such belief. Proposition 8a has long been assumed to explain strikes during the term of a collective bargaining agreement (wildcat strikes) and Proposition 8b has been assumed to explain strikes at termination of the contract. Brett & Goldberg (1979), studying wildcat strikes in the coal mining industry, reported that miners believe that wildcat strikes will prompt management to correct unsatisfactory working conditions. These results suggest that wildcat strikes may not be a spontaneous revolt but a calculated action.

Both of these propositions further assume that conditions in the work environment create the dissatisfaction or alienation that leads to industrial unrest. This assumption seems justified since many researchers have found that organizational characteristics are associated with employee attitudes (see Berger and Cummings, 1979, for a recent summary). It seems possible, then, that organizational characteristics are related to strikes.

*Size.* The size of an organizational unit (number of employees) has been shown to be associated with employee dissatisfaction (Porter & Lawler, 1965; Berger & Cummings, 1979) and strike activity (Britt & Galle, 1974; Shorter & Tilly, 1974; Brett & Goldberg, 1979). The rationale for the latter correlation is frequently given in terms of the former (Scott & Homans, 1947; Lincoln, 1978), but there is no strong empirical research identifying the organizational conditions associated with large organizational units which stimulate strikes. Brett & Goldberg (1979) argue that when other organizational characteristics are held constant, size and strike frequency are correlated because top management at large units has the same responsibilities as top management at small units, but has more repetitions of those duties, and more responsibility. Hence, the number of potential labor problems increases with the size of the work force but the capacity of top management to deal with these problems remains constant.

> *Proposition 9*: Holding other organizational characteristics constant, large organizational units will have more strikes than small ones.

*Technology.* Woodward (1965) and Eisele (1974) both reported a relationship between technology and labor relations. Industries characterized by large batch and mass technologies had poorer labor relations than industries characterized by unit or process technologies. Woodward (1970) attributed this relationship to the different control systems associated with technologies, not to the power such technologies give to certain jobs. In particular, she cites broad supervisory span of control (characteristic of large batch and mass technology) and the fact that employees resent supervisory authority more than authority imposed by the system, as factors contributing to poor labor relations.

> *Proposition 10*: Strikes may be more frequent at organizations with certain technologies because the control system associated with those technologies produces unsatisfactory conditions.

The impact of technology on strikes may also be due to the unsatisfactory or alienating characteristics of jobs associated with certain technologies. Blauner (1964) argues that workers in industries characterized by routine automated technology are alienated because they are powerless regardless of whether or not they are unionized (they lack freedom and control), their work is meaningless (their work has no relation to the whole product), they are socially isolated (they are in, but not of, society), and self-estranged (their work is a means to an end, not an end in itself). Rousseau (1978) showed that jobs typical of routine and automated technologies lack such characteristics as skill variety, task identity, task significance, autonomy, and feedback. Employees whose jobs lack these characteristics tend to be less motivated and satisfied than workers in other jobs (Rousseau, 1978; Hackman & Oldham, 1976).

> *Proposition 11*: Strikes may be more frequent at organizations with certain technologies because the jobs associated with those technologies have dissatisfying or alienating characteristics.

*Structure*. The classic study of the relationship between strikes and organizational structure is Gouldner's (1954a, 1954b) study of the attempted bureaucratization of a gypsum mine. Gouldner's findings together with the research on distressed grievance procedures (Ross, 1963; Brett & Goldberg, 1979), suggest that organizational structures that are centralized and formalized are less effective in dealing with the unsatisfactory conditions which lead to strikes than are decentralized and less formal systems.

Management tends to move decision-making authority up the management hierarchy to increase coordination of policies and practices and to avoid union whipsawing (Barbash, 1975). First-level supervisors may resolve routine problems for which they have standard operating procedures but ignore nonroutine problems over which they have no authority. In a centralized, formalized structure, nonroutine problems frequently must be started through the grievance procedure before they are seriously considered by management. While centralization and formalization of management decision making may succeed in resolving grievances in a coordinated fashion (Thompson, 1967), centralized decision-making structures easily become overloaded (Pfeffer, 1978), particularly if all decisions—routine and nonroutine—are sent up the hierarchy for resolution. The system can also become overloaded if there are standard operating procedures for only a few issues.

> *Proposition 12*: Strikes may be more frequent at organizations with centralized, formalized structures because such structures are less effective in dealing with the unsatisfactory conditions which lead to strikes than decentralized less formal structures.

Organizational structures that remove first-level supervisors' authority to resolve employee problems may also be associated with strikes if supervisors exert their remaining authority to direct the work force in an arbitrary and capricious fashion which causes friction.

The human relations movement made foremen the scapegoats for dysfunctional employee behavior (Whyte, 1962). Yet, years of supervisory training have not alleviated dissatisfaction with supervision. It seems likely that characteristics of supervisors' jobs stimulate their behavior (Fox, 1966). One of the more important characteristics may be their perceived lack of influence to direct the work force. The removal of supervisors' influence has been recognized as a major problem in implementing organizational change (Schlesinger & Walton, 1976) but has yet to be studied as a factor contributing to labor unrest.[6]

> *Proposition 13*: Strikes may be more frequent at organizations with centralized structures because first level supervisors exert their remaining authority to direct the work force in an arbitrary and capricious fashion which causes friction.

*Union Democracy*. There is an inherent tension between internal union democracy and industrial peace. The imposition of democratic procedures on unions, such as those required by the Labor-Management Reporting and Disclosure Act (regular elections with equal opportunities to nominate and speak in favor of candidates) weakens union leaders by subjecting them to competition for union office. While such competition may serve the members' interests, it may also lead to frequent strikes (Bok & Dunlop, 1970). Union leaders, aware that they must satisfy the interests of a majority of the membership to retain their offices, may press demands that they know management will not accept in order to show the membership that they are trying to represent their interests. They may also hope that management's rejection of these demands will trigger a strike and that the strike will increase their control in the union.

Social psychological research shows that groups faced with crises or external threats coalesce, though they may not coalesce around their leadership. In the Robbers Cave experiments, intergroup conflict brought a significant change in the status relations within groups, including leadership changes (Sherif & Sherif, 1958). Other research points out a tendency to more centralized decision making after a crisis (Korten, 1962; Pfeffer & Leblebici, 1973) and greater influence for group leaders (Hamblin, 1958). Union leaders seeking to increase their power in the union by setting up an external threat are thus playing a risky game, since the opposition can attribute the necessity to strike to the current leaders' incompetence at the bargaining table.

The democratic requirement that union leaders defend their offices in regular elections may also lead to more grievances being taken to arbitration. A strong union leader, essentially immune from challenge, is free to dispose of employee grievances in a manner that he thinks will best serve the interests of the entire membership, dropping those grievances that are without merit or that will cost more to take to arbitration than they are worth. If, however, that union leader is subject to political challenge, he may press to arbitration any grievance filed by a union member who has the political strength to prove troublesome if his grievance is not pushed through the entire system. If a large number of grievances are taken to arbitration as a result of such political pressures, the resolution of grievances may become substantially delayed, leading to strikes because employees want their grievances settled promptly. Taking grievances to arbitration for political reasons may also contribute to employee distrust of the grievance procedure, since the union is likely to lose a high proportion of unmeritorious grievances.

> *Proposition 14*: Internal union democracy will result in more strikes and grievances because democracy weakens the leadership which then uses strikes and/or grievances to gain or protect its influence.

*Summary*. Factors influencing the effectiveness of an industrial relations system (Figure 1) may result from the system's relationship to its social environment—the power held by one or more parties to the relationship or the conditions under which two of the parties interact. Behavioral research propositions developed in the preceding section illustrate how community social structure and job power may support employee collective action against their employer. Other propositions indicate how characteristics of the work organization, such as size, technology, and organizational structure may contribute to unsatisfactory or alienating conditions or prevent prompt relief from such conditions, and so stimulate employees to strike. The final proposition illustrates how characteristics of the union organization may contribute to strikes.

## COOPERATIVE UNION-MANAGEMENT RELATIONS

*The Nature of Cooperative Union-Management Relationships*

In an adversarial system of union-management relations, the union's role is to gain concessions from management during collective bargaining and to preserve those concessions through the grievance procedure. The union is an outsider and critic. In a cooperative system, the union's role is that of a partner, not a critic, and the union becomes jointly responsible with management for reaching a cooperative solution. Thus, a cooperative system requires that union and management engage in problem solving, information sharing, and integration of outcomes.

Cooperative systems exist within the traditional, adversarial system as additional structures for decision making. Sometimes cooperative systems develop as a result of the recognition by both union and management that some of the issues over which they bargain in a traditional fashion might be more easily resolved through cooperative problem solving. Other cooperative systems develop because management wishes union collaboration in instituting organizational change.

In the United States, cooperative systems are voluntary. During World Wars I and II, and again in the early 1970s, the federal government encouraged the establishment of joint labor-management production committees (Gold, 1976), but it has never required unions and managements to form cooperative relationships as it does collective bargaining relationships, and it sets no norms for them.[7]

*Structural Models.* There are two basic structural models for cooperative relationships. In the single channel model as in the collective bargaining model, the union represents the employees (see Figure 2). In the dual channel model, the union represents the employees in the collective bargaining system but the employee representatives in the cooperative system are elected separately from the union.

The same four parties participate in cooperative systems as an adversarial system but their roles are somewhat different. The government's role is usually limited to supplying funds or expertise to support experimental cooperative projects. The role of the union is expanded in the single channel model, since it represents employees in both the collective bargaining and cooperative system. The role of the employees is expanded in the dual channel model, since employee representatives—who may or may not be union members or leaders—participate in the cooperative system. In a sense, management's role is diminished in both models, since by entering into a cooperative system management frequently agrees to discuss issues that previously have not been discussed with the union.

*Figure 2.* Models of adversarial and two modes of cooperative union-management relationships

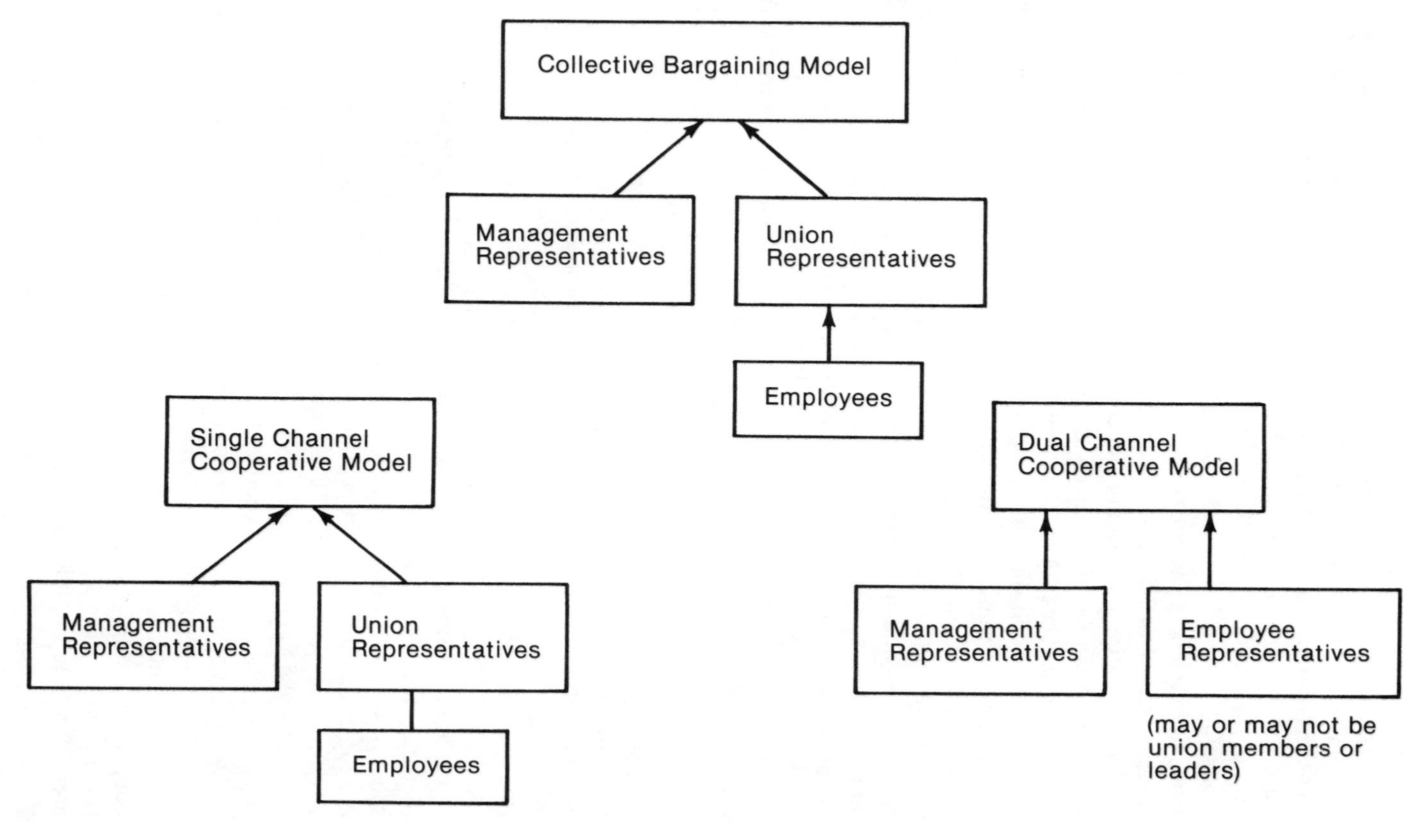

*Problems Associated with Cooperative Systems.* In the adversarial system, after union and management agree to a contract, management must interpret the contract. Management may do so in consultation with the union but if it does not, or if the union believes management's interpretation is wrong, the union can challenge that interpretation through the grievance procedure. In contrast, there are seldom formal organizational mechanisms for implementing cooperative agreements. Management's unilateral implementation of cooperative agreements violates the system's model of full collaboration in reaching decisions and joint ownership of those decisions. A similar violation of the cooperative model occurs if the union presses a grievance against the cooperative agreement.

There is, in addition, a blurring of functional distinctions in cooperative relationships, creating major problems for both labor and management representatives. In the single channel model, union and management representatives must play two roles simultaneously—one adversarial, one cooperative—and one role always has the potential of undermining the other (St. Croix, 1975). The union leaders' problems are exacerbated in the dual channel model, since the union has no legitimate control over the employee representatives in the cooperative system. Some union leaders fear that a successful dual channel cooperative system will undermine the union (Gold, 1976).

Another problem is determining which issues belong in which system (Kochan, Lipsky & Dyer, 1974). Cooperative systems appear to be more suitable for the resolution of integrative issues in which the parties ultimate goals do not conflict. Distributive issues, which involve the allocation of scarce resources, are better dealt with in the adversarial system. When an issue cannot be clearly characterized as integrative or distributive, the parties will have difficulty deciding in which system to treat that issue.

Yet another problem with many cooperative efforts stems from the fact that they are frequently introduced by government or third-party social scientists. While some theories of organizational design suggest that an interorganizational system will restructure in response to such an environmental event, Pfeffer (1978) argues that the environmental event will only stimulate the formalization of an interorganizational relationship which already exists informally. Pfeffer's argument, applied to a labor-management relationship, implies that cooperative systems must grow out of successful traditional systems. Indeed, a successful traditional relationship would seem to be a necessary precondition for an effective cooperative relationship. Parties that are unable to deal with issues in a traditional relationship without regularly resorting to economic warfare, are unlikely to be able to develop a successful cooperative relationship. Successful cooperation requires that the parties set aside their economic power, accept each other's legitimacy, and discuss issues in a trusting manner. Parties that deal with issues in a traditional relationship by recognizing the other's legitimate interests and by searching for outcomes which minimize each other's loss, have a basis on which to develop a cooperative relationship.

*The Effectiveness of Cooperative Systems.* The effectiveness of a cooperative union-management relationship should be evaluated on two different levels. Initially, the goal approach can be used to determine whether or not a cooperative relationship has achieved its purpose (assuming that the parties are able to articulate a joint purpose which is reasonably specific). Thus, a cooperative relationship set up to improve the quality of work life probably could not be evaluated but one set up to improve employee satisfaction with the work itself, safety, or employee perceptions of participation in decision making, could. Basic evaluation techniques and questions—such as whether a solution was effectuated, how it was effectuated, whether the solution was enduring, and whether the parties were satisfied with it—can be used to determine whether or not the cooperative structure achieved its purpose.

The major question about cooperative relationships, for scholars, is not whether they have, in general, achieved their purposes, but what their impact has been on the traditional relationship; this is a systemic evaluation question. It is not clear whether effective cooperative relationships spill over and make collective bargaining easier or whether employees will lose their regard for the union and collective bargaining when they are being represented in a cooperative relationship. Systemic research that focuses on cooperative issues, actors and processes, is needed to answer this question.

*Behavioral Research on Cooperative Systems*

A set of propositions for organizational behavior research on cooperative relationships is proposed in this section. The propositions identify relationships between issues, actors, processes, and effectiveness in cooperative relationships.

*Issues.* Those issues which are important to the union and typically are dealt with by collective bargaining can only be removed to a cooperative structure that is compatible with collective bargaining, i.e., a single-channel model. The argument for removing such issues to a dual-channel model is that more progress will be made in a structure which is differentiated from the collective bargaining structure (Driscoll, 1979). It is unlikely, however, that union leaders will grant enough power to the employee representatives in a dual-channel model to permit such a system to function effectively with respect to collective bargaining issues. For if the dual-channel model succeeds in resolving traditional union-management issues, the employees may see no further need for the union.

*Proposition 15*: Single-channel models are more effective structures than dual-channel models for issues related to intermediate outcomes such as strikes or distressed grievance procedures, which are a direct function of the union-management relationship.

*Proposition 16*: Single-channel models are more effective structures than dual-channel models to search for solutions to issues that are high priority for the union, e.g., wages, job security, safety.

On the other hand, single-channel models may not be suitable structures for the discussion of motivational issues, such as job design. Unions have long been suspicious of management's motives for proposing job design and typically reject management's arguments that the change will benefit employees by making their jobs more satisfying. Instead, unions see in job redesign a threat to job security since management is unlikely to share the benefits of the increased productivity that accompanies such changes in employee satisfaction (Donahue, 1976; Abel, 1975).[8] Moreover, unions seem to fear that participation in a system that is focused on the resolution of problems and implementation of solutions outside of collective bargaining will result in a diminution of the union's role as critic and worker advocate (Bluestone, 1975).

*Proposition 17*: Dual channel models are more effective structures than single channel models to discuss and resolve problems related to worker motivation.

*Actors*. Walton & McKersie (1965) argue that individuals who are effective at collective bargaining may not be effective at problem solving, since problem solving requires many behaviors that are directly counter to those used in bargaining (like openness). However, if a collective bargaining relationship is effectively containing the parties' conflict, it is likely that the representatives in that system will be able to develop an effective problem solving relationship. Effective collective bargaining would seem to require a willingness to search for alternatives and compromise both of which are also characteristics of effective problem solving.

*Proposition 18*: An effective single-channel model existing within a successful collective bargaining system will be composed of union and management representatives who are the same as the collective bargaining representatives.

If a collective bargaining relationship has been ineffective in containing conflict the parties may be willing to try a problem solving relationship as a last resort. In this case union and management representatives in the cooperative system should be different from those in collective bargaining so that the cooperative system is not burdened with the adversarial experiences of the bargaining system. The representatives in the cooperative system should have a strong mandate to search for solutions and a commitment from the bargaining representatives to consider their proposals. However, Strauss, Miles & Snow (1974) warn that having different people representing the union in each system may generate problems of communication and coordination of activities. Dual representation might also generate intraunion leadership struggles.

*Proposition 19*: Effective single channel models existing within an ineffective collective bargaining system will be composed of representatives who are different from the collective bargaining representatives.

Research experience with job design and organizational development indicates that it is easiest to change an organization when the people affected by the change are involved in designing it (Alderfer, 1967; Coch & French, 1948; Lawler, Hackman & Kaufman, 1973). In order to capitalize on the motivating aspects of participative decision making, labor representatives to a cooperative system concerned with motivational issues should be employees not union leaders.

*Proposition 20*: Labor representatives to effective dual channel models will be elected from among the employees who are to be affected by the organizational change, not from the union leaders.

*Processes*. Because the issues proposed to be most conducive to resolution in the single channel model are so intimately related to collective bargaining issues and processes, a cooperative system that is advisory to the collective bargaining system is likely to be most effective. While the cooperative system may actually resolve issues, it lacks the fully developed structure of the collective bargaining system. Cooperative systems sometimes lack procedures for employees to express their approval or disapproval of a cooperative resolution, while the collective bargaining system usually provides for a contract ratification vote. Cooperative systems typically lack a method for employees to challenge the implementation of the cooperative resolution, while the collective bargaining system contains a grievance procedure. On the other hand, a cooperative system that provided structures for approval of and challenges to its resolutions would pose a substantial threat to the collective bargaining system.

> *Proposition 21*: Effective single channel models will serve in an advisory capacity to the collective bargaining system.

In contrast to single-channel models, the issues proposed to be most conducive to resolution in dual-channel models are not central to collective bargaining. Thus, this form of cooperative system should not be advisory to the collective bargaining system but should have the freedom to resolve issues and implement these resolutions. If it does not have such power, employees may regard its processes as meaningless, and be unwilling to participate. The risk of an irresponsible dual-channel model may be controlled by a shelter agreement[9] which allows the union to become the final arbitrator of what goes on in the cooperative system; but also allows the union to step back and let the dual-channel cooperative system experiment with new procedures.

> *Proposition 22*: Effective dual channel models will have the power to resolve issues and implement solutions.

*Summary.* The conditions surrounding dual channel models for the resolution of motivational issues are particularly stringent. The likelihood of finding or developing a traditional system which has the maturity to support these conditions may be very low. Thus, the possibility that motivational issues cannot be resolved in a unionized organization other than by unilateral management action must be considered. Further, a cooperative system of any sort in a nonunion organization may violate the National Labor Relations Act (Jackson, 1977).

Research comparing single- and dual-channel cooperative systems is not available. There have been numerous experiments with both models in the United States (Gold, 1976), and the dual-channel models are law in several European countries (Furlong, 1977), but there seems to be little more than case study materials with which to evaluate the utility of cooperative labor-management systems. This situation may be improved when the results of the projects sponsored by the National Quality of Work Center and the Institute for Social Research become available.

Preliminary reports of these projects indicate the difficulty that the sponsors have had in trying to initiate cooperative projects (Lawler, 1976), the fragility of those projects (Goodman, 1976), and the potential for changes initiated by the committee to impact on productivity (Macy & Mirvis, 1976) and safety (Goodman, 1976). These reports also document the extraordinary amount of time cooperative committees need to accomplish their task (Macy & Mirvis, 1976).

## CONCLUSION

Unionized settings provide ample opportunities for behavioral research. Three areas for such research have been discussed: union organizing, the effectiveness of traditional union management relationships, and the effectiveness of more cooperative approaches to labor-management problems. Other areas of industrial relations are also appropriate for behavioral research. While there has been substantial behavioral research on union democracy, the issues are by no means resolved. Collective bargaining, too, offers an array of researchable problems (see chapter by Kochan in this volume). Organizational behavior researchers who are reluctant to study industrial relations problems may find unionized settings to be excellent sites for basic research on both psychological and sociological issues. Research designed to take advantage of the particular characteristics of a unionized setting frequently will provide generalizable findings. (See, e.g., Lieberman's, 1956, study of attitude change, and Haire's, 1955, study of perception.) Unionized settings would also seem to provide excellent sites for researchers who want to move their boundary role and conflict resolution research out of the laboratory and into the field or their interorganizational research away from exchange relationships.

While some organizational behavior researchers have capitalized on the characteristics of unionized settings and others have recognized that unionization is a factor which must be incorporated into their study design most organizational behavior researchers either avoid unionized settings or ignore the role of the union. Hammer (1978) points out the danger in ignoring the role of the union in research on traditional organizational behavior issues. Her study of employees' attitudes and perceptions based on expectancy theory, demonstrates that the collective bargaining contract not only specifies outcomes, but also impacts on performance-outcome contingencies. Including union membership or unionization as a variable in future theorizing and research, as Hammer suggests, may be technically difficult since each study would have to be done in multiple sites. Recognizing that union membership or unionization is a factor relevant to a research issue, however, is technically feasible in any study. Organizational behavior researchers routinely identify characteristics of their research settings that might limit the generalizability of their findings. Considering the impact of unionization or union membership on one's results and conclusions is simply good science.

Unionized settings are excellent sites for behavioral research, too, because they provide a variety of dependent variables—representation and union leadership election results, as well as the results of strike and contract ratification votes. Data on many of these variables are collected by government agencies and are available to researchers. Behavioral scientists should be able to use these archived data as dependent variables but typically will have to collect data on the independent variables at the unionized site.

Getting access to a unionized research site may be the most formidable task for organizational behavior researchers. Organizational behavior has a poor reputation among American unions.

> Organizational behavior appears to the union commentators as an efficiency ideology disguised as an organizational science on which trade unionism and collective bargaining are largely tangential, redundant, irrelevant, and friction inducing (Barbash, 1975, p. 9).

Organizational behavior always seems to the unions to be developing management methods to keep workers motivated, satisfied, and non-union. Taylor argued that with scientific management, unions would be unnecessary. The human relations movement advocated keeping workers satisfied so they will not need unions. Advocates of participative decision making have argued that direct individual influence is more responsive to the problems of modern work life than the collective influence that can be achieved through union representation. To unions these are thinly disguised methods of exploitation (Barbash, 1975) and they are understandably suspicious of organizational behavior research projects. Yet, many union leaders are interested in research which focuses on their problems, as opposed to what they perceive to be management's problems. Hence, there are ample opportunities for sensitive organizational behavior researchers to do research on unions and in unionized settings.

## FOOTNOTES

* I would like to thank Stephen Goldberg, Hervey Juris, and Thomas Kochan for their helpful comments at various stages of the development of this chapter.

1. The above description, as well as the one that follows, pertains to private sector employees who come under the National Labor Relations Act. While public sector employees are not covered by the NLRA, the laws governing them tend to be similar to the NLRA.

2. Approximately 20 states have banned union security arrangements for private sector employees by statutory and constitutional provisions of varying scope (Meltzer, 1977). Most states prohibit union security arrangements for public employees.

3. *Uncertainty* has not been widely associated with *dissatisfaction* in the industrial-organizational psychology literature. Expectancy theories of satisfaction (Graen, 1969; Vroom, 1964) posit that a completely *uncertain condition* (i.e., a random relationship between performance and outcome) does not contribute to dissatisfaction. The research on job stress (Kahn, Wolfe, Quinn, Snoek & Rosenthal, 1964; Caplan, Cobb, French, Harrison & Pinneau, 1975), however, shows that people whose jobs are ambiguous (uncertain) because there is a lack of clarity about work objectives, colleagues' expectations, and the scope and responsibility of the job, experience low job satisfaction and high job related tension.

4. *Collective bargaining* is used here to indicate those interactions between the parties in the industrial relations systems which occur during the negotiation of a contract. *Contract administration* covers those interactions between the parties during the life of the collective bargaining agreement.

5. Butler (1976) concluded that relative deprivation does not account for strike rates in the British and American coal mining industries. Unfortunately, he studied one of the occupational groups that Kerr & Siegal identified as forming a socially isolated, homogeneous mass. Hence, Butler's study does not test Proposition 6.

6. Gouldner (1954a, 1954b) in his classic study of the bureaucratization of a gypsum mine discussed the changed role of mine foremen in terms of formalization, not influence.

7. The social democratic governments of several European countries have legislated cooperative union-management systems. Germany, for example, has detailed laws requiring dual channel cooperative systems at the factory level, called "works councils," and codetermination on company boards of directors (Furlong, 1977). Scanlon plans are another interesting form of union-management cooperation, which are not discussed in this chapter. (See Mills, 1978.)

8. Note this union position assumes that which most organizational behavior researchers would not agree, e.g., that job redesign affects productivity through increases in job satisfaction.

9. In the Quality of Work projects, union and management typically design a shelter agreement which specifies the relationship between the cooperative project and the traditional system. These agreements allow the parties to move issues from the cooperative system to the traditional system and vice versa, allow each party to escape from the cooperative system, and provide guarantees that no worker will lose wages or work because of organizational changes promulgated by the cooperative system (Lawler, 1976).

## REFERENCES

Abel, I. W. (1975) Reported in Executive Summary Industrial Productivity Workshop. Washington, D.C.: NSF-RANN, October 1975.

Alderfer, C. P. (1967) "An organizational syndrome." *Administrative Science Quarterly*, *12*, 440–460.

Allport, F. H. (1962) "A structuronomic conception of behavior: individual and collective. *Journal of Abnormal and Social Psychology*, *62*, 3–30.

Anderson, John (1977) "A comparative analysis of local union democracy." Institute of Industrial Relations, University of California at Los Angeles, Working Paper #9.

Barbash, J. (1967) *American unions: structure, government and politics*. New York: Random House.

———. (1975) "The union as a bargaining organization: some implications for organizational behavior." Proceedings Industrial Relations Research Association.

Benson, J. K. (1975) "The interorganizational network as a political economy." *Administrative Science Quarterly, 20,* 229–249.

Berger, C. J. & Cummings, L. L. (1979) "Organizational structure, attitudes and behaviors." In B. Staw (ed.), *Research in Organizational Behavior.* Greenwich, Conn.: JAI Press.

Berkowitz, L. (1961) *The psychology of aggression.* New York: Wiley.

———. (1962) *Aggression: a social psychological analysis.* New York: McGraw-Hill.

———. (1972) "Frustrations, comparisons, and other sources of emotional arousal as contributors to social unrest." *Journal of Social Issues, 28,* 72–92.

Bieri, J. (1968) "Cognitive complexity and judgment of inconsistent information." In R. Abelson, E. Aronson, W. McGuire, T. Newcomb, M. Rosenberg, and P. Tannenbaum (eds.), *Theories of cognitive consistency: A sourcebook.* Chicago: Rand McNally.

Blauner, R. (1964) *Alienation and freedom.* Chicago: University of Chicago Press.

Bluestone, I. (1975) "Speech." American Psychological Association Annual Meeting. Chicago.

Bok, D. C. & Dunlop, J. J. (1970) *Labor and the American community.* New York: Simon and Schuster.

Brett, J. M. & Goldberg, S. B. (1979) "Wildcat strikes in bituminous coal mining." *Industrial and Labor Relations Review, 32,* 465–483.

Britt, D. & Galle, O. (1974) "Structural antecedents of the shape of strikes: a comparative analysis." *American Sociological Review, 39,* 642–651.

Brickman, P. (1974) "Role structures and conflict relationships." In P. Brickman (ed.), *Social conflict.* Lexington, Mass.: Heath.

Butler, R. J. (1976) "Relative deprivation and power: a switched replication design using time series data of strike rates in American and British coal mining." *Human Relations, 29,* 623–641.

Campbell, A., Converse, P., Miller, W. & Stokes, D. (1964) *The American voter.* New York: Wiley.

Caplan, R. D., Cobb, S., French, J. R. P., Jr., Harrison, R. V. & Pinneau, S. R., Jr. (1975) "Job demands and worker health: main effects and occupational differences." HEW Publication No. (NIOSH) 75–160. (USG-PO Catalog No. HE 20.7111:J57. USG-PO stock No. 1733–00083) Washington, D.C.: U.S. Printing Office.

Coch, L. & French, J. R. P., Jr. (1948) "Overcoming resistance to change." *Human Relations, 1,* 512–532.

Crosby, F. (1976) "A model of egotistical relative deprivation." *Psychological Review, 83,* 85–113.

Derber, M., Chalmers, W. E. & Edelman, M. T. (1961) "Assessing union-management relationships." *Quarterly Review of Economics and Business, 1,* 27–48.

Dollard, J., Doob, L., Miller, N., Mower, O. & Sears, R. (1939) *Frustration and aggression.* New Haven: Yale University Press.

Donahue, T. (1976) "Speech before International Conference on Trends in Industrial and Labor Relations." Reported in *World of Work, 1,* 6.

Driscoll, J. W. (1979) "Working creatively with a union: lessons from the Scanlon plan." *Organizational Dynamics,* Summer, 61–80.

Dunlop, J. T. (1958) *Industrial relations systems.* Carbondale, Il.: Southern Illinois University Press.

Edelstein, J. D. & Warner, M. (1975) *Comparative union democracy.* New York: Wiley.

Eisele, C. F. (1974) "Organization size, technology, and frequency of strikes." *Industrial and Labor Relations Review, 27,* 560–571.

Evan, W. M. (1976) "Organization theory and organizational effectiveness: a preliminary analysis." In S. L. Spray, *Organizational effectiveness*. Kent, Ohio: Kent State University, Comparative Administration Research Institute.

Fox, A. (1966) "Industrial sociology and industrial relations Royal Commission on Trade Unions and Employers' Association." London: Her Majesty's Stationery Office.

Furlong, James (1977) *Labor in the boardroom*. Princeton, N.J.: Dow Jones Books.

Getman, J. G., Goldberg, S. B. & Herman, J. B. (1976) *Union representation elections: law and reality*. New York: Russell Sage Foundation.

Gold, C. (1976) "Employer-employee committees and worker participation." NYSSILR Key Issues Series #20.

Goodman, P. S. (1979) *Assessing organizational change: the Rushton quality of work experiment*. New York: Wiley Interscience.

Gouldner, A. W. (1954a) *Wildcat strike*. Yellow Springs, Ohio: Antioch Press.

———. (1954b) *Patterns of industrial bureaucracy*. Glencoe, Il.: The Free Press.

Graen, G. (1969) "Instrumentality theory of work motivation." *Journal of Applied Psychology Monograph, 53, 2,* Pt. 2.

Greer, C. R. & Martin, S. A. (1978) "Calculative strategy decisions during union organizing campaigns." *Sloan Management Review, 19,* 61–74.

Hackman, J. R. & Oldham, G. R. (1976) "Motivation through the design of work: test of a theory." *Organizational Behavior and Human Performance, 16,* 250–279.

Haire, M. (1955) "Role perception in labor management relations: an experimental approach." *Industrial and Labor Relations Review, 8,* 204–216.

Hall, R. H. (1972) *Organizations: structure and process*. Englewood Cliffs, N.J.: Prentice Hall.

Hamblin, R. L. (1938) "Leadership in crises." *Sociometry, 21,* 322–335.

Hammer, T. H. (1978) "The role of the union in organizational behavior: a study of relationships between local union characteristics and worker behavior and attitudes." *Academy of Management Journal, 4,* 560–577.

Hamner, W. C. & Smith, F. J. (1978) "Work attitudes as predictors of unionization activity." *Journal of Applied Psychology, 63,* 415–421.

Herman, J. B. (1976) "Cognitive processing of persuasive communications." *Organizational Behavior and Human Performance, 19,* 126–147.

Jackson, C. C. (1977) "An alternative to unionization and the wholly unorganized shop: a legal basis for sanctioning joint employer-employee committees and increasing employee free choice." *Syracuse Law Review, 28,* 809–846.

Kahn, R. L., Wolfe, D. M., Quinn, R. P., Snoek, G. D. & Rosenthal, R. A. (1964) *Organizational stress*. New York: Wiley.

Katz, D. & Kahn, R. L. (1978) *The social psychology of organizations*, 2nd ed. New York: Wiley.

Kerr, C. & Siegal, A. (1954) "The interindustry propensity to strike—an international comparison." In A. Kornhauser, R. Dubin, and A. Ross (eds.), *Industrial Conflict*. New York: McGraw-Hill.

Kochan, T. A. (1979) "How American workers view labor unions." *Monthly Labor Review,* 102, 4, 23–41.

Kochan, Lipskey & Dyer (1974) "Collective bargaining and the quality of work: the views of local union activists." *Proceedings,* Industrial Relations Research Association, 150–162.

Kornhauser, A., Dubin, R. & Ross, A. (1954) (eds.), *Industrial conflict*. New York: McGraw-Hill.

Korten, David C. (1962) "Situational determinants of leadership structure." *Journal of Conflict Resolution, 6,* 222–235.

Kuhn, J. W. (1961) *Bargaining in grievance settlement, the power of industrial work groups.* New York: Columbia University Press.

Lawler, E. E. (1976) "Union management collaboration in job and organization design." Speech, Academy of Management.

———, Hackman, J. R. & Kaufman, S. (1973) "Effects of job redesign: a field experiment." *Journal of Applied Social Psychology, 3,* 49–62.

Lazarsfeld, P. F., Berelson, B. & Gaudet, H. (1968) *The people's choice,* 3rd ed. New York: Columbia University Press.

Levine, S. & White, P. E. (1961) "Exchange as a conceptual framework for the study of interorganizational relationships." *Administrative Science Quarterly, 5,* 583–601.

Lieberman, S. (1956) "The effects of changes in roles on the attitudes of role occupants." *Human Relations, 9,* 385–402.

Lincoln, J. R. (1978) "Community structure and industrial conflict: an analysis of strike activity in SMSAs." *American Sociological Review, 43,* 199–220.

Losada, M. (1978) "Consumer attitudes and behavior: study of the lead lag relationship by means of Fourier synthesis." Institute for Social Relations.

Macy, B. A. & Mirvis, P. H. (1976) "A methodology for assessment of quality of work life and organizational effectiveness in behavior economic terms." *Administrative Science Quarterly, 21,* 212–226.

McGuire, W. J. (1968a) "Selective exposure: A summing up." In R. P. Abelson, E. Aronson, W. J. McGuire, T. M. Newcomb, M. J. Rosenberg & P. H. Tannenbaum (eds.), *Theories of cognitive consistency: A sourcebook.* Chicago: Rand McNally.

———. (1968b) "Theory of the structure of human thought." In R. P. Abelson, E. Aronson, W. J. McGuire, T. M. Newcomb, M. J. Rosenberg & P. H. Tannenbaum (eds.), *Theories of cognitive consistency: A sourcebook.* Chicago: Rand McNally.

McGuire, W. J. (1968c) "The nature of attitudes and attitude change." In G. Lindsey and E. Aronson (eds.), *Handbook of Social Psychology,* Vol. 3, Reading, Mass.: Addison-Wesley.

Meltzer, B. (1977) *Labor law.* Boston: Little Brown.

Mills, D. Q. (1978) Labor-management relations. New York: McGraw-Hill.

Newcomb, T. M. (1956) "The prediction of interpersonal attraction." *American Psychologist, 11,* 575–586.

Olsen, Mancur, Jr. (1965) *The logic of collective action.* Cambridge, Mass.: The Harvard University Press.

Pepitone, A. (1968) "The problem of motivation in consistency models." In R. P. Abelson, E. Aronson, W. J. McGuire, T. M. Newcomb, M. J. Rosenberg & P. H. Tannenbaum (eds.), *Theories of cognitive consistency: A sourcebook.* Chicago: Rand McNally.

Pfeffer, J. (1978) *Organizational design.* Arlington Heights, Il.: AHM Publishing Corporation.

——— & Leblebici, H. (1973) "The effect of competition on some dimensions of organizational structure." *Social Forces, 52,* 268–279.

Porter, W. & Lawler, E. E. (1965) "Properties of organization structure in relation to job attitudes and job behavior." *Psychological Bulletin, 64,* 23–51.

Quinn, R. P. & Staines, G. L. (1978) "The 1977 quality of employment survey: Section two sampling methods." Survey Research Center, University of Michigan.

Ross, A. M. (1963) "Distressed grievance procedures and their rehabilitation." Proceedings of the Sixteenth Annual Meeting, National Academy of Arbitrators, BNA.

Rousseau, D. (1978) "Characteristics of departments, positions, and individuals: contexts for attitudes and behavior." *Administrative Science Quarterly, 23,* 521–540.

Sayles, L. (1958) *Behavior of industrial work groups.* New York: Wiley.

Schein, E. H. (1965) *Organizational psychology*. Englewood Cliffs, N.J.: Prentice-Hall.

Schlesinger, L. A. & Walton, R. E. (1976) "Work restructuring in unionized organizations: risks, opportunities and impact on collective bargaining." Industrial Relations Research Association Proceedings.

Schmidt, S. & Kochan, T. (1978) "Interorganizational relationships: patterns and motivation." *Administrative Science Quarterly, 22,* 220–234.

Schriesheim, C. A. (1978) "Job satisfaction, attitudes towards unions and voting in a union representation election." *Journal of Applied Psychology, 63,* 548–552.

Scott, J. F. & Homans, G. C. (1947) "Reflections on the wildcat strikes." *American Sociological Review, 12,* 278–287.

Sears, D. O. & Freedman, J. L. (1967) "Selective exposure to information: A critical review." *Public Opinion Quarterly, 31,* 194–213.

Sherif, Muzafer & Sherif, Carolyn. (1958) *An outline of social psychology*. New York: Harper.

Shorter, E. & Tilly, C. (1974) *Strikes in France: 1830–1968*. London: Cambridge University Press.

Simmel, G. (1950) *The sociology of George Simmel* (trans. K. H. Wolff). New York: Free Press.

St. Croix, C. (1975), Reported in Executive Summary Industrial Productivity Workshop. Washington, D.C.: NSF-RANN, October 1975.

Stagner, R. & Rosen, H. (1965) *The psychology of union-management relations*. Belmont, CA.: Wadsworth.

Strauss, G. (1977) "Union government in the U.S.: Research past and future." *Industrial Relations, 16,* 215–242.

———, Miles, R. E., & Snow, C. C. (1974) "Implications for industrial relations," in G. Strauss, R. E. Miles, C. C. Snow and A. Tannenbaum, *Organization Behavior* IRRA Series.

Tannenbaum, A. (1969) "Unions." In J. G. March (ed.), *Handbook of Organizations*. Chicago: Rand McNally.

——— & R. L. Kahn. (1958) *Participation in union locals*. Evanston, Il.: Row Peterson.

Thompson, J. (1967) *Organizations in action*. New York: McGraw-Hill.

Turk, H. (1973) "Comparative urban structure from an interorganizational perspective." *Administrative Science Quarterly, 18,* 37–55.

*United Mineworkers Journal,* December 16–31, 1976.

Van de Ven, A. H., Emmett, D. C. & Koenig, R., Jr. (1975) "Frameworks for interorganizational analysis." In A. Negardi (ed.), *Interorganization theory*. Kent, Ohio: Kent State University Press.

Vroom, V. (1964) *Work and motivation*. New York: Wiley.

Walton, R. E. & McKersie, R. B. (1965) *A behavioral theory of labor negotiations*. New York: McGraw-Hill.

Weick, K. E. (1969) *The social psychology of organizing*. Reading, Mass.: Addison-Wesley.

Woodward, J. (1965) *Industrial organization: theory and practice*. London: Oxford University Press.

Woodward, J. (1970) *Industrial organization: behavior and control*. London: Oxford University Press.

Wortman, C. B. & Brehm, J. W. (1975) "Responses to uncontrollable outcomes: an integration of reactance theory and the learned helplessness model." *Advances in Experimental Psychology*. New York: Academic Press.

Yuchtman, E. & Seashore, S. E. (1967) "A system resource approach to organizational effectiveness." *American Sociological Review, 32,* 891–903.

# INSTITUTIONALIZATION OF PLANNED ORGANIZATIONAL CHANGE*

Paul S. Goodman and Max Bazerman**

CARNEGIE-MELLON UNIVERSITY

Edward Conlon***

GEORGIA INSTITUTE OF TECHNOLOGY

## ABSTRACT

The concept of institutionalization, which concerns the process by which organizational change is sustained, is examined. Basic definitional attributes and a two-phase model of institutionalization are presented. Then factors in the literature which affect the degree of institutionalization are reviewed and related to the model. Some of the factors include: the organization's reward system, transmission mechanisms, group forces, and diffusion processes.

**Research in Organizational Behavior, Volume 2, pages 215–246**

**ISBN: 0-89232-099-0**

## INTRODUCTION

The 1970s have witnessed a rapid proliferation of planned organizational interventions (Goodman & Lawler, 1977). The goals of these efforts have been to increase productivity for the organization and to improve the working life of the employee. One of the major issues in these interventions is whether the change effort can be sustained over time. The focus of this chapter is on the concept of institutionalization which concerns the process by which changes in social systems are maintained over time.

Institutionalization is one of the critical concepts in organizational change. Failure to institutionalize new behaviors clearly detracts from the effectiveness of the organizational change. Our view of the organizational literature is that there is little systematic conceptual or empirical work on this topic. The goal of this chapter is to develop a theoretical framework and to identify the factors that contribute to institutionalization. Our analysis will be limited to planned organizational change where the focus is on altering organizational structure (e.g., role relationships, reward systems, technology) or organizational process (e.g., communication, decision making). Interventions primarily oriented to changing individual behavior through some form of training (e.g., sensitivity training) are excluded.

## A SELECTIVE REVIEW

Organizational change is a central issue in organizational theory. Unfortunately, most of the discussions of organizational change provide few insights into the theoretical issues or the processes of change. The primary mode of examining organizational change has been to outline general phases of change, describe intervention techniques, or review research findings.

The work by Lewin (1951) and Schein (1969) has focused on the processes of unfreezing, changing, and refreezing from a psychological perspective. This conceptualization is useful to the extent that it helps to organize our thinking about change processes. It fails, however, to provide much insight into why refreezing or institutionalization does or does not occur. Still another approach uses gross variable classifications and implies causality through the use of a "flowchart" to illustrate a framework for change (Beer & Huse, 1972; Friedlander & Brown, 1974). These approaches are quite general, provide little insight into critical processes such as institutionalization, and rarely generate any testable hypotheses.

The organizational change literature can also be grouped by the intervention techniques discussed by, for example, Friedlander & Brown (1974) and Katz & Kahn (1978). While this approach is instructive in delineating alternative techniques for change, it tells very little about the theoretical issues underlying the change process.

There is also an empirical literature of organizational change. One possibility is that a theory of planned organizational change processes might be developed inductively from these empirical studies. Unfortunately, the quality of these studies is poor (White & Mitchell, 1978). They are devoid of the methodological rigor that would allow drawing generalizations about change processes. The studies that attempt a rigorous assessment of the effectiveness of planned organizational change tend to focus only on short-term results. The problem of institutionalization of change however, demands a longitudinal assessment of the change effort over time. In the few studies of institutionalization, the principal focus has been on determining whether the change has persisted rather than on why it has persisted (Miller, 1975; Seashore & Bowers, 1978).

Discussions of institutionalization and other processes of change can be found in literatures outside the area of planned organizational change (e.g., Parsons, 1951; Homans, 1961; Berger & Luckman, 1966; Buckley, 1967; Meyer & Rowan, 1978). For example, the functional position (cf. Parsons, 1951) ties persistence of social structure to the functional prerequisites of social systems. Homans (1961) uses the concept of institutionalization to differentiate transactions based on social exchange and transactions embedded in social structure. Berger & Luckman (1966), working from an ethnomethodological position, conceptualize institutionalized acts as those behaviors that acquire shared meaning through a process of reciprocal typification. Institutionalized acts are repeatable by any member of the social system without a significant change in the meaning of those acts for others in the social system.

While these efforts provide alternative views of institutionalization, they are developed at a fairly general level, they do not identify factors affecting variation in levels of institutionalization, they do not focus on the process of institutionalization, and most important, the level of explanation is not easily applicable to the topic of planned organizational change.

Given the theoretical significance and the current state of the literature, the first step will be to develop a theoretical framework that can be used to organize our current body of knowledge, and more importantly, to identify the direction for future research. We begin with a definition and then distinguish institutionalization from other concepts in the literature. Second, the main contribution of the chapter, a two-phase model of institutionalization will be presented. Finally, we use the conceptual model to organize the factors from the empirical literature that affect institutionalization.

## BASIC DEFINITIONS AND CONCEPTS

*Institutionalized Act*

Our conceptualization of institutionalization focuses on specific behaviors or acts. An *institutionalized act* is defined as a behavior that persists over time, is performed by two or more individuals in response to a common stimulus, and exists as a social fact. Behavior as a social fact means that the behavior exists external to any individual, is part of a social reality, and is not dependent on any particular individual. An institutionalized act is then a social construction transmitted across generations of organizational members. An institutionalized act is also a behavior performed by multiple individuals given a common stimulus. The act is not only a social construction but also occurs in a social context. Individuals in the social context have knowledge of their own performance of the target behavior and that others are performing the same behavior for a similar rationale. Lastly, persistence, an important characteristic of an institutionalized act, means that the act will be evoked over time. These three characteristics—*persistence*, *performance by multiple others*, and *its existence as a social fact*—are the defining characteristics of an institutionalized act.

Institutionalized acts vary in degree. That is, they vary in the degree to which they are accepted as a social fact, are performed by multiple others, and persist over a long time period. An act not highly institutionalized would not be evoked in common with others exposed to a common stimulus, would not be exhibited over long time periods, and, if maintained, would probably require some direct form of control. That is, it would be performed in response to some direct reinforcement mechanism rather than being embedded in social reality. Two major concerns of this chapter are the definition of different levels of institutionalization and the identification of the causes of different levels of institutionalization.

The following concepts may further institutionalization.

*1. Institutionalization as an Act or Process.* Our primary definition focuses on institutionalization as a specific act or behavior. The concept of process in this analysis will focus on the dynamic mechanisms which influence whether a behavior will persist, be performed by multiple individuals, and will exist as a social fact. Three major processes will be delineated in our two-phase model of institutionalization: acquisition, which concerns how beliefs about the new work behavior are formed; reward allocation, which concerns the type and nature of reinforcement schedules maintaining the new behavior; and transmission, which concerns the mechanism by which new organizational members are socialized into the new work behavior (cf. Zucker, 1977).

2. *Institutionalization as an Organizational or Individual Phenomenon.* An institutionalized act is an organizational phenomenon, a social construction of reality that exists independent of any individual. While the emphasis here is on the social versus the individual level of behavior, it is impossible to conceptualize institutionalization without recognizing the individual level of analysis. The processes of acquisition, reward allocation, and transmission are directed at individuals. Further note that the focus here is on the behavior of individuals in social organizations, not on social institutions (e.g., educational, religious, economic) per se (Meyer & Rowan, 1978).

3. *Institutionalization and Motivation.* The definition of institutionalization alludes to some forces which evoke and sustain institutional acts; delineation of the character of those forces should further clarify the discussion of institutionalization.

The motivational component in institutionalization is mediated by the process of reward allocation. There are four principal classes of forces that affect the level of institutionalization. First is the simple allocation of external rewards or punishments. In this case, the behavior is adopted and continued in order to receive rewards or to avoid punishments. It is the contingency between the behavior and allocated rewards that motivates the institutionalized act. A second social influence is internalization. In this case, the individual adopts and maintains a behavior because it is intrinsically satisfying and congruent with that individual's value scheme. That is, the individual adopts and maintains the behavior to achieve congruence between his behavior and his value system. Identification represents a third class of forces (Kelman, 1958). In this case, the individual accepts influence in order to maintain a satisfying relationship with another individual or with a group. Lastly, a behavior will be adopted and maintained when it is perceived as a social fact. In this case, the individual complies with the behavior because it is accepted as social knowledge which facilitates meaning and predictability in social action. It is a social convention not supported through any direct rewards. These four classes, although not exhaustive, represent the major forces that affect institutionalization. An important qualification is that none of these forces are necessary conditions for institutionalization. A single force or any combination of forces might bear on any institutionalized act. Also, it might be expected that at different levels of institutionalization different classes of forces would be important. For example, at low levels of institutionalization, instrumental rewards or internalization might be necessary to maintain the behavior. Highly institutionalized acts, on the other hand, may be accepted and maintained simply as social facts, that is, as social knowledge passed from one generation to another.

It is important to remember that these motivational forces must be understood within the context of the definition of an institutionalized act. That is, internalization may lead to persistence of individual behavior, but may not be sufficient to cause institutionalization given the definition of an institutionalized act as behavior which persists over time, is performed by two or more individuals, and exists as a social fact. Internalization contributes to institutionalization when multiple individuals find the behavior congruent with their value systems, are aware that others perform the same behavior, and consider the behavior appropriate for a specific group or social organization.

4. *Institutionalization and Persistence.* One defining characteristic of institutionalization is persistence of behavior over time. Persistence in the context of planned organizational change implies recurrent responses evoked by a cue. Persistence is not an all-or-nothing phenomenon. There are clearly degrees of persistence which can be identified in terms of response rates over time. Persistence can be described as the probability of evoking an act given a particular cue and the functional form of that response rate over time. Behavior that is evoked at the same rate at each time period represents persistence in a steady state. If the response rate were to decline over time, we would say that the level of persistence (and of institutionalization) would have declined.

There is no a priori standard to determine how long a behavior must last before it is institutionalized. Instead, an institutionalized act may be measured in terms of degree; that is, the degree to which a behavior persists is a measure of the degree of institutionalization.

Another issue related to persistence concerns the level of specificity between a cue and act. It could be argued that a specific cue should evoke a specific act. Another point of view holds that a cue will evoke a common class of acts. Our conceptualization of persistence assumes the latter view. That is, persistence is defined in terms of the probability of evoking a behavior from a common class of responses rather than the probability of evoking a specific response. This position permits some evolution and modifications of institutionalized acts; there is a zone of acceptable responses. The concept of a zone is derived from the notion of a stimulus generalization gradient.

To illustrate this idea in one recent organizational intervention (Goodman, 1979), the behavior of communicating between work shifts was introduced to improve organizational effectiveness. Initially each member of a work group would communicate with his counterpart each day during the change of shifts. A year into this intervention, the workers would communicate only if there was a problem. Two years into the program a crew representative served as the communicator. In this example, the specific communication behaviors changed but the general behavior of communicating to coordinate interdependent work activities persisted.

*Related Concepts*

The definition of institutionalization may be sharpened by contrast with other concepts. A selective set of concepts is enumerated below for this purpose.

*Commitment.* Commitment may be variously defined. For convenience we use the one discussed by Salancik (1977). Commitment is the binding of the individual to behavioral acts. The acts are performed in relation to beliefs, attitudes, organizations and other social objects. For example, one might hold beliefs about the necessity to conserve energy. Commitment in this context would refer to behavioral acts such as buying a small car, reducing fuel usage, shutting off lights, etc. These acts bind the individual to the social object of energy conservation. The degree of commitment is a function of the explicitness or deniability of the act, the revocability of the act, whether the act was adopted by personal choice or external constraints, and the extent to which the act is known by others (Salancik, 1977).

Commitment and institutionalization are similar in that they both focus on behavior. Both concepts also relate to resistance to change. Once there is high commitment to a particular act or an act becomes institutionalized, the likelihood of changing that act diminishes. The two concepts differ in that commitment refers to a psychological process while institutionalization refers to the constructions of social facts. A single individual can make a commitment; institutionalization, on the other hand, requires the behavior of two or more individuals. Institutionalization also implies transmission of acts across generations of group members; this is not true in the case of commitment. An institutionalized act also persists over time; commitment to an act occurs at one point in time. It may lead to persistence, but it is not defined by persistence. Commitment can precede institutionalization, but it should not be considered a necessary condition.

*Group Conformity.* Group conformity is the yielding of individual behavior to group forces. That is, uniformity in behavior can be evoked through information provided by or through compliance to group pressures. Conformity in this definition requires that there is some conflict between the individual and group positions. The more the individual yields to the group position, the greater the conformity. The Asch (1956) and Crutchfield (1955) experiments are classic tests of conformity.

Conformity parallels institutionalization in that both concepts treat the uniformity of behavior and both assume that behavior is embedded in a social context. Conformity, however, generally develops from group pressure—a direct negative sanction. While there are a number of different forces (e.g., internalization) that might facilitate institutionalization, direct social control is only one of these forces. A second difference is that institutionalization refers to persistence in behavior by multiple others. The definition of conformity is not based on any notion of persistence.

*Norms.* Norms are pre- or proscriptions about behavior. They are statements about ranges of behaviors an individual should or should not perform (Blake & Davis, 1964). There are many similarities between norms and institutionalized acts. Both are social facts which exist independently of any individual. Also, both include statements about appropriate or required behavior commonly held by others and supported by some social context (Jackson, 1966). In one sense, norms can be thought of as the product of the institutionalization process. The major difference between these two concepts is one of emphasis. Much of the literature on norms focuses on the structural characteristics of these concepts. The concern here is how new forms of behavior are developed and maintained over time.

*Diffusion.* Diffusion refers to the extension and adoption of a new work behavior in a social system (cf. Rogers & Shoemaker, 1971). That is, it concerns the spread of innovation from one setting to another. In planned organizational change, diffusion refers to the spread of the change effort from one target area to another (cf. Zaltman, 1973). Institutionalization and diffusion are different but interdependent concepts. Diffusion includes the concept of institutionalization. Diffusion cannot occur completely without institutionalization. That is, innovation must not only be adopted in a new social setting; it also must persist. Institutionalization can occur without diffusion. Again, the two concepts differ in emphasis. The literature on diffusion focuses primarily on whether the innovation will be adopted in a new social setting; the theory of institutionalization concerns whether the behavior will persist.

## PHASES OF INSTITUTIONALIZATION

This section examines the two-phase model of institutionalization. We begin with the individual phase and then move to the structural phase. To facilitate this analysis, we assume that a new form of behavior has not been introduced. There is no structure or social reality relative to this behavior. Although the focus here is on an organization or subsystem, no assumptions are made about the existence of well-defined groups that may be transmitting social facts pertinent to the new focal behavior. That is, we start by asking how a new form of behavior becomes institutionalized rather than how an existing institutionalized act gets transmitted across generations. This position is important because it permits the analysis to start at the individual level.

### *Phase One—Individual Level of Analysis*

The analysis begins as a new behavior is introduced into an organizational setting. First, we consider whether an individual decides to adopt and to continue the new behavior. Following this, we examine the effect of multiple individuals adopting and continuing the new behavior. As multiple individuals become aware of each other performing the new behavior, the analysis must shift to phase two.

*Decision to adopt.* The decision to adopt concerns the initial adoption of the new behavior. We can understand this decision by reference to some of the concepts in expectancy theory (cf. Mitchell, 1974). Here we are using this theory as a way to think through the choice process rather than embracing the formal assumptions of that model.

The decision to adopt is related to the following factors.

1. *The perceived ability of the target person to perform the new form of work behavior.* The issue is whether the individual perceives that he or she is capable of performing the behavior in question. In a study of a Scanlon plan installation, Goodman & Moore (1976) reported that individuals who felt capable of making suggestions, a key behavior in that plan, did so; others, because of their perceptions of their own ability or of their work, reported that they could not make suggestions and did not. This belief then forms from the interaction among the proposed behaviors, perceptions of self, and the work environment.

2. *The perceived relationship between the new behavior and resultant outcomes.* In order for an individual to adopt a new behavior, there must be some prior belief that the behavior will be rewarding. If the target population for change does not perceive rewards flowing from the behaviors, it is unlikely the behavior will be adopted (Goodman, 1979).

3. *The valence of the outcomes.* Valence, a critical factor in the adop-

tion decision, refers to the attractiveness of the outcomes. It is assumed that the individual makes a comparison between the new work behavior and some alternative behavior. The choice should reflect the alternative with the highest expected value (or valence). This choice is made within the context of limited rationality (March & Simon, 1958). That is, the individual approaches the adoption decision as a limited sequential processor of information. The level of awareness of outcomes is incomplete and varies amoung individuals. Also, the mechanisms to weigh and combine outcomes are at best crude approximations of the theoretical mechanisms for determining expected value.

The level of valence of any outcome is a function of the individual's needs and the amount of available outcomes. The types of outcomes (rewards or punishments) utilized in the adoption decision are another issue. Extrinsic rewards, those mediated through some external source, are probably the dominant type of reward in the adoption decision. Pay, recognition, and approval can all affect the decision to adopt a new work behavior. Intrinsic rewards, those mediated internally, may be used, but their effect is probably weaker in the adoption decision than in the continuation decision. The adoption decision is made on the expectation of rewards, not on the experience of rewards. It is probably easier to assign a valence to an expected amount of money than to an expected feeling of accomplishment. A third type of reward is identification. In this case, the individual adopts a behavior in order to maintain a satisfactory relationship with the person(s) requesting that behavior (Kelman, 1958). The outcomes do not follow from the behavior but rather from the relationship between the target person and the individual(s) requesting the change.

We have identified different types of outcomes for two reasons. First, the adoption decision and the continuation decision probably incorporate different types of rewards. Another reason for identifying the type of reward is that the reward utilized in the adoption decision might affect the continuation decision. Different types of rewards are evoked by different cues. For example, if the decision to adopt is based primarily on identification, then the availability of the source of identification should affect the decision to continue. Absence of the identification stimulus or cue should decrease the probability of continuation.

The two processes important in the decision to adopt are acquisition and commitment. Prior to the actual decision to adopt, the individual forms beliefs about the nature of that behavior and the reward contingencies associated with that behavior. The acquisition process concerns the acquisition of these beliefs. In the context of planned organizational change, the individual may adopt beliefs about new behaviors because of communications from a change agent or relevant other and/or because the individual has performed similar behaviors in the past and generalizes his beliefs about that behavior to the newly introduced behavior. The credibility, trustworthiness, etc., of the communicator and the similarity of the new work behavior to past behaviors determines the contents of the beliefs about the behavior and its reward contingencies (cf. Oskamp, 1977). The nature of the beliefs, then, affects the adoption decision which, in turn, can bear on the level of institutionalization (Goodman & Moore, 1970).

There are other contextual factors which bear on the adoption decision, for example, the number of learning sources. The greater the number of sources for learning about behavior-reward contingencies relevant to the new work behavior, the higher the social validity of the behavior. While we cannot predict the direction of the adoption choice, we can expect the behavior to be more stable the higher the social validity. A related hypothesis concerns the level of congruence among the learning sources. The greater the level of congruence the more likely it is that the new work behavior will be adopted. In the case of the identification process, the greater the number of congruent influences with whom the target person identifies, the more likely is the adoption of the behavior. The greater the incongruity of sender expectations, the less likely it is that the new work behavior will be adopted.

The commitment process shapes the decision to adopt and can affect the level of institutionalization of the new behavior. Schein (1969) has suggested that the context for this decision is important in understanding the outcomes of change. He argues that the amount of freedom or control the individual has over the adoption decision affects the character of the institutionalization. At one extreme, the target person is captive and does not voluntarily accept the target role. The source of the change agent's power is position rather than expertise. At the other extreme, the target person volunteers into the change project, is free to terminate this relationship, and defers to the change agent as a function of expertise. This continuum for adoption may be useful in understanding the continuation decision. The hypothesis is that adoption decisions originating from the "control" end of the continuum would be unstable and highly contingent on the presence of the controller, while decisions originating at the "freedom" end of the continuum would probably enhance commitment to the new behavior and increase the probability of continuation of that behavior.

*Decision to Continue.* The persistence of a new behavior in the individual phase occurs for two reasons. First, a reevaluation of the adopted behavior is *not* cued. Second, if a reevaluation of the adopted behavior is cued, then the attractiveness of the selected behavior will determine whether the behavior will be continued.

*Factors not evoking the decision to continue.* The decision to continue will *not* be evoked, and hence behavior will persist, under the following conditions.

1. *There is a congruence between the expected outcomes and the actual outcomes.* The decision to adopt was based on expectations of certain outcomes. If those expectations are realized, then the adopted behavior might be expected to continue.

The types and schedule of extrinsic rewards will determine whether the decision to continue will be evoked. If the mediator of those rewards is visible over time to the target person, and if rewards are distributed in some predictable manner that is congruent with expectations, then the behavior should continue without any conscious reevaluation. What is critical here is whether the extrinsic rewards flow in some predictable manner. It may be that over time there is a discrepancy between the level of rewards expected prior to the adoption decision and the actual level of rewards. But, if the discrepancy grows slowly and the allocation of rewards remains fairly predictable and is perceived as equitable, then the behavior may persist anyway. The expectation level is dynamic and should adjust.

Identification (Kelman, 1958) is another force that can sustain behavior. As long as the target person derives satisfaction from his relationship with the sender of change, then the new work behavior should persist. That is, if the experienced satisfaction is congruent with the expected satisfaction, then the decision to continue will not be evoked. The identification process can last beyond the decision to adopt.

Internalization (Kelman, 1958) can also affect the institutionalization of the new behavior. In this case, the individual incorporates the new behavior into his value system. The new behavior is performed not as a means but as an end. Where positive outcomes of performing the behavior are internally mediated (intrinsic), the behavior should persist without any conscious reevaluation. The actual outcomes are consistent with the individual's value system.

2. *The level of commitment can lead to persistence.* As shown above, the level of commitment is increased when the adopted behavior is selected freely (versus coercively), explicit (i.e., not easily deniable), and publicly known (Salancik, 1977). Given a high level of commitment, some stability of the selected behavior might be expected because of the resistance to change that behavior.

The first part of this examination of the decision to continue points out that there are forces which will preclude the reevaluation of the adoption decision. As long as these forces prevail which create congruency between actual outcomes and expectations, the adopted behavior will persist. We argue that, in this environment, there is no conscious reevaluation of the adoption decision. It is even possible that the individual will persist in performing the new behavior although the original rationale for performing the behavior no longer exists. This may occur for the following reasons. Individuals as information processers can attend to only a limited array of cues. The decision to adopt generates a commitment which further limits the range of cues the individual attends. Changes may have occurred in the cost/benefit ratio, but the individual still persists in performing the adopted behavior.

*Factors evoking the decision to continue.* The decision to continue can be analyzed in terms of two processes. First, there are cues that cause the individual to reevaluate the adoption decision. Second, faced with the evaluation, the individual organizes a set of information, evaluates, and then decides whether to continue the behavior. There is no literature that identifies directly those factors which cue reassessments of ongoing behavior, although it would be relatively simple to list the potential candidates. Instead of generating such a list, we identify some two broad classes of cueing factors that serve to illustrate the process: inconsistency and new alternatives.

1. *Inconsistency.* When adopted, behavior was expected to result in some set of valued outcomes. Following actual performance of the behavior, however, information may arise to suggest that one or more of the initial expectations was incorrect, or that the outcomes were differentially valued from expectations. This inconsistency between prior expectations and the resulting information will increase the likelihood of revoking the decision to continue.

2. *New alternatives.* One correlate of introducing planned change programs in organizations may be the tacit encouragement to incumbents to experiment with new behaviors and seek "better" modes of performance. Hence, alternatives to the adopted behavior may be generated and evoke the decision to continue.

The decision to continue may be cued repeatedly following adoption; hence, overt persistence of a behavior may actually result from a number of serial reassessments of the behavior. The effectiveness of any particular factor in cueing a decision to continue depends on the future rewards that the individual perceives will accrue. For example, information about the nonaccrual of a valued outcome (i.e., a contradiction) is more likely to trigger a reevaluation than is information about a neutral outcome. The same argument could be made for unexpected outcomes and new alternatives.

Once the decision is evoked, the process of deciding anew whether to continue the behavior is similar to the decision to adopt. The individual remembers or forms beliefs about his capacity for performing a behavior, the connection between the behavior and its rewards or punishments, and the valence of rewards and punishments. The question is *whether the adopted behavior will dominate some identified alternative*. Again, our view is of a decision maker with limited rationality (March & Simon, 1958). We do not see the decision maker engaged in elaborate search routines for alternative behavior or elaborate evaluations of alternative outcomes. Indeed, the availability of information, enhanced by the saliency or recency of an event, may affect whether the new work behavior will continue.

While the process of deciding is the same, the content of the beliefs differs between the adoption and continuation decisions. In the interval between adoption and evocation of the decision to continue, the individual has had the opportunity to reformulate the critical beliefs through *feedback, direct reinforcement,* and *observation of others' behaviors*. For example, observing that others continue that behavior, even in the light of personally discrepant outcomes, may facilitate institutionalization. These three learning mechanisms are probably more powerful shapers of behavior than the processes of communication and generalizing from past experience, which underlie the formation of the beliefs in the decision to adopt (Goodman & Moore, 1976). The differences in the learning processes for the two decisions may mean that the character of the beliefs is more explicit in the decision to continue than in the decision to adopt. The less ambiguity, the easier it is to process the beliefs. Easier processing may lead to great convergence of and predictability between beliefs and behavior.

*Phase Two—Structural Level*

The conceptualization of institutionalization has focused on the decision to adopt and the decision to continue. We view these two decisions, at the individual level of analysis, as critical determinants of the nature of the institutionalized act. We now turn our attention to the structural level, where an institutionalized act is performed by multiple others, exists as a social fact and persists over time.

This section charts a transition between the individual and structural levels. Whereas the focus has been on the individual as decision maker, it is clear that decisions to adopt or to continue occur in the context of groups and organizations. As we turn our attention to others and their role in the adoption and persistence of new forms of behavior, we can build the bridge between the individual and structural level. For institutionalization to take place, the behavior must be part of the social structure.

Three conditions are important in the transition from individual behavior to institutionalization at the structural level. First, individuals must perceive relevant others, given some common stimulus, performing the new behavior. Relevant others would be the members of some defined social organization. Second, there must be a common belief that it is appropriate for members of this social organization to perform that behavior. The concept of "appropriateness" evolves from the social organization, not from the valence of the behavior for the individual. Third, there must be common belief that the social organization will sanction performance of the new behavior (Katz & Kahn, 1978).

The above conditions describe the transition between the two phases in the model: the individual and the structural. In the following section we will examine three factors which affect these conditions. These factors are the physical setting, the social organization's norms and goals, and the cohesiveness of the social organization.

1. *Physical setting.* The character of the physical setting affects the level of interaction. A physical setting which constrains interaction will limit the visibility of others' behavior. In this case, the individual will not be able to learn about relevant others' behaviors—one of the conditions for institutionalization. In the absence of knowledge of others' behaviors, it is unlikely that the individual will develop beliefs about "appropriateness" conditions and "group sanctions" conditions discussed above. A physical setting which facilitates visibility of others' behaviors creates a necessary condition for the development of beliefs about appropriateness and group sanctions.

A corollary of the influence of the physical setting is the nature of the communication system. Since the critical variable is communication among individuals, the nature of the communication system may compensate for the character of the physical setting in creating awareness of others' behaviors.

2. *Social organization norms and goals.* The relationship between the new form of behavior and the norms and goals of the social organization affects whether the new behavior will be perceived as "appropriate" by group members. Previously, we treated new work behavior in a descriptive manner. Here we examine how a normative label is attached to the behavior.

Social organizations are also collections of norms. Norms define appropriate behaviors for group members. They identify behaviors which are functional for organizational goals, and they provide predictability for social interactions. The degree of congruence between the new work behavior and existing norms affects whether the new behavior will be perceived as appropriate. The process by which this occurs is hypothesized as follows. A norm for appropriate behavior exists in the organization. New work behavior is introduced which is similar to and congruent with the existing norms. Following a similarity gradient, group members generalize the appropriateness "label" to the new behavior (Breer & Locke, 1965).

The perception of behavior as "appropriate" is illustrated by the following example. Assume there is a work group that embraces norms about intragroup cooperation. Assume that this group has little interaction with other groups. A planned organizational change program is introduced to bring about intergroup cooperation. Individuals begin some form of intergroup cooperation and become aware that relevant others are engaged in this behavior. The hypothesis is that, to the extent that intergroup cooperation facilitates the group's goals, it is more likely the group will deem this behavior as appropriate for its members. Similarly, the greater the perceived congruence between inter- and intragroup cooperation, the more likely intergroup cooperation will be considered appropriate group behavior.

*3. The cohesiveness of the social organization.* A highly cohesive group is one which is very attractive for its members. In such groups individuals identify strongly with the group and are willing (by the process of identification) to perform behaviors to maintain that satisfactory relationship. Similarly, groups that are highly cohesive have the resources and mechanisms to enforce compliance with group norms. Groups low in cohesiveness, on the other hand, do not have effective sanctioning devices. The degree of cohesiveness contributes then to the third condition for institutionalization: the belief that sanctions (positive or negative) will follow performance (or nonperformance).

Given awareness of others' performance, the belief in appropriateness, and the expectation of sanctions, the focal behavior becomes part of the social structure. As an institutional act the behavior is then part of social reality, independent of the behavior of individuals.

It is important to recognize that the distinction between the individual and structural level is arbitrary, made primarily to facilitate presentation. The decisions to adopt and to continue occur in a social context. Either before or after the decision to adopt, the individual becomes aware of others performing the new work behavior. Over time, the individual (while deciding whether to continue) learns whether the behavior is considered appropriate by group members and what sanctions are attached to the new behavior. The phases described separately above are in reality occurring simultaneously.

The existence of an institutionalized act does not guarantee its persistence, however. Other mechanisms need to be evoked to maintain the institutionalized act. Transmission refers to the socialization process by which institutionalized acts are passed on from old to new organizational members. Although the process of socialization has been widely examined (cf. Schein, 1968), there is little information about this process in the context of planned organizational change. It is not within the scope of this chapter to examine the merits of various socialization processes. The point is that some transmission process is necessary for persistence. Also, the degree of institutionalization will determine, to some extent, the type and character of the transmission process. If behavior is highly institutionalized, the necessity for elaborate or extensive transmission will not be as great as if the behavior were not widely performed, not viewed as appropriate by all members in the group, and not effectively sanctioned.

The second mechanism critical for persistence is the reward allocation process. Compliance as a basis for allocating rewards or punishments and identification are two powerful maintenance processes. The success or failure of these mechanisms will be affected by the degree of institutionalization. Behavior that is highly institutionalized requires little maintenance. Less institutionalized behavior requires greater application of compliance or identification mechanisms.

## FACTORS AFFECTING DEGREE OF INSTITUTIONALIZATION OF PLANNED ORGANIZATIONAL CHANGE

This section reviews a series of factors that affect the institutionalization of planned organizational change. It is hoped that this review will be valuable both to those designing organizational interventions as well as to those diagnosing the success or failure of planned organizational interventions.

Selected empirical studies, mostly field investigations, will be reviewed. We draw also on the findings of a few pertinent laboratory experiments that are related to the process of institutionalization. The review is selective in the following ways. First, we examine only those studies related to the problem of sustaining change over time. Unfortunately, many of the documented studies of planned organizational change have only a short-term focus and must, therefore, be eliminated. Second, there are studies reporting changes over time (cf. Kimberly & Nielson, 1978) but offering no explanation as to why the change was stable; these are similarly neglected. Case studies are included which focus on the issue of institutionalization in sufficient detail to provide some insights into this process. Finally, only studies published in the major organizational psychology, sociology, and change journals are reviewed.

Two categories of factors affecting institutionalization are discussed. Factors that affect the decisions and mechanisms described in the theoretical section are examined first. These include, at the individual level, factors affecting the decision to adopt and to continue. At the structural level, factors such as transmission are studied. The second category is the organizational context itself, both internal and external. The enumeration of these factors is limited to the data sources identified above.

*Reward Allocation Systems*

1. The type of reward seems to bear on the degree of institutionalization. While there is a large literature on rewards and reinforcement schedules, there are few empirical studies on planned organizational change that identify the relative influence of different types of rewards (e.g., intrinsic, identification, internalization). Those field studies that have identified institutionalized behavior generally point to the existence of internally mediated rewards. Generally the greater the autonomy, control, and responsibility the workers experience in the new work organization, the greater the level of institutionalization (cf. Walton, 1975). Goodman (1979) reported institutionalization of changes in work safety practices due to an intervention that provided workers greater authority and responsibility in decisions regarding safety. The attractive feature of internally mediated rewards is that the rewards are built into the performance of the desired behavior and are thus highly contingent on performance.

Some studies have reported that the removal of negative outcomes can facilitate the institutionalization of new work behaviors. An intervention in a coal mine produced structural changes that minimized criticism workers might experience for practicing certain safety procedures. In the absence of these sanctions, the new safety practices were institutionalized (Goodman, 1979).

Combining different types of reward systems can also affect the degree of institutionalization. In a study of a Scanlon plan (Moore & Goodman, 1976), which has remained intact over a four-year period, several types of rewards were in operation. The company-wide bonus was paid fairly regularly. Many of the workers strongly identified with the plant manager who supported the plan. Also, the workers enjoyed some of the intrinsic aspects of the plan, such as increased responsibility for guiding plant activities. In general, the combination of different rewards might be expected to encourage persistence. Two well-publicized interventions (Walton, 1978; Goodman, 1979) which did not persist were characterized by a paralled emphasis on intrinsic factors and the failure to develop a company-wide bonus system. It is reasonable to conclude that the reliance on intrinsic factors such as greater responsibility and authority may not be sufficient incentive. It can be assumed that the regular payment of a company bonus has a strong impact on workers' perceptions of the legitimacy and viability of the change effort. Perceptions of lower levels of legitimacy and viability can be seen to contribute to the failure to institutionalize in these two experiments.

There are, unfortunately, no studies of how various combinations of rewards affect the institutionalization of behavior. This is a viable direction for future research.

2. Discrepancies between expected and actual rewards also affect the process of institutionalization. The theoretical section argued that the individual is faced with the decision to adopt. If the new form of work behavior is adopted, it will persist until a discrepancy is perceived between expected and actual rewards. This discrepancy will evoke the decision to continue. If rewards fall short of expectations over time, the behavior will not persist. The new behavior will lose its legitimacy. In a study of an intervention in a bank, Frank & Hackman (1975) report that some of the expected rewards (greater variety in work) never materialized. In addition, some jobs were already enriched, so that it was unlikely that any additional rewards were possible. In this case, the planned interventions (e.g., autonomous work teams) never became institutionalized.

3. Shifting expectation levels also affect institutionalization. An intervention promises certain rewards. These rewards influence both the decision to adopt and the decision to continue. In practice, the process is more complex. Let us assume that the expectation level of rewards is higher than those rewards available prior to the intervention. Following the intervention, the actual rewards may exceed the expected rewards. Level of aspiration theory predicts that expectations will be even higher for future rewards (Lewin, 1935). At some point, the expectations for the rewards may outstrip the actual level. In reviews of several experiments with new work organization, Walton (1975) and others (cf. Goodman, 1979) have reported initial satisfaction with the new-found rewards. However, over time the level of enthusiasm, and institutionalization, is seen to decline. The workers may be experiencing a discrepancy between expected and actual rewards. A corollary explanation may be understood through the concept of adaptation level. Prior to the intervention let us assume the adaptation level (Helson, 1964) for a given reward is lower than the expected reward. Initially if the expected reward (e.g., level of job variety) is allocated, it should be highly valued. Over time, however, we would expect the adaptation level for variety to shift upward. The consequence should be that the attractiveness of a given level of variety should decline. This in turn might lower the level of institutionalization.

4. A related problem is that there are many unanticipated consequences of organizational interventions, some of them negative. In a review of work organization experiments, Goodman & Lawler (1977) note recurrent conflicts that were unanticipated by the organizational partici-

pants. Many of these interventions, for example, created greater role conflict for first-line supervisors and often for middle management. These unanticipated problems have been noted by others (cf. Walton, 1975; Frank & Hackman, 1975) to increase the costs of participation and to work against persistence.

*Sponsorship*

Withdrawal of sponsorship. Most programs in planned organizational change have some sponsor in the management hierarchy. This individual plays an important role in legitimizing the project, providing a flow of rewards, monitoring and controlling the new behaviors, and providing support in times of crisis. One of the consistent findings from this limited literature is that once sponsorship is withdrawn from the project, the new work behaviors decline. The institutionalized act is evoked less frequently in the absence of the sponsor and fewer members perform that behavior.

The withdrawal of sponsorship can follow from common organizational practices rather than be inherent to the change project. For example, Crockett (1977) reports a major organizational intervention in the State Department in which substantial changes were observed to persist for years. However, when the initiator of the project, a political appointee, left office, the organization reverted to its traditional form. The new administrator was not sympathetic to the values and structure of the change program. As support and legitimization for the program decreased, the degree of institutionalization declined. A similar effect was reported by Walton (1978) when the sponsors of the famous Topeka experiment left the organization. In some cases, the sponsor left temporarily (Frank & Hackman, 1975); in other cases (Walton, 1975; Miller, 1975), the sponsors focused attention on other organizational matters. In all cases, however, the persistence of the new structures declined.

A study by Scheflen, Lawler, & Hackman (1971) of an incentive program to reduce absenteeism provides some insights into one class of reasons why sponsorship might be withdrawn. In this project, the researchers introduced, with controlled levels of participation, a group incentive to curb absenteeism. Over the years, the results were very good for the high participation groups that designed the incentive system (versus those for some control groups which did not participate in the design of the experimental system). Despite years of lower levels of absenteeism in the participation groups, management cancelled the incentive plan in some of these groups. Following cancellation, absenteeism increased. The researchers argued that management's rejection of the program, even in light of favorable results, was due to its lack of involvement in the planning of the program. Although top management had given its support, those middle managers responsible for the administration of the incentive plan were not involved in its development. The concept of multiple sponsorship emerges from this study. While it is clear that withdrawal of top management support means the end to any program, this study would indicate that withdrawal of middle-management sponsorship and other sponsors can also effectively end institutionalization.

The nature of the sponsorship can affect the degree of institutionalization. The sponsor's roles include legitimizing the new project, providing a flow of rewards, monitoring or controlling the behaviors, and/or providing support in times of crisis. If the primary role of the sponsor is monitoring and controlling behavior, the behavior will not be highly institutionalized. In this case, if the sponsor left the organization or focused his attention elsewhere, the degree to which the behavior is institutionalized should decline. By the same token, if the sponsor's primary role is to legitimize or support the new work behavior in times of crisis, then the absence of the sponsor should not affect the degree of institutionalization to the same extent. In this latter case the sponsor is not continuously maintaining the new behavior.

### *Transmission*

*1. New Organizational Members.* Socialization is one of the major steps in institutionalization. After an act is institutionalized, it is still necessary to transmit information about appropriate behaviors to new members. None of the studies under review treat the relative effectiveness of different transmission mechanisms. The major finding is the absence of transmission mechanisms. Failure to transmit information about behavior to new members should decrease the level of institutionalization.

Miller's follow-up study (1975) of Rice's (1953) intervention in an Indian textile mill found that the groups that maintained the structure initiated by Rice sixteen years previously had relatively low turnover. On the other hand, the work groups with greater mobility did not exhibit the same level of institutionalization. The problem, of course, is not simply mobility, but rather the resultant failure to develop specific transmission mechanisms to socialize new leaders and members. Walton (1975) and Goodman (1979) report similar results. Walton noted that new members were not informed of their rights and obligations in the new work system. The existence of unsocialized members weakens the level of institutionalization as members differ in their perceptions of appropriate member behavior. Several laboratory experiments (cf. Jacobs & Campbell, 1961) concerning the persistence of group norms with changing group membership report similar phenomena.

While the necessity seems obvious, it is interesting that effective transmission mechanisms have been neglected in some well-known organizational interventions. One reason for the omission is that there is a natural tendency to focus on the "front-end" of an intervention—getting it started—rather than on the mechanisms to keep it going.

2. *New Organizational Roles.* Organizational interventions lead to the development of new roles. The long-run success of the intervention depends on the degree to which occupants of those roles can be trained. Many of the roles are complicated and require a sustained socialization effort. Transmission involves then not only the socialization of new members but also the training of old members to insure that the appropriate behavior will persist. Mohrman, Mohrman, Cooke, & Duncan (1977), in a study of survey feedback in a school district, reported that the training was not extensive enough to institutionalize the major roles. One consequence was that many of the schools within the district did not continue many of the new work behaviors after the project's first year of operation. Lacking sufficient personnel, the viability of the project was threatened. Another study (Goodman, 1979) describes a major training program in the first year of the intervention which led to the development of a new set of roles. However, in the second and third years of the project, there were fewer systematic training experiences. The failure to maintain the training process over time in order to *renew* role behaviors contributed to some degree to the decline of the change project.

*Group Forces*

The theoretical section argued that the social organization or group plays an important role in the institutionalization process. As individuals become more aware that others are performing the same behaviors, that the behaviors are considered appropriate by the group, and that the behaviors are sanctioned by the group, the new behaviors are more likely to become institutionalized. While there are many studies of the development of groups as norm-sending bodies and on group cohesiveness (cf. Davis, 1969), only a few of the studies reviewed here address the issue of group forces as they bear on institutionalization.

The level of interaction among members of the target group seems to determine whether the group will play a major role in the institutionalization process. The more meetings among target group members, the greater is the number of interactions and the greater the identity within the target group (Walton, 1975; Goodman, 1979). Identification serves as one force to maintain the new behavior. The development of group identification also permits the group to be a dispenser of rewards and punishments.

Minimizing competition within the target group by changing the evaluation system from an individual to a group standard (Lesieur, 1958; Goodman, 1979) is another way to strengthen the group forces.

*Feedback*

Feedback is another factor affecting the degree of institutionalization. The effectiveness of feedback for promoting change has been demonstrated by various studies (cf. Nadler, 1977) of the survey feedback approach to organizational development. These studies do not, however, examine particular behaviors or the institutionalization of these behaviors.

*1. Feedback and the Institutionalization of Behavior.* Although there are many studies of the effect of feedback on performance, relatively few have examined institutionalization as a dependent variable. In a laboratory simulation, Conlon (1978) found feedback concerning individually valued outcomes (i.e., pay) to be more strongly related to the degree of institutionalization than feedback concerning organizationally valued outcomes (i.e., quality).

*2. Level of Feedback Aggregation.* In using feedback to evoke changes, the appropriate level of aggregation for the feedback is an important variable. For example, if an intervention requires cooperative behavior within a work group, is feedback about the performance of specific individuals appropriate? Berkowitz, Levy & Harvey (1957), in a study of task group behavior in a military setting, found that group feedback produced greater task-oriented behavior and more group pride than did individualized feedback. It is reasonable to assume that motivation will increase persistence. Although no data are available, we hypothesize that the amount of cooperation required for the task may determine what is appropriate feedback. For increasing persistence on highly cooperative tasks, group feedback may be effective, while for divisible, noncooperative tasks, individual feedback may be more appropriate.

*Commitment*

The nature of the commitment process in the adoption decision can increase the resistance of new behavior to change. Individuals are more likely to persist at behaviors to which they are committed. Of the four ways cited by Salancik (1977) to produce commitment, volition and publicity are the most relevant to planned change contexts.

*1. Volition and Planned Change.* Volition refers to the degree to which an individual perceives free choice in making a decision. Free choice may apply to one's choice to participate in a change program (i.e., volunteering) or one's ability to determine the context or nature of new behavior (i.e., the design of the intervention). When individuals are asked to volunteer in an organizational experiment, they are likely to exhibit greater commitment than when a program is imposed. The act of volunteering means the actor has made a free choice. Similarly the voluntary decision not to participate leads the decision maker to resist adoption of the new work behaviors (Goodman, 1979).

There is more research on the second area of volition: designing the nature of a work change. Scheflen, Lawler & Hackman (1971) state that participation in designing new work behaviors induces a higher level of commitment among participants than would accrue without participation. Indeed, some studies suggest that participatively-introduced change may be more durable than others (cf. Seashore & Bowers, 1978; Scheflen et al., 1971). To the extent that the participants feel that they have been responsible for the selection and content of a new work behavior, we would expect greater commitment and persistence.

2. *Publicity and Planned Change.* Publicity refers to the extent to which others know about the performance of a particular act (e.g., decision to adopt). The greater the publicity, the greater the commitment. In the Scheflen et al. (1971) study, the participants had group discussions to formulate a plan for reducing absenteeism. Individuals who publicly indicated their approval of the plan in these discussions became committed to adopting and to continuing the new behavior. In the Rushton mine quality of work study (Goodman, 1979), some behaviors became institutionalized and others did not. One institutionalized behavior, communication between crews, was characterized by high visibility (publicity). The adoption of this behavior was visible to individuals both internal and external to a work crew. The noninstitutionalized behavior—job exchange—occurred underground and was visible, at most, only to the work crew. The visibility of these adopted behaviors may have affected the degree to which individuals felt bound to continue the behavior. For the communication behavior, it would be more difficult for individuals to later devalue the behavior since it had been publicly adopted. For the job switching behavior, only the crew could know about prior adoption, perhaps making devaluation easier.

*Diffusion*

Diffusion refers to the extension and adoption of new forms of work behavior in a social system. The diffusion process can play a major role in facilitating institutionalization. Most change efforts focus on specific target groups. Often the target is a specific work group or a plant. It is unlikely that the change effort would include a total organization. Given the narrower focus of most change efforts, there will always be a larger formal system surrounding the target group. Walton (1975) has argued that the special treatment given to these groups creates a star-envy phenomenon which generates pressures to destroy the change effort. Diffusion serves to spread the new form of work innovation throughout the system and thus to counter the invidious comparisons between treatment and nontreatment groups and to legitimate the new work innovation.

Walton (1975) has done some of the most careful analyses of diffusion in the context of planned organizational change. In an examination of eight work restructuring experiments, the effect of diffusion on institutionalization is documented—failure to diffuse the innovation to new systems weakens the level of institutionalization in the original target system. In the Rushton mining experiment (Goodman, 1979), new forms of work organization were introduced into some work sections but not others. An attempt to diffuse the work innovations into the rest of the mine failed. Over time new forms of work organization practices in the original experimental sections began to decline due to the loss of legitimation in the larger system.

*Internal Contextual Factors*

The factors discussed thus far directly affect the processes of institutionalization. Internal factors refer to the organizational context surrounding the change effort.

*1. The Congruence of New Forms of Work Organization with Existing Organizational Structure Affects the Level of Institutionalization.* Congruence, or consistency, refers to the fit between the intervention structure and the organization's values and structure. The finding that cuts across a variety of studies is that the greater the consistency, the higher the level of institutionalization.

Seashore & Bowers (1978) explain the level of institutionalization by the congruence between the organizational change and the values and motives of the individual participant. In their investigation, where the changes were more congruent with the values and motives of the employees, a high level of institutionalized resulted. Mohrman et al. (1977) found in a study of a school system that the change activities persisted in those schools which had prior experience with these activities and in those schools where the intervention structure was congruent with the existing authority system. Similar findings have been reported in the relationship between experimental plants and corporate headquarters. Fadem (1976) suggests the greater the discrepancy between the experimental settings and corporate policies, the less likely the project will be institutionalized.

The explanation for the relationship between consistency and institutionalization is the following. The intervention is introduced in the target group. The new work behavior is accepted; the issue is not one of resistance to change. The greater the discrepancy between the target group and the larger organizational context, the greater the opportunity for tension. The tension arises because the norms in the target system diverge and hence challenge the norms in the larger system. The target group is a deviant group. Given this situation, we would expect forces to be generated from the larger system to bring the target group back into line. A number of studies (cf. Walton, 1975; Goodman, 1979) provide some support for this explanation.

*2. Character of the Boundary Conditions.* Since the target group organization operates in a larger system, there is a need to manage the boundary relationship between both groups. The issue here is not one of consistency. The boundary representative of the target group acts as a buffer against pressures from other organizational groups, and extracts whatever resources it can obtain from the larger system. Alderfer (1976) has hypothesized that the openness of boundary conditions are curvilinearly related to desirable characteristics of successful change programs. Miller's analysis of the persistence of change in Indian textile mills argues that institutionalization was facilitated when the buffer conditions were effectively managed. When the boundary representatives were not able to buffer external pressures (i.e., demands from higher authorities), the internal functioning of the group, and hence the new form of work behavior, dissipated.

*3. Intergroup Dependencies.* Planned organizational change takes place in a web of interdependencies. The character of these interdependencies can affect the level of institutionalization. Frank & Hackman (1975), in an investigation of a change effort in a bank, noted the breakdown in a related work group hindered the persistence of the intervention in the target population. The failure of a data processing department to produce the necessary information led to the downfall of the intervention strategy. Goodman (1979) reports a similar finding. Although many of the new forms of work behavior were adopted in the mining section, lack of cooperation from other work areas in the mine contributed to the eventual decline of this intervention.

*External Contextual Factors*

These factors refer to events outside the organization that affect the institutionalization process within the organization. The focus is on the environment as it affects the persistence of new forms of work organization.

*1. Nature of the Environment.* Miller (1975) found differential degrees of institutionalization in different weaving groups in the Indian textile mill.

One factor explaining these differences was the nature of the marketplace for the different groups. Those groups with the lowest degree of institutionalization operated in a very competitive market where the level of quality was a critical factor in affecting sales. This demand for high quality led to a great deal of pressure on the work group. The pressure over time led to the replacement of the new forms of organization with more traditional organizational arrangements. For other work groups the market for its goods was less competitive and the company held a more secure niche. In this situation the new forms of work organization initiated by Rice in the 1950s remained in place. Walton (1978) also documented the importance of the organization's environment. He reported that as production pressures increased in the Topeka plant, there were few group meetings, more suboptimization of the shift level (at the expense of the following shift), and lower levels of quality, which had once been a source of pride. Each of these consequences worked against the institutionalization of new forms of work organization. For example, fewer meetings leads to less group identification, which is a powerful force in maintaining institutionalized behavior.

2. *Union-Management.* There is relatively little empirical work on the effect of unionization on planned organizational change. This is unfortunate, because in any attempt to restructure work where union members are the target population, the union will be a major determinant of the success of that change effort.

Kochan & Dyer (1976) developed one of the best conceptual frameworks to understand organizational change in the context of labor-management relations. One of their basic arguments is that the company and union management have conflicting goals and both parties use power to achieve their goals. The scene is one of conflict rather than cooperation. Most planned organizational interventions are built on cooperation. There is an inherent conflict then between the nature of labor-management relationships and the assumptions underlying most planned organizational interventions.

From the modest number of empirical studies on planned organizational change in a union-management context (Lewicki & Alderfer, 1977; Driscoll, 1978; Herman & Macy, 1977; Goodman, 1979; and from the Kochan & Dyer model) we can identify the following characteristics of labor-management relations that bear on institutionalization.

(a) If both parties perceive the new forms of work organization as facilitating the attainment of their respective goals, the level of institutionalization should increase. In the Rushton mine quality of work experiment, the union participated to improve safety while management was more interested in increasing productivity. As long as the experimental program led to achievement of both goals, both parties remained committed to the intervention (Goodman, 1979).

(b) The greater the congruency between the values and structure of the intervention with the institutions of collective bargaining, the higher the level of institutionalization. In the Rushton experiment, the payment of overtime and grievance procedures were different from the provisions in the collective bargaining agreement and different work groups operated with different procedures. This led to opposition within the work force which worked against the institutionalization of new work behaviors (Goodman, 1979).

(c) The more successfully labor and management work out traditional labor-management issues, the more likely they can maintain commitments over time to new forms of work organization. At issue is a spillover effect that has appeared in a number of change programs. Here failure to solve traditional issues such as grievances spills over into relationships which require joint problem-solving behavior. Adversary behaviors drive out cooperative behaviors and new forms of work behavior decline (Goodman, 1979). The separation of collective bargaining relationships and the planned organizational change relationships seem necessary to maintain the integrity of both institutions.

(d) Certain planned organizational change efforts can lead to increased conflict within the union, which can decrease levels of institutionalization. A number of current labor and management quality of work experiments lead to increased interaction between labor and management outside the traditional collective bargaining procedure. These interactions are sometimes seen as management's attempt to coopt union officials. This leads to increased tension between union leaders and members which can lead to the union's withdrawal of support from the change effort, and hence the decline of new forms of work behavior. Another tension may arise when innovations in the target organization are at variance with the values and policies of the international union. In this case the local arrangements may conflict with the national collective bargaining practices which would lead the international to work against the newer forms of work organization and, therefore, the viability of the change effort.

## CONCLUSION

Organizational change is a central concept in organizational theory. Current experiments in new forms of work organization have emphasized the need to increase our understanding of organizational change processes. This paper has elaborated on one of the central concepts in organizational change: institutionalization. Although the process of sustaining change over time is an obvious ingredient of any organizational change, there have been few attempts to conceptualize institutionalization or to empirically examine its antecedents or processes. The goal of this paper was to develop a conceptual framework of institutionalization. Basically the focus was to delineate this concept in order to increase our understanding of change processes. No formal theory has been developed. Rather we have tried to specify the construct space of institutionalization. The level of specification is in sufficient detail to enhance our understanding of institutionalization and to pave the way for systematic empirical studies testing hypotheses derived from this framework.

Our approach was to define institutionalization as an act and process and then to differentiate it from other concepts. Most of our conceptualization focused on the two-phase model which traces the development of an institutionalized act from a series of individual-level decisions (i.e., to adopt and to continue) to its existence as part of social structure. A careful examination of this phase model provides insights into those factors that facilitate or inhibit the persistence of new forms of work behavior. Although there is no systematic literature on institutionalization of planned organizational change, we were able to identify a set of factors which seems to influence the level of institutionalization. Some of the factors, such as the nature of reward systems, transmission mechanisms, and group forces affected the persistence of new forms of work behavior. The organizational context surrounding the planned organizational intervention, as well as characteristics of the organization's environment also contributed to the institutionalization of new forms of work behavior. Although the empirical studies were not formal tests of institutionalization, the critical factors identified in these studies were congruent with our two-phase model of institutionalization.

One objective of this paper is to call for a change of focus in the organizational change literature. Some of this literature is characterized by very general theorizing. Others, primarily those involved in the practice of changing organizations, have developed personal testimonies or case reports of interventions. Currently, cataloguing intervention techniques or advocating certain techniques over others is common practice in the literature. Unfortunately, none of these approaches is going to increase our understanding of organizational change. We need to identify the critical processes in organizational change and then delineate these processes, develop hypotheses and systematically test these hypotheses. Hopefully our approach to institutionalization may begin the development of a better understanding of organizational change.

## FOOTNOTES

* This Chapter was supported in part by ONR Contract #00014-79-C-0167.

** Now at the University of Texas at Austin.

*** The authors would like to thank Robert Atkin for his helpful comments on an earlier draft of this paper.

## REFERENCES


Alderfer, C. (1976) "Change processes in organizations." In M. Dunnette (ed.), *Handbook of industrial and social psychology*. Chicago: Rand McNally College Publishing.

Asch, S. E. (1956) "Studies of independence and conformity: A minority of one against a unanimous majority." *Psychological Monographs*, 70 (9, Whole No. 416).

Beer, M., & Huse, E. F. (1972) "A systems approach to organizational development." *Journal of Applied Behavioral Science*, 8(1), 79–101.

Berger, P., & Luckman, T. (1966) *The social construction of reality: A treatise in the sociology of knowledge*. Garden City, NY: Doubleday.

Berkowitz, L., Levy, B., & Harvey, A. (1957) "Effects of performance evaluations on group integration and motivation." *Human Relations, 10,* 195–208.

Blake, J., & Davis, K. (1964) "Norms, values, and sanctions." In R. Feris (ed.), *Handbook of Modern Sociology*. Chicago: Rand McNally.

Breer, P., & Locke, E. (1965) *Task experience as a source of attitudes*. Homewood, IL: Dorsey.

Buckley, W. (1967) *Sociology and modern systems theory*. Englewood Cliffs, NJ: Prentice Hall.

Conlon, E. J. (1978) "On the persistence of behavior in planned change contexts: Some effects of instrumental feedback" (Working Paper No. M-78-1). Unpublished manuscript, Georgia Institute of Technology.

Crockett, W. (1977) "Introducing change to a government agency." In P. Mirvis & D. Berg (eds.), *Failures in organizational development: Cases and essays for learning*. New York: Wiley-Interscience.

Crutchfield, R. S. (1955) "Conformity and character." *American Psychologist, 10,* 191–198.

Davis, J. H. (1969) *Group performance*. Reading, MA: Addison-Wesley Publishing.

Driscoll, J. (1978) "Change strategies for union-management cooperation: The Scanlon Plan." (Working Paper). Unpublished manuscript, Massachusetts Institute of Technology.

Fadem, J. (1976) "Fitting computer-aided technology to workplace requirements: An example." Paper presented at the 13th annual meeting and technical conference of the Numerical Control Society, Cincinnati, March 1976.

Frank, L. L., & Hackman, J. R. (1975) "A failure of job enrichment: The case of the change that wasn't." *Journal of Applied Behavioral Science*, *11*(4), 413–436.

Friedlander, F., & Brown, L. (1974) "Organization development." *Annual Review of Psychology, 25,* 313–341.

Goodman, P. S. (1979) *Assessing organizational change: The Rushton quality of work experiment*. New York: Wiley-Interscience.

———, & Lawler, E. E. (1977) "New forms of work organization in the United States." Monograph prepared for the International Labor Organization, Geneva, Switzerland.

———, & Moore, B. (1976) "Factors affecting acquisition of beliefs about a new reward system." *Human Relations, 29*, 571–588.

Helson, H. (1964) *Adaptation-level theory*. New York: Harper & Row.

Herman, J. B., & Macy, B. A. (1977) "Labor-management relationships in collaborative quality of working life projects." Paper prepared for the Quality of Working Life Assessment Conference, University of Michigan, Ann Arbor, Michigan, July 1977.

Homans, G. (1961) *Social behavior: Its elementary forms*. New York: Harcourt Brace.

Jackson, J. (1966) "Structural characteristics of norms. In I. Steiner & M. Fishbein (eds.), *Current studies in social psychology*. New York: Rinehart and Winston.

Jacobs, R. C., & Campbell, D. T. (1961) "The perpetuation of an arbitrary tradition through several generations of a laboratory microculture." *Journal of Abnormal and Social Psychology, 62,* 649–658.

Katz, D., & Kahn, R. L. (1978) *The social psychology of organizations* (2nd ed.). New York: John Wiley & Sons.

Kelman, H. (1958) "Compliance, identification and internalization: Three processes of attitude change." *The Journal of Conflict Resolution, 2,* 51–60.

Kimberly, J. R., & Nielson, W. R. (1978) "Organization development and change in organization performance." In W. L. French, C. H. Bell, Jr., & R. A. Zawacki (eds.), *Organizational development: Theory, practice, and research.* Dallas: Business Publications.

Kochan, T. A., & Dyer, L. (1976) "A model of organizational change in the context of union-management relations." *Journal of Applied Behavioral Science, 12,* 59–78.

Lesieur, F. G. (Ed.). (1958) *The Scanlon Plan: A frontier in labor-management coordination.* New York: Wiley.

Lewicki, R., & Alderfer, C. (1977) "The tensions between research and intervention in intergroup conflict." In P. Mirvis & D. Berg (eds.), *Failures in organizational development: Cases and essays for learning.* New York: Wiley-Interscience.

Lewin, K. (1935) *A dynamic theory of personality: Selected papers.* D. K. Adams & K. E. Zenor (trans.). New York: McGraw-Hill.

———. (1951) "*Field theory in social science*" D. Cartwright (ed.). New York: Harper.

March, J., & Simon, H. (1958) *Organizations.* New York: Wiley.

Meyer, J. W., & Rowan, B. (1977) "Institutionalized organizations: Formal structures as myth and ceremony." *American Journal of Sociology, 83*(2), 340–363.

Miller, E. J. (1975) "Socio-technical systems in weaving, 1953–1970: A follow-up study." *Human Relations, 28*(4), 349–386.

Mitchell, T. (1974) "Expectancy models of job satisfaction, occupational preference and effort: A theoretical, methodological and empirical appraisal." *Psychological Bulletin, 81,* 1053–1077.

Mohrman, S. A., Mohrman, A. M., Cooke, R. A., & Duncan, R. B. (1977) "A survey feedback and problem solving intervention in a school district: 'We'll take the survey but you can keep the feedback.' " In P. Mirvis & D. Berg (eds.), *Failures in organizational development: Cases and essays for learning.* New York: Wiley-Interscience.

Nadler, D. A. (1977) *Feedback and organizational development: Using data-based methods.* Reading, MA: Addison-Wesley Publishing.

Oskamp, S. (1977) *Attitudes and opinions.* Englewood Cliffs, NJ: Prentice Hall.

Parsons, T. (1951) *The social system.* New York: Free Press.

Rice, A. K. (1953) "Productivity and social organization in an Indian weaving shed: An examination of the socio-technical system of an experimental automatic loomshed." *Human Relations, 6*(4), 297–329.

Rogers, E. M., & Shoemaker, F. F. (1971) *Communication of innovations.* New York: The Free Press.

Salancik, G. (1977) "Commitment and the control of organizational behavior and belief." In B. Staw & G. Salancik (eds.), *Directions in organizational behavior.* Chicago: St. Clair Press.

Scheflen, K., Lawler, E., & Hackman, J. (1971) "Long term impact of employee participation in the development of pay incentive plans." *Journal of Applied Psychology, 55,* 182–186.

Schein, E. H. (1968) "Organizational socialization and the profession of management." *Industrial Management Review*, Winter 1968, 1–16.

———. (1969) "The mechanisms of change." In Bennis, Benne & Chin (eds.), *The planning of change* (2nd ed.). New York: Holt, Rinehart and Winston.

Seashore, S. E., & Bowers, D. G. (1978) "Durability of organizational change." In W. L.

French, C. H. Bell, Jr., & R. A. Zawacki (eds.), *Organization development: Theory, practice, and research.* Dallas: Business Publications.

Walton, R. E. (1975) "The diffusion of new work structures: Explaining why success didn't take." *Organizational Dynamics,* Winter 1975, pp. 3–21.

———. (1978) "Teaching an old dog food new tricks." *The Wharton Magazine,* Winter 1978, pp. 38–47.

White, S. E., & Mitchell, T. R. (1978) "Organizational development: A review of research content and research design." In W. L. French, C. H. Bell, & R. A. Zawacki (eds.), *Organizational development: Theory, practice, and research.* Dallas: Business Publications.

Zaltman, G. (1973) *Processes and phenomena of social change.* New York: John Wiley and Sons.

Zucker, L. G. (1977) "The role of institutionalization in cultural persistence." *American Sociological Review, 42,* 726–743.

# WORK DESIGN IN THE ORGANIZATIONAL CONTEXT[1]

Greg R. Oldham

UNIVERSITY OF ILLINOIS

J. Richard Hackman

YALE UNIVERSITY

## ABSTRACT

This chapter focuses on factors in the organization that can compromise the implementation and persistence of work redesign. In the first half of the chapter we discuss ways that three organizational systems—the technological system, the personnel system, and the control system—can constrain the implementation of substantial and meaningful changes in the work itself. We then explore ways that several organizational practices—training practices, career development practices, compensation practices, and supervisory practices—can reduce the likelihood that the effects of work redesign will persist in an organization. Throughout the chapter we examine possible

**Research in Organizational Behavior, Volume 2, pages 247–278**

**ISBN: 0-89232-099-0**

strategies for improving the "fit" between redesigned work and organizational systems and practices. Since there is relatively little research on the aforementioned topics, much of the chapter is devoted to exploring the dynamics of these phenomena and to suggesting some research directions for generating more systematic knowledge about them.

## INTRODUCTION

People bring energy, talent, and personal needs or wants to their jobs. Jobs make demands on people and offer opportunities for people to use their talents and seek satisfaction of their needs. The goodness of the "fit" between the characteristics of people and the properties of their jobs is one factor that affects both work productivity and worker satisfaction.

When work is redesigned, the person-job fit is altered. We are particularly interested in those cases when the intent is to "enrich" the person-job relationship—that is, to make the work more meaningful for the job-holder, to provide him or her with increased personal responsibility for managing the work, and to increase the job-holder's knowledge of the results of the work activities (Hackman & Oldham, 1976, in press).

Research has shown that, under appropriate circumstances, work redesign can enhance both organizational productivity and the personal satisfaction of job-holders (cf. Katzell & Yankelovich, 1975; Katzell, Bienstock & Faerstein, 1977). Yet our observations of work redesign programs suggest that attempts to change jobs frequently run into—and sometimes get run over by—other organizational systems and practices, leading to a diminuation (or even a reversal) of anticipated outcomes.

Sometimes it turns out that the changes actually made in the work are far less radical or far reaching than those originally contemplated by the managers or consultants who initiated work redesign, and not large enough to make measurable differences in employee attitudes and behavior (Frank & Hackman, 1975). Other times substantial changes in the work actually are installed, but their effects fade over time (Walton, 1977).

The "small change" and "vanishing effects" phenomena call into question the efficacy and permanence of planned organizational change through the redesign of tasks and jobs. How are we to understand these phenomena? One possibility, of course, is simply that tasks do not make much of a difference in people's behavior and attitudes (cf. Salancik & Pfeffer, 1977). Another, and one we find more plausible, has to do with the ways that work design and other organizational systems and practices *interact*. The "small change" effect, for example, often develops as managers begin to realize that radical changes in work design will necessitate major changes in other organizational systems as well. When managers discuss possible "enriching" changes in jobs, a frequent comment is something like, "Yes, but we couldn't do that, because..." And then follows a description of how the contemplated change would require revision of corporate personnel systems, or the technology of the work unit.

In most cases the concerns are valid: large changes in one organizational system (in this case, the work itself) invariably require alterations in related systems. These alterations are anxiety-arousing for the people involved, or expensive to make, or contrary to organization-wide policies—or frequently, all of the above. So numerous small compromises are made from the "ideal" work design to minimize the disruptiveness and cost of the changes. The net effect, in many cases, is a project that meddles with the work rather than redesigns it. The changes are safe, feasible, inexpensive—and ineffectual.

Sometimes—especially in decentralized and low-technology organizational units—relatively substantial alterations are actually made in how the work is designed. When this happens other organizational systems and practices invariably are affected, and occasionally are thrown into disarray. People don't like the way they are paid anymore. First-line managers feel stripped of their authority and unpleasantries develop between them and the employees. Organizational control systems no longer work.

What we have is an aberration in one system that creates difficulties for many surrounding systems and for the people who manage those systems. Either the surrounding systems must accommodate the new way the work is being done or vice versa. Our observations suggest that the innovation "wins" relatively infrequently. Instead, the innovation is modified, slowed, or redefined in such a way as to be less of a problem for the surrounding systems. It is almost as if a foreign substance were introduced into the body: once its presence is recognized, the white blood cells arrive and attempt to render it impotent. The defenders usually prevail, given enough time, and the result is the "vanishing effects" phenomenon noted above.

Previously we have suggested that work redesign could be an excellent point of entry for broad-scale programs of organizational change (e.g., Hackman, 1975). The idea was that tasks, once changed, tend to stay changed. And because redesigning the work does create a "ripple effect" on surrounding systems, parts of the organization that previously were not amenable to change might become so following changes in tasks and jobs. By starting with work redesign and gradually folding in changes in related systems to make them consistent with the enriched work, major reorientations in how organizations function could be achieved.

That set of ideas has turned out to work better in theory than in practice. Clearly we underestimated both the difficulty of carrying out significant changes in the work itself, and the degree to which changes in tasks wind up being altered by surrounding organizational systems, rather than vice versa. Our present view, informed by numerous cases in which even well-executed work redesign interventions have failed to have a broad or persisting impact, is that one must deal with the design of work and with surrounding organizational systems simultaneously, or very nearly so.

This is precisely what sociotechnical systems theorists have advocated for years (e.g., Davis & Trist, 1974). It is clear to us now, as it should have been before, that the case for changes with multiple foci is a strong one, and that redesigned jobs may be at considerable risk unless they are congruent with other organizational systems and practices. If one adopts this view, then it becomes critical to understand *which* systems and practices most strongly compromise the implementation and persistence of work redesign, and to determine *how* these effects come to pass.

In this chapter we present some thoughts on these issues, and propose some new directions for research on work design that we believe could increase understanding about them. In the first half of the chapter we explore ways that existing organizational systems constrain the *implementation* of work redesign. Then, in the second part of the chapter, we examine ways that certain organizational practices can chip away at the *persistence* of changes that have been made in the design of work. While organizational systems and practices also affect the *diffusion* of work redesign within and across organizations, we restrict our attention here to matters that have to do with getting substantial changes in jobs made in the first place, and maintaining those changes over time. Readers interested in systemic effects on the diffusion of organizational changes are referred to Kimberly (in press) and Walton (1977).

Our chapter will necessarily be more speculative than most others in this series. While there are numerous case studies and tales of woe regarding problems in the implementation and persistence of work redesign, there is almost no systematic research on the topic. Therefore, we are unable to point out many contradictory findings in the research literature, or to propose tests of competing hypotheses about the phenomena we address, or to evaluate which lines of existing research seem more or less fruitful. Instead, we must suffice with an attempt to construe the phenomena we are addressing in researchable terms and to suggest some directions for research on these phenomena that may help in understanding them better.[2]

## CONSTRAINTS ON THE IMPLEMENTATION OF WORK REDESIGN

As noted above, sometimes it is not possible to alter the design of work substantially enough to make meaningful differences in the person-job relationship. What differentiates between those organizational circumstances in which substantial changes in the design of work can be made, and those when only small changes in the work itself are feasible? The answer, we suggest, has primarily to do with the properties of three organizational systems: the technological system, the personnel system, and the control system. As will be seen below, each of these systems has direct influences on how work is designed, and can limit the degree to which jobs can be changed and enriched.

Specifically, the technology of the organization has powerful effects on the specific tasks that must be performed, and how those tasks are arranged and sequenced. The personnel system can constrain flexibility in determining which employees can do what aspects of the work, and limiting the possibility of combining tasks into large and meaningful modules. And the control system can constrain both the scope of jobs and the procedures that are used to perform them. Under some circumstances, these systems may render work redesign wholly infeasible—for example, when the technology is both fixed and expensive—when the personnel system enforces adherence to a detailed and specific set of job descriptions, and when work processes are prescribed and enforced by an elaborate and inflexible control system. Work redesign is almost sure to fall victim to the "small change" effect under such conditions—unless it is possible to redesign these constraining systems simultaneously with the redesign of the work itself.

### *Technological System*

The technology of an organization can constrain the feasibility of work redesign by limiting the number of ways that jobs within the technology can be designed. In certain kinds of technologies, for example, it simply is not possible to build meaningful amounts of autonomy, variety, or feedback into the jobs (Slocum & Sims, 1978).

A well-worn but illuminating example of such a technology is the automobile assembly line. Employees working on the line have little control over work pace—this is controlled by the line itself. Moreover, both the size of the work unit for which employees are responsible and the variety of skills needed to complete the work of the unit are severely limited for technological reasons. What remains in many assembly line technologies are fractionalized, segmented jobs—jobs that must remain that way as long as the technology is the way it is (Blauner, 1964; Walker & Guest, 1952).

Perhaps the key to understanding how technology limits the characteristics of jobs is the concept of employee *discretion*. Rousseau (1978) has argued that one of the most important features of a technological system is the degree to which discretionary behavior is required of the human operator. When few degrees of discretion are required or allowed by the technology, then work procedures are by definition highly standardized and structured. Under such conditions, employees' jobs are usually segmented and routinized, and contain little variety, autonomy, identity, and significance for the people who perform them. In essence, the technology usurps many of the most desirable features of the work.

Recent research has documented that technologies which permit few degrees of discretion are in fact empirically associated with simple and routine jobs. For example, Rousseau (1977) used Thompson's (1967) classification scheme to categorize technologies according to the degree of discretion permitted. The technological categories were: *long-linked* (exemplified by assembly line systems), *mediating* (where inputs are first sorted into groups and each group subsequently is subjected to prescribed treatments), and *intensive* (where customized techniques are applied to an input based on feedback from that input). Results showed that assembly line technologies were associated with low levels of several job characteristics (as measured by the Job Diagnostic Survey) relative both to national norms for the instrument (Oldham, Hackman & Stepina, 1979) and to job characteristic scores obtained for the mediating and intensive technologies. These results provide empirical support for the idea that certain technological systems do limit the presence of certain motivating characteristics of jobs.

Thus, any effort to redesign work in a technology that permits little employee discretion (e.g., long-linked systems) is probably doomed to failure from the outset because of the mechanics of the system itself. The system has been designed to accommodate only segmented, routinized jobs. The only redesign activities that are likely to be feasible, then, are those that involve relatively small changes in the work itself (e.g., giving employees some choice of tools). However, as we have suggested earlier, this usually amounts to meddling with the work rather than enriching it—and the effects are likely to be neither substantial nor long-lasting.

These arguments suggest that if work is to be meaningfully redesigned in an organization either: the technology must be of the type which provides at least moderate employee discretion (e.g., mediating or intensive); or the whole technology must be changed to be compatible with the characteristics of enriched work.

The latter route was taken, apparently with success, when the management of Volvo planned its new automobile assembly plant at Kalmar

(Gyllenhammar, 1977). It was decided early in the planning process to create enriched, challenging jobs at the new plant. The idea, however, was to design these jobs within a traditional conveyor line technology. As planning progressed, Gyllenhammar and his associates realized that the traditional technology and the innovative jobs were inherently incompatible. It became clear that if nontraditional, enriched jobs were to be created, a nontraditional technology also would be required. This realization led to cancellation of plans for the conveyor line and, instead, the development and installation of a technology permitting large amounts of employee discretion (e.g., moveable automobile carriers).

The cost of altering technologies can be very high and in many cases will be prohibitive. Volvo's nontraditional technology, for example, entailed an initial investment that was ten percent higher than would have been the case if the traditional technology had been installed. Unless such an investment is made, however, the possibility for substantial and meaningful redesign of employee jobs is sharply reduced. Indeed, in many cases it will be advisable *not* to try to enrich work for traditional technological systems, but instead to find ways to manage people as effectively and humanely as possible within those systems.

To aid in such decision making, and to increase understanding about the extent and dynamics of technological influences on the design of jobs, better measures are required to assess the interdependencies between technological systems and the jobs that exist within them. The recent work of Rousseau (1977) is a major step in this direction. But additional conceptual and empirical progress must be made before it will be possible to identify clearly those technological systems that are (and are not) appropriate for work redesign activities, and to specify what is required if enriching changes in jobs can be made within a given technological system.

*Personnel System*

In the interest of having a clear, fair, and concrete basis for recruiting, selecting, and placing people on jobs, elaborate personnel systems often are developed in organizations. These systems often result in a set of fixed job descriptions that specify exactly who is to do what. As will be seen below, such job descriptions can introduce rigidity into how the work itself is designed, and limit flexibility in changing the duties of people who hold specific jobs.

Typically, a job description consists of a written statement that describes the basic duties and responsibilities of the job-holder (Wexley & Yukl, 1977). Job descriptions can be very elaborate and detailed. Consider, for example, "functional" job descriptions, which may specify all of the following (Fine & Wiley, 1974):

1. *The action the employee is expected to take in the job.* "A task statement requires a concrete, explicit *action* verb. Verbs which point to a process (such as *develops, prepares, interviews, counsels, evaluates,* and *assesses*) should be avoided or used only to designate broad processes, methods, or techniques which are then broken down into explicit, discrete action verbs" (pp. 7–8).

2. *The result the employee is expected to accomplish.* "The purpose of the action performed must be explicit so that (1) its relation to the objective is clear and (2) performance standards for the worker can be set" (p. 8).

3. *The tools the employee is expected to use.* "A task statement should identify the tangible instruments a worker uses as he performs a task; for example, telephone, typewriter, pencil/paper, checklists, written guides, and so forth" (p. 8).

4. *The instructions the employee is expected to follow.* "A task statement should reflect the nature and source of instructions the worker receives. It should indicate what in the task is prescribed by a superior and what is left to the worker's discretion or choice" (p. 8).

While an explicit, well-developed job description containing the information listed above can be useful to management in developing performance criteria and in determining selection, placement and training needs and practices, it also may impede the implementation of work redesign. In particular, descriptions that specify precise actions, tools, and instructions for the job-holder provide so much detail, and so limit flexibility, that it may be impossible to meaningfully alter the design of any given jobs within a large, interdependent work system. In essence, work redesign is a difficult proposition because either the whole job description apparatus would have to be changed if any subset of jobs within that system were changed; or arrangements would have to be made for the foçal jobs to be outside the traditional job description system.

These alternatives are likely to be resisted by numerous individuals who have a vested interest in maintaining the system as it exists—such as managers who created the system and who may be charged with maintaining and enforcing it. Moreover, fixed job descriptions sometimes are the product of years of union-management negotiations, and are enforced with a legal, binding contract. Because of these conditions, both parties are likely to have vested interests in the descriptions. The very possibility of altering them clearly challenges these interests and creates the necessity for additional negotiations between management and the union.

In one instance, the administration of a university was convinced that work redesign was desirable for numerous secretarial jobs throughout the university. However, the personnel system of the university included a set of rigid secretarial job descriptions that had been accepted by both personnel officers and the employees' union. The implementation of the work redesign program would have involved substantial revisions in the secretarial job descriptions and salary schedules. This, of course, would have required new union-management negotiations concerning the job descriptions as well as additional discussions about salary for any revised job descriptions. After considering the possibility of these additional negotiations, the administration scrapped the entire work redesign project.

All of this suggests that having some "fuzziness" and slack present in job descriptions can be helpful in carrying out work redesign. If this slack is not present, and if the job descriptions cannot be circumvented or changed in major ways, then three possibilities exist. First, the redesign project will not be implemented at all (as in the university example given above). Second, a work redesign program will be implemented that is consistent with the explicit job descriptions. The probable result here is a relatively small change in the work content, producing few desirable results. A third possibility is that "fuzziness" will somehow be introduced into the job descriptions, prior to or simultaneously with the work redesign. It is in this latter situation that favorable results are most likely.

Research is now needed to determine the most effective *process* for introducing slack into job descriptions. In particular, we need to know about processes that will facilitate the loosening of descriptions without prompting excessive conflict between labor and management. Does the most effective process involve the union in an advisory capacity or in a fully collaborative role? What are the appropriate contributions by personnel officers in redesigning job descriptions? What are the consequences of job description changes *alone* on employee behavior and attitudes? Is it best to introduce changes in job descriptions prior to or simultaneously with work redesign? Answers to questions such as these should greatly enhance our understanding of the interdependencies between job descriptions and the design of work—and reduce substantially the number of work redesign activities that fail because they run afoul of existing personnel systems.

### *Control System*

A third potential barrier to the implementation of work redesign is the organization's control system. By *control system* we refer to any "mechanical" system in place in the organization that is designed to control and influence employee behavior in an impersonal, impartial, and automatic fashion (Reeves & Woodward, 1970). Control systems include budgets and cost accounting systems, production and quality control reports, and attendance measuring devices.

Most control systems share certain structural properties (Lawler & Rhode, 1976):

1. They collect, store and transmit information in the form of abstract measures of reality. Usually they deal with information in the form of quantitative measures intended for use by trained personnel.

2. The information is stored and transmitted in a specific form and with a specific frequency.

3. The summarized information is distributed to a predetermined group of people. This group may or may not include all the members of the organization. (Lawler & Rhode, 1976, p. 6).

The existence of a control system with such properites obviously allows management to coordinate the activities of different jobs and departments. That is, redundancy of duties and tasks can be avoided by carefully specifying and measuring the behaviors of employees in different areas or jobs. In addition, control systems provide a basis for taking corrective action when employee behavior or work outcomes do not conform to standards.

However, control systems also tend to limit the complexity and challenge of jobs located within the system. Because it is important to pinpoint accountability, control systems often specify in considerable detail exactly who is to do what specific tasks—thereby restricting the autonomy in employees' jobs. Also, control systems often rigidify and standardize the work, so that performance indices can be developed and applied to all employees and work activities within the system.

An example of the effects of control systems on work design is provided in a study of the purchasing department in a large organization (Lawler & Rhode, 1976). A financial control system was designed both to provide information about work outcomes to higher management, and to prevent fraud and theft. In this situation, the control system required that employees who handled payments to vendors not talk with those vendors, and specified that each employee handle only a few of the activities necessary to pay the vendors. Thus, the jobs necessarily were routine and highly repetitive in order to meet the criteria for a "good" control system. The result in this particular case was low employee satisfaction and work performance.

Any subsequent attempt to redesign work in the purchasing department described above (or in other departments with rigid control systems) would probably result in only small changes in the work itself. For one thing, many control systems simply cannot tolerate substantial increases in the complexity of the work done by individual job-holders, or in the level of autonomy people have to manage their own work processes and procedures. Moreover, it may be difficult to introduce enriched job-based feedback for employees (a common change made when work is redesigned) and still have the control system function as intended. The reason is that data collected as part of the control system typically are supplied to staff personnel and line managers for use in managing unit performance. To redesign the work so that feedback comes naturally to the person who is doing the work (e.g., by having the jobholder do his or her own testing and inspection, or by placing that individual in direct contact with the "client" of the work) would throw many control systems into disarray. This is particularly likely if the *form* of the naturally-occurring feedback were not readily quantifiable or if it varied from job to job within the organizational unit.

To avoid the "small change" trap when work is redesigned in organizations where rigid control systems are already in place, it often is necessary to alter the control systems themselves as part of the work redesign activity. Unfortunately, significant redesign of control systems (and in particular "loosening" of them) is unlikely in many instances. In some cases, tampering with a control system may even be illegal. Certain government agencies, for example, require organizations to engage in specific, strict quality control activities. A contract may be awarded to an organization (or withheld from it) because of the organization's quality control procedures. The costs to an organization of loosening its procedures in such circumstances would usually (and justifiably) be considered unacceptably expensive.

In addition, there may be large internal costs associated with altering or scrapping a control system. Establishing a good control system often involves a large initial investment, perhaps including the purchase of computer hardware and development of sophisticated programs to assess unit productivity and management performance. Altering such a system could involve the costly development of new control system technology in addition to the person-hours required to set-up the technology. Finally, whenever there is an organizational system in place, there are personnel whose own jobs depend on the maintenance of that system; control systems are no exception. Attempts to substantially alter a control system (particularly if the idea is to "loosen" a technically sophisticated system) may encounter substantial resistance on the part of staff who have a personal and professional interest in the preservation and further refinement of the system in essentially its present form.

In summary, it appears that implementation of work redesign activities that involve *substantial* changes in jobs may often require simultaneous alteration of existing control systems. If such alterations are not feasible (e.g., for legal, economic, technical or political reasons), then the magnitude of the changes that can be made in the design of the work may be considerably restricted—and work redesign may not be advisable.

Unfortunately, there is almost no research available in the literature on the relationship between control systems and the design of work, so it is impossible at present to know just how serious a problem control systems typically pose for work redesign activities. Indeed, we do not even have descriptive data on how the characteristics of control systems and the properties of jobs within them are empirically related across organizational units. Research on the interdependencies between control systems and the design of work seems to us well warranted, both to further our understanding of the *systemic* properties of organizational units, and to provide practical guidance about the nature and magnitude of changes that feasibly can be made in organizational units that employ various types of controls over work and workers.

*Summary: Constraints on Implementation*

In this section we have suggested that rigidities built into an organization's technological, personnel, and control systems often can prevent the installation of meaningful changes in how work is designed. Attempts to redesign work within existing rigid boundaries typically result in small changes in the work itself—the kind that are likely to produce few desirable outcomes.

The organizational systems discussed above not only serve as a constraint in getting meaningful changes in jobs made, but also are part of the reason why the jobs may *need* to be changed. That is, each of the three systems reviewed tends to influence how jobs are structured, and we have seen that the type of job design that is consistent with a traditional technological system, personnel system, or control system often tends to involve work that is routinized and simplified rather than complex and challenging.

So work redesign may be especially difficult to carry out successfully under precisely the circumstances when it is most needed to improve work motivation and the personal satisfaction of employees. The alternatives in such circumstances are three. The first is to decide not to redesign the work, and to look to other devices for improving the functioning of the organizational unit. The second is to proceed with work redesign despite the constraints posed by the technological, personnel and/or control systems. This alternative risks, on the one hand, succumbing to the "small change" pitfall (if the systemic constraints are not overcome) and, on the other, throwing the existing systems into disarray (if substantial changes somehow do get made, but are inconsistent with the functioning of the established systems). And the third alternative is to redesign the organizational systems themselves, either prior to or simultaneously with redesign of the work, so that they can accommodate and support employees' work on the enriched jobs. This alternative is obviously not an easy, risk-free or inexpensive undertaking; but in many cases it may be the only way to proceed if the risk of small change and the risk of organizational disarray are to be simultaneously avoided when work is redesigned.

It must be emphasized that the implications drawn above are little more than informed speculation. The hard fact of the matter, as noted earlier, is that we have few research data to use either in determining the magnitude of the incongruence that can exist between work redesign and existing organizational systems, or for developing strategies for dealing with that incongruence. To allow firmer and more trustworthy conclusions to be drawn regarding the phenomena we have been addressing, research is needed on the following questions.

1. When plans for work redesign fail to be implemented, or when implementation results only in small changes in the work itself, what are the factors that compromise the plans for change that initially were laid? Careful survey research on work redesign projects could help ascertain the degree to which the technology, the personnel system, and the control system do have the constraining effects posed here—or could show that other aspects of the organization are far more significant in determining when significant changes in the work itself are, and are not, actually made.
2. What are the empirical relationships between the properties of the systems discussed here and the characteristics of the jobs that exist within them? Is it true that elaborate technologies, personnel systems, and control systems tend to be associated with relatively routinized work? What specific job characteristics show the greatest covariation with different aspects of these organizational systems? As noted above, research on the relationships between technology and the design of jobs is already well under way (cf. Rousseau, 1977). But this is not the case for either personnel systems or control systems, and cannot be so until ways of measuring the key properties of those systems are developed—which is yet another challenging research task.
3. How do work redesign activities develop when initiated in various systemic contexts? Once data are available regarding the first two questions posed, it will be possible to conduct both comparative case studies and action research (using quasi-experimental methods) to trace what happens when work redesign interventions are made under organizational circumstances that vary in a priori "favorableness" to job changes. Ultimately, such research should help in the development of *conceptual* understanding of the relationship between organizational systems and the design of work. It should also provide a basis for

planning and executing organizational changes involving work redesign that are much more differentiated and far less speculative than has been possible here.

*The Congruence between Organizational Practices and Redesigned Work*

Getting substantial changes made in the work can be difficult enough, as noted in the previous section; but that is only half the story. Once such changes are in place, questions about the congruence of the enriched work with "standard" organizational policies and practices come to the fore. As noted in the introduction to this chapter, even well-conceived and well-executed changes in jobs can fail to have lasting effects when they are inconsistent with the way the organization is managed.

Both the aspects of the organization that are addressed and the dynamics of their effects differ as we turn from questions about the magnitude of job changes made to questions about the persistence of their effects. Previously we focused on intact, entrenched organizational systems; now we emphasize on-going managerial practices.

Although almost any policy or practice of management at least potentially can affect how people react to enriched work, we have selected four organizational practices that appear most likely to lead to the "vanishing effects" phenomenon described earlier: (1) training practices, (2) career development practices, (3) compensation practices and (4) supervisory practices.

*Training Practices*

By training practices, we mean instructional processes initiated by the organization to improve the job-relevant knowledge, skill, or attitudes of organization members. Training is a very popular device for attempting to improve the motivation and productivity of employees. Yet the benefit of time and money spent on training programs appears to depend substantially on how the work of the trainees is designed. Indeed, we will see that sometimes training (widely viewed as an inherently valuable activity) can actually make things worse rather than better.

The irony is that training often is provided when it isn't much needed and is eliminated in precisely those circumstances when it could have real benefits. Consider, for example, employees who work on simple, routinized jobs. Training is unlikely to have beneficial effects for these individuals, since the requirements of the job usually can be mastered very quickly without any special instruction. (O'Toole, 1975, has estimated that employees can learn most routine jobs within two weeks, simply by proceeding to do the work.) Because training is objectively unnecessary in such circumstances, it may be experienced by employees as an attempt by management to gain even more control over their on-the-job behavior. The result is likely to be no improvement in work performance (the employees knew all they needed to know to do the job already) and heightened feelings of frustration and disillusionment with management. Yet when managers note performance problems on simple jobs, training is one of the most popular and widely used techniques for attempting to correct those problems (Hackman & Oldham, in press, Chapter 2).

On the other hand, training activities sometimes are completely eliminated after the work has been enriched (Hackman & Oldham, in press). The belief, apparently, is that work redesign will solve all problems of job performance and that employees will informally provide one another with help in gaining any new knowledge or skill that may be required. These are very optimistic assumptions. The actual consequences of work redesign sans training, in many cases, is an increase in the *motivation* of employees to work effectively (because of the improved design of the work itself) but a decrease in their *capability* to do so (because new skills are required that they do not presently hold). We know of one example of this in a large transportation organization. The job of reservationist was enriched such that it required a variety of skills and abilities and a good deal of autonomy on the part of the incumbents. However, the organization neglected to provide employees with sufficient information about how they should go about completing the new, enriched tasks. The result was a group of frustrated reservationists who were not able to take advantage of the new opportunities the enriched job provided.

It appears, therefore, that training programs can substantially affect the persistence of changes caused by work redesign. Two types of training may be especially useful in avoiding such problems. The first is "technical" training, to ensure that employees have the knowledge and the skills necessary to execute their enriched tasks competently. If work redesign has been successful, then employees will care more than previously about performing well. They should experience self-reward when they find they have done well, and feel displeased with themselves when they fail. A good technical training program for employees on enriched jobs can increase the likelihood that their work experiences are characterized more often by self-reward than by displeasure with their performance.

The second type of training that often is needed when work is redesigned has to do with the management of interpersonal relationships and decision-making processes. When work is designed in accord with the dictates of the scientific management approach, employees have little objective need to coordinate and negotiate with others to get the work done or to make decisions about work processes or scheduling. All such matters are decided by management and specified in clear detail for those who actually perform the work. On enriched jobs, however, where a great deal of decision making and coordination may be required, the prior work experiences of the employees may have provided them with few chances to exercise or hone their skills carrying out such activities. So even if the employees are competent to execute the technical aspects of their enriched work, problems may develop because of insufficient knowledge and skill about how to *manage* their new and expanded work responsibilities. Training about such matters should be welcomed by the affected employees and could have substantial benefits on work performance, on employee attitudes, and on the social climate of the work unit.

As yet, there have been few studies of the interactions between training activities and work redesign. While a great deal is known about how to design good training programs for specific purposes (Bass & Vaughan, 1968; Goldstein, 1974), there is still much to learn about the focus and timing of training programs that will be most helpful to employees whose work is being enriched. Two research questions seem particularly pressing. First, we need to know whether or not technical and/or interpersonal training will actually enhance work performance and employee satisfaction on enriched jobs—and, if so, how such training can best be structured. (So far as we know, that question has yet to be addressed in the context of work redesign activities.) Second, little is known about the most appropriate timing of training activities when work is redesigned. Should training for work on enriched tasks be done before the jobs are actually changed? That would help increase the "readiness" of employees for their expanded responsibilities but the training might not have its intended effects because the trainees do not yet experience a real need for the knowledge provided and skills taught. Or should training come after job changes are made? Presumably the experienced need would then be present but psychological and behavioral dysfunctions might already have appeared because needed knowledge and skills were not immediately at hand when work began on the redesigned jobs. If we wish to better understand the interactions between training practices and the design of work, and if we seek to minimize the risk of "vanishing effects" that stem from insufficient or inappropriate knowledge and skill, then research on questions such as those posed above would seem well warranted.

*Career Development Practices*

Career development, as used here, refers to the process by which a synthesis is worked out between employee aspirations and the opportunities for mobility that are present in the work environment. Ideally, this synthesis will result in the fulfillment of both individual and organizational objectives (VanMaanen & Schein, 1977). Specific organizational practices having to do with career development include job rotation, various promotional systems, and workshops on life planning and career development.

It is becoming clear that how people respond to their jobs is strongly affected by the stage of their career and their tenure on the job (Katz, 1978 and Katz, Chapter 3 of this volume). It appears, for example, that a person's responsiveness to motivational opportunities that are present in enriched work may be diminished both for very new employees (who may have their hands full just getting settled into the organizational routine) and for employees who have a great deal of experience on the job (who may have adapted more or less permanently to the existing properties of their jobs).

In this section, we look at the other side of the coin—namely, how career development practices may affect the success and persistence of work redesign activities. Rather than focus on individual responsiveness to job characteristics, we examine whether career development practices are appropriately responsive to the experiences individuals have on enriched jobs. As will be seen below, the effectiveness of a work redesign program may be significantly compromised if career development programs do not help individuals respond and adjust satisfactorily to their new, on-the-job experiences, problems, and aspirations. We will address the issue separately for three types of employee reactions to enriched work: those of the "overstretched" employee, those of the "fulfilled" employee, and those of the "growing" employee.

*The Overstretched Employee.* We have suggested previously that work redesign may not be appropriate for certain people—such as employees with weak needs for personal growth, or with knowledge and skill that are not appropriate for the demands of the job (Hackman & Oldham, 1976). These individuals may find enriched work threatening and may balk at being "pushed" or stretched too far by the work. When employees are overstretched, adverse consequences may appear both for the persons involved and for their employing organizations (cf. Blood & Hulin, 1967; MacEachron, 1977; Turner & Lawrence, 1965). Examples include an increase in personal anxieties, psychological or behavioral withdrawal from the job, and various counterproductive activities that express employees' displeasure with the newly enriched work.

Typical career development practices (e.g., life-planning workshops and promotional schemes) that are geared to the upwardly mobile employee may be completely out of place in this situation—and produce few desirable outcomes. If management is interested in retaining overstretched employees, alternative career development practices that are responsive to their special situation may be in order. One such approach assumes that the growth aspirations of over-stretched employees can and will be "rekindled" by the work itself. The idea is that after experiencing challenging work, employees may begin to desire it and respond positively to it. Under this assumption, then, some form of employee counselling might be all that is required to help overstretched employees begin to take advantage of the opportunities available in the enriched work. Later, after the enriched work had been mastered and the initially overstretched employees were comfortable with it, then discussions of further career opportunities would be initiated.

A second approach assumes that the overstretched employee is *not* likely to grow to meet the demands of enriched work. In this case, creating alternative, downward career paths that lead to simpler jobs more consistent with the employees' needs may be in order. This might involve creating new, nonchallenging jobs or transferring employees presently in lower level jobs to create additional openings for work on those jobs; in either case, the downward movement of the overstretched employee would require downward transfers to be legitimized. This would be difficult in many organizations because of strong norms against any downward movement. To overcome this norm, it might be necessary to begin by moving downward obviously competent employees who would prefer less demanding work (Hall, 1976).

Which assumption is correct—that employees will, or will not, gradually come to value and respond positively to enriched work that initially is psychologically threatening to them? There is, at present, little research evidence on the topic, and that which does exist is inconsistent (e.g., Andrisani & Nestel, 1976; Brousseau, 1978; Hackman, Pearce & Wolfe, 1978; Kohn & Schooler, 1976; Hall, Goodale, Rabinowitz & Morgan, 1978). Further research is needed on this question, as it is on strategies for structuring and legitimizing opportunities for downward transfers for overstretched employees on enriched jobs who remain that way even after a period of support and counselling.

*The Fulfilled Employee.* A second type of response to work redesign is that of the fulfilled individuals who are basically satisfied with the responsibilities and challenges of their newly enriched jobs. Fulfilled persons perform well at work, but have no particular desire to move upward in the organizational hierarchy. Instead, they are pleased with their jobs and want to retain the amount of responsibility they presently have. Career development practices designed for the upwardly mobile employee may be inappropriate for the fulfilled individual. Indeed, such practices run the risk of overstretching these employees. As suggested earlier, this can have adverse consequences—especially if the employee is performing well on his or her current job.

What are the career development practices that may be appropriate for fulfilled employees? What would one do to retain current levels of challenge and responsibility (and thereby avoid the risk of overstretching the employees), yet not create conditions where stagnation may emerge? Two types of developmental practices may be especially appropriate in these circumstances. The first is traditional job rotation. In this practice individuals are periodically rotated through jobs where new learnings and skills can be obtained, yet which require little permanent or additional responsibility. Rotation, then, essentially involves short-term movement with employees eventually returning to their regular positions.

A second possibility is the formation of *lateral* career paths (Schein, 1978). These paths allow employees to move into different functional areas (e.g., manufacturing and finance) at approximately the same horizontal level in the organization. While these paths provide opportunities for movement, movement takes place without increases in responsibility. Thus, these paths should be attractive to an employee who is basically content with the level of responsibility in his or her current position.

As with the practice of downward transfer discussed earlier, there may be a stigma attached to job rotation and lateral career paths because, historically, employees who have moved anywhere but upward have been viewed as "failures." This stigma is not likely to be an easy one to reshape, but formal and public policies that directly legitimize (and even reward) lateral movement may ultimately create a positive image for such practices.

*The Growing Employee.* Another possible response to work redesign programs is movement by employees into a "growth cycle." In this condition, employees are so stimulated by the enriched nature of their work that they seek even higher levels of responsibility and additional opportunities for on the job learnings. After a period of time, then, employees who initially were challenged and stimulated by an enriched job may find that the job now provides insufficient opportunities for continued growth. If action is not taken for employees in this situation, stagnation and disillusionment may result. Because they are no longer being stretched by their work, these employees may feel that their careers are at a standstill. Further, they may come to believe that the organization has little interest in providing the kinds of opportunities they seek and therefore begin to look to other organizations for more challenging work.

Organizations might reduce the likelihood that human resources will be wasted in this fashion by installing career development practices tailored to meet the needs of the growing employee. Such practices should provide individuals with increasing responsibility and with opportunities to continue their growth and development within the organization. One such practice involves establishing hierarchical career paths that allow individuals to move upward within their function to levels of leadership and authority (Schein, 1978). These paths should provide persons new growth opportunities simultaneously with increases in responsibility for organizational outcomes. Another practice would require that special assignments be offered to the "growing" employee. These would be short-term, challenging jobs filled with opportunities to exercise authority and creativity. Because opportunities for hierarchical mobility will always be limited by the relatively small number of top positions in organizations, some ingenuity in designing stimulating short-term assignments would seem well worthwhile in organizations where significant numbers of employees have demonstrated increased (rather than diminished) desires for further growth and development after work redesign.

The career development practices suggested in this section serve to indicate the diversity of activities that may be called for if (as predicted) it turns out that different employees respond differently to the enrichment of their work. It would be informative to have research findings in hand to help in matching particular types of employee responses with particular kinds of career development activities. Perhaps of particular use would be research on the following three topics. First, can employees in fact be partitioned according to their dominant response to enriched work (i.e., overstretched, fulfilled, and growing), and what are the major antecedents and consequences of these response patterns? Or is the typology useful only for heuristic purposes? Second, how malleable are employee needs and career aspirations as a function of work experiences and opportunity structures in organizations (Kanter, 1977)? Once "set" by experience (perhaps early in life) do needs and aspirations remain mostly unchanged until opportunities for their expression arise? Or do work experiences constantly shape and reshape what people want and need in their careers? As noted above, evidence on these issues presently is scattered and inconsistent. Finally, how can alternative, nontraditional (and nonhierarchical) career paths be established and legitimized within organizations? How can the stigma that is attached to all-but-upward job changes be ameliorated to allow increased flexibility in achieving good fits between people and their jobs? Such questions will be difficult to research but they are important if we are to significantly advance present understanding of the interactions between how work is designed and how careers evolve within organizations.

*Compensation Practices*

How people are paid for their contributions to the organization also can adversely affect the persistence of work redesign activities and can neutralize some of the beneficial effects of enriched jobs. Three aspects of compensation arrangements are dealt with here: the absolute level of pay desired by employees after their jobs have been redesigned; the form of payment (i.e., contingent vs. noncontingent); and the focus of the compensation system (i.e., on individual employees, work groups, or larger organizational units).

*Absolute Level of Pay.* One of the controversial issues connected with the practice of work redesign is its possible impact on employees' demands for pay. On the one hand, some commentators have argued that enhancing employees' responsibilities through work redesign usually results in demands for more money, because people simply *expect* higher pay for greater responsibility (Lawler, 1977). Alternatively, it has been suggested that enriching the work content only rarely leads to demands for higher pay (Walters & Associates, 1975). The argument here is that responsibility at work and pay demands are basically independent of one another; indeed, that an improved job may provide sufficient psychic rewards, that more material rewards would be motivationally superfluous. We know of no systematic research that has contrasted these viewpoints. Furthermore, we have observed cases where additional money is demanded after work redesign and cases where it is not. It appears that under some circumstances employees do view pay and responsibility as interconnected, while other times they seem to perceive pay and work content as unrelated. What is clear is that if some (or all) employees in a work unit become dissatisfied with pay levels, undesirable consequences are likely. We have demonstrated previously, for example, that pay dissatisfaction can distract the attention of employees from enriched work and orient their energy instead toward coping with this more pressing problem (Oldham, Hackman & Pearce, 1976). The result may be relatively how levels of motivation and performance on enriched jobs.

If such consequences are to be avoided, some improvement in the level of pay offered employees following work redesign may be required. One approach to the problem is to give all employees in the affected work unit substantial pay increases as a sign that the newly enriched jobs are important and that management is serious about the changes being made (Lawler, 1977). Alternatively, savings that result from the work redesign program might be shared with employees in the work unit on a proportional basis (Walters & Associates, 1975). Either of these approaches seems likely to reduce pay dissatisfaction that results from work redesign and thereby increases the chances that behavioral changes prompted by the enriched jobs will persist. At present, however, there are no studies in the literature that compare the effects of the "flat increase" and the "gain sharing" strategies for raising compensation levels following work redesign.

Also at issue is the timing of changes in compensation arrangements for employees. One possibility is to design and announce a new pay plan prior to (or simultaneously with) changes in the work itself. Alternatively, several plans might be developed prior to the changes but held in reserve until (and unless) signs of pay dissatisfaction appear following work redesign. The first approach heads off possible problems with pay level before any damage is done and provides to employees a sign of management's good faith before the changes themselves are made. The advantage of the second approach is that the pay plan that eventually is introduced can be tailored to the particular compensation problems that emerge. Again, there presently is no research that examines the consequences of these alternative approaches.

*Form of Payment.* When jobs are enriched, how should people be paid? Is it advantageous to use salaries, hourly wages, or some type of incentive or bonus for good performance? There is a good deal of controversy about the matter. On the one hand, some commentators (e.g., Deci, 1971) have argued that contingent rewards (e.g., bonuses) may be inappropriate for tasks that are intrinsically meaningful and interesting. The reasoning is that the employee paid on a contingent basis may conclude that he or she performed the task *because of* the external reward and, therefore, that the task must not be very satisfying or interesting in and of itself. According to this view, a bonus system could change an employee's perception of the reasons for his or her behavior and ultimately diminish the motivational benefits of enriched work. Several laboratory investigations support this basic position (see Deci & Porac, 1978; and Staw, 1976 for reviews).

Advocates of Deci's position argue that noncontingent rewards (e.g., salaries or hourly wages) are most appropriate for and supportive of enriched work. These rewards are seen as allowing individuals to experience all the benefits of redesigned work. Moreover, salary plans provide employees with freedom and flexibility and treat individuals as mature adults. In this sense, salary systems encourage responsibility and trustworthiness among employees, which is compatible in spirit with most work redesign programs.

A contrasting position is that contingent pay plans (e.g., bonuses and piece-rate systems) are perfectly appropriate for enriched jobs. The notion is that the rewards available from the pay plan and those available from the work itself are additive. Thus, work motivation should be maximized when employees are paid contingently for performance on enriched tasks. There also is evidence that supports this view (e.g., Arnold, 1976; Hamner & Foster, 1975; Wyatt, 1934). For example, the Hamner & Foster research contrasted the effects of three pay systems (i.e., no pay, noncontingent pay, and contingent pay) for people working on a meaningful task. Results showed that individuals performed best when paid on a contingent basis.

What are we to conclude from these seemingly contradictory findings? Is it more appropriate to use noncontingent or contingent payment systems for enriched work? The answer, we believe, may depend on two factors. The first factor is the degree to which it is possible to measure the output of enriched jobs. For many such jobs, this would involve measuring whole work units rather than small segments of the work. Simple quantity of output may not be the most appropriate output measure. As we have previously suggested, quality of the work performed is more likely than work quantity to be enhanced through work redesign (Hackman & Oldham, 1976). To the extent that managers and employees agree about what work outcomes should be measured (and how they should be measured), contingent pay systems may be indicated. When there is no agreement or when measurement is impossible, noncontingent systems (e.g., salary) may be more appropriate and more motivationally effective.

A second factor is the level of trust between management and employees (Lawler, 1971). Contingent reward systems may be incompatible with enriched work if employees perceive the systems as attempts by management to control and manipulate their behavior on enriched jobs. In such cases any motivational advantages of contingent systems may be more than offset by suspicion of them; salary plans may be more appropriate. If, on the other hand, there is high trust in the organization, employees are more likely to believe that the plans will be administered fairly; they may see them as a fair and appropriate means of sharing in the gains generated by high work productivity. In such circumstances, contingent payment systems would seem fully congruent with enriched, challenging work.

*Focus of the Compensation System.* If a contingent reward system is used to compensate people for their work on enriched jobs, then a decision must be made about whether to administer the system on the basis of individual, group, or organization-wide performance. We concur with Lawler (1977) that the appropriate focus of the system depends on the amount of interdependence that characterizes the design of the work itself. When employees are basically independent of one another, it is most reasonable to use an individual reward plan. Employees are paid for their own performance not that of others in the organization. There are numerous jobs for which employees work quite independently (e.g., salespersons) and where individual pay plans have been used with success. On the other hand, when work has been designed for a group of employees, individual plans may not be appropriate. In such cases, it may be difficult to measure individual contributions to the group product and therefore difficult to reward individuals differentially. Moreover, individual incentives usually do not reward employees for cooperation and teamwork, which often are essential for effective group performance (see Hackman, 1978).

So, for group work design, some type of group incentive system usually should be most effective. Individuals would share rewards for completion of the group task not for completion of their own part of the task. Such group incentive schemes have been used successfully in many sociotechnical work redesign projects (see Davis & Trist, 1974).

Finally, unit- or organization-wide incentive plans (e.g., Scanlon plan) may be most effective when cooperation is necessary (or desired) among all employees in the organization, or when the contributions of individuals or groups to organization-wide performance cannot easily be disentangled. In these plans, bonuses based on measures of unit or company performance are given to all employees. When they are functioning properly, the better the organization functions, the better off are the employees. It is to the advantage of employees to produce more, to cooperate with others, and to adopt new procedures and technologies (Lawler, 1977).

*Conclusion: Compensation Practices.* It is clear that there are major interdependencies between how work is designed and how people are paid. These interdependencies have to do with the *level* of pay that employees find satisfactory after their work responsibilities have been expanded, with the *form* of payment (i.e., contingent vs. noncontingent), and with the *focus* of the compensation system (i.e., individual vs. group vs. larger organizational units). Yet solid research findings are not available to help answer key questions that have been addressed in our discussion of compensation practices for enriched jobs.

Here are some of the issues that seem to us to be especially in need of systematic research.

- What factors influence the beliefs and attitudes of employees about their pay after job responsibilities have been expanded or enriched? When will people feel that a "better job" is compensation in itself? When will they feel that they are being "exploited" by doing more work for the same pay?
- At what time should attention be given to compensation arrangements in a work redesign project? Before the changes, simultaneously with them, or after work has begun on the enriched jobs? What are the special benefits and risks associated with each of these alternatives?
- What factors moderate the effects of contingent vs. noncontingent

compensation arrangements for work on enriched jobs? Do the factors we have suggested (measurability of work outcomes and level of employee-management trust) most strongly determine when contingent payment will be effective, or are there other more powerful factors that must be accounted for?
- How important is the level of interdependence among employees in determining the consequences of individual-focused (vs. group- or unit-focused) compensation arrangements? What other variables must also be considered in understanding the impact of pay plans with different foci on employees who perform enriched jobs?

We have generated hypotheses about all of these questions in preceeding pages but throughout we were forced to rely on speculation, on our own experience and observations, and on a smattering of research findings which often were uncontrolled case reports. Given the apparent importance of compensation practices in determining when the effects of work redesign will prosper (and when they will disappear over time) more systematic research on questions such as those posed above strikes us as being of very high priority.

*Supervisory Practices*

The final set of practices to be considered deals with the behaviors of first-line supervisors toward employees whose jobs have been redesigned. Research has demonstrated that enriching employees' jobs can cause serious strains in the relationship that exists between them and their supervisors (Alderfer, 1967) and numerous case reports have shown that such relationship problems can lead to an early and unanticipated demise of even well-conceived change projects.

These strains may be rooted in changes in the *role* of the supervisor that accompany the redesign of rank-and-file jobs. In many cases, autonomy, decision-making responsibility, discretion, and quality control activities are removed from the job of the supervisor and assigned to subordinates as part of the enrichment process. Such shrinkage of the responsibilities of the supervisor may result in a substantial (and not necessarily constructive) change in the supervisor's behavior. This apparently is what transpired in a study of the effects of work redesign on telephone operators' reactions to their work (Lawler, Hackman & Kaufman, 1973). Many of the responsibilities of first-line supervisors were assigned to the operators, which resulted in supervisors experiencing large amounts of "free time" after the changes had been made. Most of the supervisors chose to use this time to supervise the operators rather closely as they worked on their newly enriched jobs. Postchange assessments showed few changes in employee motivation or satisfaction—but a substantial *decrease* in operator perceptions of the respect and fair treatment they received from their supervisors. The authors attributed the failure of the project to generate improvements in motivation or satisfaction to this deterioration in supervisor-subordinate relationships.

The pattern of results obtained in this study is consistent with the more general notion that reducing a supervisor's autonomy and power often prompts overcontrolling, rules-minded and excessively critical behaviors on the part of the supervisor (Kanter, 1977). Such behaviors can more than offset any positive changes in employee motivation and satisfaction resulting from the enrichment of the work itself.

How might such counterproductive behaviors by supervisors be avoided when work is redesigned? The usual approach, of course, is to send the affected supervisors off to a training program where they would learn how to behave in ways that are constructive and supportive of their subordinates. The problem is that such programs have been shown to be largely ineffective in generating lasting changes in managerial attitudes and behaviors (Campbell, Dunnette, Lawler & Weick, 1970).

What may be required, then, is to redesign the supervisor's job, so that providing support for subordinates in performing their enriched tasks becomes a natural part of the supervisor's own responsibilities. Among the tasks that could be built into the supervisor's job are the following (cf. Walters & Associates, 1975):

- Gathering data for charting trends and forecasts in work volumes and workforce needs
- Training employees in their new responsibilities, and counselling with them about both work-related problems and career opportunities
- Helping subordinates set performance goals, and reviewing with them their performance in attaining those goals
- Providing increased openness of communication both upwards (i.e., sharing employee concerns with higher management) and downwards (i.e., sharing with employees information about changes in organizational objectives and policies)
- Developing and testing with subordinates innovations in methods and procedures for executing and coordinating the work
- Working on aspects of the work context (e.g., compensation, control systems, opportunity structures, equipment and space) that may be causing dissatisfaction or impeding employees' work
- Managing the evolution of the job enrichment process itself.

The focus of these activities is to support subordinates in performing their work effectively; the list does not include such traditional supervisory activities as direct monitoring of subordinates' behaviors, checking their work, or serving as disciplinarian. Many of the tasks listed require that the supervisor turn his or her attention upward and outward in the organization—managing the organizational context so that it facilitates high subordinate motivation and effective work performance. Thus, if the supervisor is to perform these tasks well, his or her own job must contain considerable responsibility, discretion, and control. These qualities would have to be pushed down to the supervisor from higher levels of organizational management and the net effect would be to enrich the supervisor's job, just as previously had been done for the subordinates' jobs.

Such changes in supervisory jobs should prompt more supportive and effective behaviors by supervisors for at least three reasons. First, supervisors now would have the *power* to help their subordinates in meaningful ways. For example, the supervisor might have some significant influence over the pay system in his or her unit or the career development practices that are available to employees. As we have seen, attention to organizational practices such as these can enhance the likelihood that work redesign changes will persist. Second, because of his or her new responsibilities the supervisor should have *less free time* to closely supervise. In fact, there might be so many time constraints and demands that supervisors would be forced to give even more freedom and discretion to their employees. Finally, because the supervisor feels that he or she is now an integral member of the management team, he or she is likely to have a *greater investment* in the success of the redesign effort. This should result in behaviors that directly support the redesigned work—especially if supervisors have been trained in the skill of helping (rather than bossing) their subordinates, and if they have been personally involved in the redesign of their subordinates' jobs.

The idea that supervisory work redesign can create appropriate behavioral patterns in supervisors is supported by a study by Davis & Valfer (1966). These researchers investigated the impact of a supervisory work redesign program on both the behavior of the supervisors and the performance of their units. Supervisors were given authority and responsibility for controlling all operational and inspection functions required to determine final acceptance of the products or services assigned to their work groups. After this change, most supervisors began to rotate workers and to provide training to facilitate worker interchangeability. Moreover, some supervisors assigned inspection activities entirely to employees. Supervisors found that they had less free time than before and they felt less dependent on their superiors. The net effect was an improvement in work quality and stable levels of productivity.

Further studies along the lines of the Davis & Valfer research would be very helpful in understanding how the behavior of supervisors can support (or counteract) the effect of enriched work on rank-and-file employees. Also useful would be simple descriptive studies (including survey research) to provide documentation of what changes typically occur in the jobs and the behaviors of supervisors when their subordinates' work is redesigned. Especially informative would be case studies rich with descriptive detail of how supervisors react to the changes that are made when job enrichment is carried out in their work units. Such research surely would yield numerous new hypotheses about the person and the role of the supervisor in work redesign projects—hypotheses that are both more detailed and conceptually richer than the speculations we have set forth here.

## WORK REDESIGN AND THE ROLE OF MIDDLE MANAGERS

We have examined in this chapter several organizational systems and practices that may interact with the redesign of work—sometimes amplifying the effects of enriched work, sometimes counteracting them. In the first part of the chapter, we identified three organizational systems that can constrain the installation of large meaningful changes in the work itself: the technological system, the personnel system, and the control system. When attempts are made to redesign work in an inhospitable systemic context, we argued, the most likely outcome is a set of small changes that can have few noticeable or lasting effects. Work redesign will fail in such circumstances simply because the work itself cannot be changed substantially enough to make a difference in how people behave at work.

Next we highlighted four organizational practices that may compromise the "staying power" of changes that result from enriched work: training practices, career development practices, compensation practices, and supervisory practices. When these practices support the kinds of attitudinal and behavioral changes that are brought about by work redesign, then the effects of the job changes should be both strong and lasting. When organizational practices are incongruent with the new attitudes or behaviors, however, then the benefits of enriched work are unlikely to persist over time—the "vanishing effects" phenomenon we discussed at the start of the chapter.

Throughout the chapter we have lamented the paucity of research findings that relate directly to the phenomena we have been addressing and we have posed numerous research questions that could aid in understanding the relationship between work redesign and those systems and practices that characterize the organization in which it takes place. Yet there is a more general hypothesis, never stated explicitly, that underlies the very organization of the chapter. It is that the intact *systems* that characterize an organization (and in particular, the technological, personnel and control systems) have a substantial effect on the magnitude of changes that are made when work is redesigned but not much of an impact on the persistence of the effects of work redesign. The management *practices* of an organization, on the other hand, are viewed as having a substantial impact on persistence of work redesign effects but as having little influence on the magnitude of the job changes that are made. This general hypothesis, like most of the smaller ones scattered throughout the chapter, awaits research test.

Regardless of the validity of our organizing hypothesis, it does seem to us indisputable that both the "small change" and "vanishing effects" phenomena are properties of organizations as social systems. And for this reason, it seems necessary to stop construing work redesign as a short-term, limited focus "fix" for specific attitudinal and behavioral problems observed among rank-and-file workers. Instead, it appears more appropriate to view changes in how work is structured as involving alterations in how the social system as a whole functions and how such changes will affect and be affected by other aspects of that system.

One implication of this view is that we need much better understanding than we presently have of the role of *middle management* when jobs and work systems are redesigned. In many organizations, middle managers, much more than supervisors or top managers, have responsibility for the organizational systems and practices that we have identified as critical to the potency and persistence of changes resulting from the redesign of work (cf. Oldham, 1976). It is the middle manager, for example, who is most likely to be in a position to alter control systems or to initiate a change in the work technology; it is he or she who may be able to revise compensation practices or to redesign the job of a subordinate manager.

The role of the middle manager in the work redesign process is not an easy one. He or she must not only take responsibility for assessing existing organizational systems to see whether plans for change should proceed but must also worry about bringing other managerial practices into congruence with the revised work system after jobs are changed. Moreover, while all of this is going on, the middle manager will need to deal with the fact that responsibilities are being pushed down in the organization—i.e., first-line responsibilities to the rank-and-file, second-line and staff responsibilities to first line—and that eventually the design of his or her own job may be affected, not necessarily for the better, by the change process as it continues to unfold.

Presently, little is known either about how middle managers involve themselves in the process of work design, or what kinds of behaviors and managerial strategies are more likely to facilitate rather than impede effective change in how jobs are structured. Research is needed, then, not only on the interdependencies between work redesign and organizational systems and practices but also on the key role of middle managers in system-wide change processes that involve alterations in the design of work. Until we have better understanding of how interdependent and sometimes-conflicting organizational systems fit together when changes are made in the design of work, and how managers respond to the clashes that can occur among these systems, we are not likely to generate a very robust understanding of either the "small change" or the "vanishing effects" phenomena that prompted us to write this chapter in the first place.

## FOOTNOTES

1. Work on this chapter by the second author was supported in part by Contract N00014-75C-0269 of the Psychological Sciences Division, Office of Naval Research, to Yale University.

2. The organization of this chapter is based on two underlying assumptions. First, throughout the chapter we treat organizational systems and practices as separate and unique properties. Organizational systems will refer to formal, structural characteristics (e.g., technological systems) that are firmly entrenched in the organization itself. Organizational practices will refer to on-going managerial programs and processes that are directed at employees (e.g., career development practices). A second organizing assumption is that systems and practices have differential effects on the implementation of work redesign and its persistence in the organization. The argument is that structural systems in place in the organization are most responsible for preventing or blocking the actual installation of work redesign. On the other hand, we suggest that managerial practices primarily influence the degree to which the effects of work redesign will persist over time. Both of these arguments (i.e., that systems and practices are independent and have differential effects) are based on our observations of various work redesign projects and provide an organizing framework for the chapter; however, we have little empirical data to support these arguments and they clearly await research test.

## REFERENCES

Alderfer, C. P. (1967) "An organizational syndrome." *Administrative Science Quarterly, 12*, 440–460.

Andrisani, P. J. & Nestel, G. (1976) "Internal-external control as a contributor to and outcome of work experience." *Journal of Applied Psychology, 61*, 156–165.

Arnold, H. J. (1976) "Effects of performance feedback and extrinsic reward upon high intrinsic motivation." *Organizational Behavior and Human Performance, 17*, 275–288.

Bass, B. M. & Vaughan, J. A. (1968) *Training in industry: The management of learning*. Belmont, CA: Wadsworth.

Blauner, R. (1964) *Alienation and freedom*. Chicago: University of Chicago Press.

Blood, M. R. & Hulin, C. L. (1967) "Alienation, environmental characteristics and worker responses." *Journal of Applied Psychology, 53*, 456–459.

Brousseau, K. R. (1978) "Personality and job experience." *Organizational Behavior and Human Performance, 22*, 235–252.

Campbell, J. P., Dunnette, M. D., Lawler, E. E. & Weick, K. E. (1970) *Managerial behavior, performance, and effectiveness*. New York: McGraw-Hill.

Davis, L. E. & Trist, E. L. (1974) "Improving the quality of working life: Sociotechnical case studies." In J. O'Toole (ed.), *Work and the quality of life*. Cambridge: MIT Press.

Davis, L. & Valfer, E. (1966) "Studies in supervisory job design." *Human Relations, 19,* 339–352.

Deci, E. L. (1971) "The effects of externally mediated rewards on intrinsic motivation." *Journal of Personality and Social Psychology, 18,* 105–115.

——— & Porac, J. (1978) "Cognitive evaluation theory and the study of human motivation." In D. Greene & M. Lepper (eds.), *The hidden costs of rewards.* Hillsdale, N.J.: Lawrence Erlbaum Associates.

Fine, S. A. & Wiley, W. W. (1974) "An introduction to functional job analysis." In E. Fleishman & A. Bass (eds.), *Studies in personnel and industrial psychology.* Homewood, Ill.: Dorsey Press.

Frank, L. & Hackman, J. R. (1975) "A failure of job enrichment: The case of the change that wasn't." *Journal of Applied Behavioral Science, 11,* 413–436.

Goldstein, I. L. (1974) *Training: Program development and evaluation.* Belmont, CA: Wadsworth.

Gyllenhammar, P. G. (1977) *People at work.* Reading, MA: Addison-Wesley.

Hackman, J. R. (1975) "On the coming demise of job enrichment." In E. L. Cass & F. G. Zimmer (eds.), *Man and work in society.* New York: Van Nostrand Reinhold.

———. (1978) "The design of self-managing work groups." In B. T. King, S. S. Streufert & F. E. Fiedler (eds.), *Managerial control and organizational democracy.* Washington, D. C.: Winston & Sons.

Hackman, J. R. & Oldham, G. R. (1976) "Motivation through the design of work: Test of a theory." *Organizational Behavior and Human Performance, 16,* 250–279.

——— & Oldham, G. R. (in press) *Work redesign.* Reading, MA: Addison-Wesley.

———, Pearce, J. L., & Wolfe, J. C. (1978) "Effects of changes in job characteristics on work attitudes and behaviors: A naturally occurring quasi-experiment." *Organizational Behavior and Human Performance, 21,* 289–304.

Hall, D. T. (1976) *Careers in organizations.* Pacific Palisades, CA: Goodyear.

———, Goodale, J. G., Rabinowitz, S., & Morgan, M. (1978) "Effects of top-down departmental and job change upon perceived employee behavior and attitudes: A natural field experiment." *Journal of Applied Psychology, 63,* 62–72.

Hamner, W. C. & Foster, L. W. (1975) "Are intrinsic and extrinsic rewards additive?: A test of Deci's cognitive evaluation theory of task motivation." *Organizational Behavior and Human Performance, 14,* 398–415.

Kanter, R. M. (1977) *Men and women of the corporation.* New York: Basic Books.

Katz, R. (1978) "The influence of job longevity on employee reactions to task characteristics." *Human Relations, 31,* 703–725.

Katzell, R. A. & Yankelovich, D. (1975) *Work, productivity and job satisfaction.* New York: Psychological Corporation.

———, Bienstock, P., & Faerstein, P. H. (1977) *A guide to worker productivity experiments in the United States 1971–1975.* New York: New York University Press.

Kimberly, J. Managerial innovation. (in press) In P. C. Nystrom & W. H. Starbuck (eds.), *Handbook of organizational design.* New York: Oxford Univ. Press.

Kohn, M. L. & Schooler, C. (1976) The reciprocal effects of the substantive complexity of work and intellectual flexibility: A longitudinal assessment. Paper presented at the American Sociological Association convention, August, 1976.

Lawler, E. E. (1971) *Pay and organizational effectiveness: A psychological view.* New York: McGraw-Hill.

———. (1977) "Reward systems." In J. R. Hackman & J. L. Suttle (eds.), *Improving life at work.* Santa Monica, CA: Goodyear.

———, Hackman, J. R., & Kaufman, S. (1973) "Effects of job redesign: A field experiment." *Journal of Applied Social Psychology, 3,* 49–62.

——— & Rhode, J. G. (1976) *Information and control in organizations.* Pacific Palisades, CA: Goodyear.

MacEachron, A. E. (1977) "Two interactive perspectives on the relationship between job level and job satisfaction." *Organizational Behavior and Human Performance, 19,* 226–246.

Oldham, G. R. (1976) "The motivational strategies used by supervisors: Relationships to effectiveness indicators." *Organizational Behavior and Human Performance, 15,* 66–86.

———, Hackman, J. R., & Pearce, J. L. (1976) "Conditions under which employees respond positively to enriched work." *Journal of Applied Psychology, 61,* 395–403.

———, Hackman, J. R., & Stepina, L. P. (1979) "Norms for the Job Diagnostic Survey." JSAS *Catalog of Selected Documents in Psychology, 9,* 14 (Ms. 1819).

O'Toole, J. (1975) "The reserve army of the underemployed. II—The role of education." *Change,* June, 26–44, 60–63.

Reeves, T. & Woodward, J. (1970) "The study of managerial controls." In J. Woodward (ed.), *Industrial organization: Behavior and control.* London: Oxford.

Rousseau, D. M. (1977) "Technological differences in job characteristics, employee satisfaction and motivation: A synthesis of job design research and sociotechnical systems theory." *Organizational Behavior and Human Performance, 19,* 18–42.

———. (1978) "Characteristics of departments, positions, and individuals: Contexts for attitudes and behavior." *Administrative Science Quarterly, 23,* 521–540.

Salancik, G. R. & Pfeffer, J. (1977) "An examination of need-satisfaction models of job attitudes." *Administrative Science Quarterly, 22,* 427–456.

Schein, E. H. (1978) *Career dynamics: Matching individual and organizational needs.* Reading, MA: Addison-Wesley.

Slocum, J. W. & Sims, H. P. (1978) "A typology of technology and job redesign." Working Paper, The Pennsylvania State University.

Staw, B. M. (1976) *Intrinsic and extrinsic motivation.* Morristown, N.J.: General Learning Press.

Thompson, J. D. (1967) *Organizations in action.* New York: McGraw-Hill.

Turner, A. N. & Lawrence, P. R. (1965) *Industrial jobs and the worker.* Boston: Harvard University Graduate School of Business Administration.

VanMaanen, J. & Schein, E. H. (1977) "Career development." In J. R. Hackman & J. L. Suttle (eds.), *Improving life at work.* Santa Monica, CA: Goodyear.

Walker, C. R. & Guest, R. H. (1952) *Man on the assembly line.* Cambridge: Harvard University Press.

Walters, R. W. & Associates. (1975) *Job enrichment for results.* Reading, MA: Addison-Wesley.

Walton, R. E. (1977) "Work innovations at Topeka: After six years." *Journal of Applied Behavioral Science, 13,* 422–433.

Wexley, K. N. & Yukl, G. A. (1977) *Organizational behavior and personnel psychology.* Homewood, Ill.: Irwin.

Wyatt, S. (1934), *Incentives in repetitive work: A practical experiment in a factory.* Industrial Health Research Board Report No. 69. London: H. M. Stationery Office.

# ORGANIZATIONAL GROWTH AND TYPES:

## LESSONS FROM SMALL INSTITUTIONS

A. C. Filley and R. J. Aldag

UNIVERSITY OF WISCONSIN

## ABSTRACT

This paper discusses organizational growth and development. It reviews theoretical and empirical typologies of organization and the patterns of leadership, structure, and environmental adaptiveness which accompany the types. Development and change are considered as processes which move an organization from one generic type to another. Size is seen as a consequence of type characteristics.

## INTRODUCTION

The nature and determinants of organizational growth are the focus of this discussion. The approach taken to consider these issues differs in a

**Research in Organizational Behavior, Volume 2, pages 279–320**

**ISBN: 0-89232-099-0**

number of ways from those characteristic of much previous research and theory. First, rather than emphasize the role of differences in *degree* of particular variables, this paper will focus on differences of *kind*. Thus, for example, a choice between customized or mass-produced products as a company's preferred technology will have major impact upon the ease of growth and the requisites for organizational survival.[1]

Second, the discussion does not suppose that organizations have homologous relations with living organisms. However tempting it is to draw such parallels, the resulting anthropomorphism would be reductionism of the worst sort. Man is related to the butterfly in the sense that they are both living organisms; he is not similarly related to an organization. Reference to the organization as "it" in the following discussion merely refers to a social unit designed to achieve some purpose. Third, the discussion is not concerned with differences in size; instead, its focus is upon the growth of an organization. As Penrose (1959, p. 88) pointed out, growth is a process, size is a state. One learns little about the former by investigating the latter. Fourth, we are *not* concerned with the normative change of organizations commonly identified as organizational development; instead, change is considered descriptively as it occurs in association with other variables.

Finally, we are not concerned here with large, very complex organizations with multiple enterprises and multiple product bases. Instead, the discussion assumes the existence of a single enterprise operating with a single product base under a single body of administration. Studies of large institutions may be fruitful but hardly typify the business population. According to the 1972 Census of Manufacturers, 89 percent of the manufacturing enterprises had less than 100 employees. By its own definitions, the U. S. Small Business Administration (1977, p. 3) estimates that 97 percent of the U. S. business concerns are small businesses.

## ORGANIZATION TYPES

### *Classification of Organizations*

To date, most taxonomies of organizations have utilized single criteria such as size (Kimberly, 1976), technology (Child, 1973; Thompson, 1967; Woodward, 1965), control systems (Etzioni, 1961), prime beneficiaries (Blau & Scott, 1962), industry type, and degree of perceived environmental uncertainty (Lawrence & Lorsch, 1967).

In contrast, a variety of theoretical discussions and some empirical work have suggested the presence of complex patterns of organization which differ systematically from each other. Several of these are shown in

Figure 1. Unless the similarities are merely the result of citing common authority, and such seems not to be the case, they suggest a homology to different levels of human organizations. As indicated, three institutional types are often cited, with stability in the first type, rapid growth in the second, and consolidation and institutionalization in the third. In addition, such theories posit an order to the types. However, as we will see, while there may be factors which influence the relative probabilities of alternative movements between the types, the assumption of a deterministic progression is apparently invalid.

A number of other typologies have also been presented (summarized in Starbuck, 1965 and Feldbaumer, 1973) but these are either differences in size or else imply an anthromorphic evolution from infancy to adolescence to adulthood. Such an evolutionary view would require acceptance of the notion that organizations decline and, ultimately, die of old age. Since old age has not to the knowledge of the authors been seriously proposed as a cause of failure of particular organizations, the idea of an evolutionary imperative seems simplistic.

Explanations of organizational development and change which focus upon "abrupt, major transformations which sharply distinguish one period of organization history from another" (Starbuck, 1968, p. 113) have been labeled *metamorphosis models*. As mentioned above, such models emphasize either internal characteristics, or size, or age, or some combination of the three. Metamorphosis models stress the importance of discontinuities in explaining organization change:

> The history of an organization is divided into stages. Structural changes which occur within a given stage are de-emphasized as having lesser importance; structural changes which take place between two consecutive stages are emphasized as having greater importance. Consequently, the historical development within a single stage is cast as a relatively smooth and continuous process, but the overall development pattern is marked by sharp and discrete transitions from one stage to the next (Starbuck, 1968, p. 113).

Unfortunately, this global view of metamorphosis incorporates bases for transitions, such as size, which are of fixed order and others, such as independent types, which need not be so ordered. This stress on evolution seems to derive from a penchant for making organizations analogous to living organisms.

There is some empirical support for the argument that organizations undergo fundamental changes of character, though not necessarily for the view that such changes follow a deterministic sequence. Starbuck (1968) analyzed the history of ten firms, finding that changes were better explained in terms of metamorphosis than a smooth evolution. He noted that the significant points of change appeared to be the result of both manage-

*Figure 1.* Institutional Types

| *Author* | *Institutional Type* | *Type I* | *Type II* | *Type III* | *Relationship to Growth* |
|---|---|---|---|---|---|
| Davis (1951) | Firm | Pioneering-Owner-manager | Exploitation-Promoter | Stabilization-Professional Manager | Innovative advantage in II; Competition in III |
| Dale (1960) | Firm | — | Genius management | Systematic management | Initiators leading to crises in II; Preservation and continued success in III |
| Weber (1947) | Institutions | Traditional authority | Charismatic authority | Bureaucratic authority | Greatest efficiency in III |
| Bennett & McNight (1956) | Nation | Feudal | Dynamic politico-economic expansion | Consolidation | Rapid growth in II |
| Reisman, Glazer & Denney (1950) | Nation | Tradition-directed | Inner-directed | Other-directed | Rapid growth in II |
| Rostow & Millikin (1957) and Rostow (1960) | Nation | Traditional and Precondition for take-off | Take-off and drive to maturity | High mass consumption | Rapid growth in II |
| Hoffer (1958) | Mass Movement | Man of words as leader | Fanatic as leader | Practical man of action as leader | Rapid growth in II; Preservation in III |
| Lester (1958) | Unions | Pregrowth period | Militant turbulance | Institutional middle age | Rapid growth in II; Consolidation in III |

rial strategies pursued and external problems to be solved. In addition, metamorphic changes have been reported by Filley (1962) for five firms and by Feldbaumer (1973) for three firms. A variety of case studies also stress such major shifts in organizational structure and function (Kimberly, 1976; Dale, 1960). Shifts in patterns of organization have also been reported in several studies using Guttman scaling techniques (Carneiro & Tobias, 1963; Hall & Tittle, 1966; Udy, 1958). Although they assume unidimensionality and they are not arranged into stages or types, per se, such scalings have displayed successive additions of organizational characteristics.

In general it appears that organizations do experience shifts in their basic character, that common patterns of structure and growth are to be found in various forms of human organization, and that unlike the case with organisms the patterns need not follow each other in a prescribed order.

*An Organizational Strategy Typology*

One way to classify organizations is to identify different characteristics which cluster together more than they do in combination with other clusters. One such typology based on three strategies for organizing and for dealing with the environment has been presented by Filley. A detailed description of each type has been presented elsewhere (Filley, House & Kerr, 1976), and is merely summarized here. For convenience they are labeled a *craft type,* a *promotion type,* and an *administrative type.*

The *craft type* organization appears to be strongly influenced by a chief executive who seeks comfort and company survival objectives and who engages in duties which are largely technical rather than administrative. Policies evolve by tradition and the organization is structured into levels of power rather than an impersonal hierarchy of offices. The absence of real or expected change is associated with low staffing needs for indirect labor and morale based on minimal discrepancies between what members have and what they expect to have. The organization has low levels of risk taking and low levels of innovation outside the conventional technology of the firm. The craft type exists primarily at the mercy of a benevolent environment, since it is not adaptive to changes in its market environment. Craft firms emphasize either the making or the selling of their product or service and they use conventional methods for both processes. One would expect to find little growth in the craft firm since growth would occur as a fortuitous experience rather than as a desired and planned outcome.

The *promotion type* organization appears to be strongly influenced by a chief executive who is charismatic and the promoter of the firm's innovative advantage. Policies are fluid according to tactical requirements and all organization members are directly influenced by the promoter. Indirect labor provides technical support and personal assistance for the promoter. Work functions develop as problems warrant and morale is high as long as organization members share mutual high expectations about future outcomes. The organization is innovative and is readily redirected by the promoter. It exists and succeeds primarily to exploit some unique product or market advantage, and that advantage ends when competitors and imitators enter the field.

The *administrative type* organization is characterized by professional management and by the existence of an institutional character which exists independent of organization members. Its management seeks to adapt to its competitive situation through planning and through the use of formal organization. It is formally structured according to technical and size imperatives, and has balanced functional development. Indirect labor is present to provide planning and control, and morale is based upon a homogeneity of member values. Changes take place within the formal budgeting and planning process. Product improvement rather than product innovation is emphasized.

*Organizational Metamorphosis.* Metamorphosis in organizations is probabilistic, not deterministic. While changes from one type of organization to another may be common, there is no necessary sequence from craft to promotion to administrative. At least two imperatives for change seem to be important. First, there is a temporariness in the promotion type in a free market economy (Hoffer, 1958; Weber, 1947). Second, there may be a collective rationality in both the craft and promotion types induced by survival needs or by the desire to gain attractive objectives associated with the administrative type. For example, there may be collective pressure to move a craft type into an administrative type when incentives for growth, for greater profit, or for protection from external threat are felt by organization members. However, as will be discussed later, there are powerful forces that limit the ability of the craft executive to adopt a different organizational strategy.

Thus, while there is no completely specified sequence of types through which an organization must pass, pressures for transitions between types do exist. To that extent, something on the order of a metamorphosis may be evidenced.

*Leadership and Metamorphosis.* In order to adequately explain the differences between the promotion and administrative organization and to understand the forces mandating the temporariness of the promotion type, it is necessary to consider the characteristics of the promoter. The label "promoter" is intended to be descriptive, not derogatory. While the label "entrepreneur" would carry less of an implicit value connotation, that term has been rendered ambiguous due to its application to initiators, owners, or even managers of small enterprise. In any case, a variety of authors (Collins, Moore & Unwalla, 1964; Schumpeter, 1935; Weber, 1947; Hoffer, 1958; Charan, 1976; Kimberly, 1976) have distinguished between the person or role involved in exploiting some kind of innovative advantage (i.e., promoter or entrepreneur) and the person or role involved in institutionalizing and managing an enterprise (i.e., manager).

In particular, Weber (1947) described a charismatic leader who was perceived by subordinates as possessing exceptional qualities which set him apart from ordinary men. The charismatic image was effective as long as the leader was successful and his presence was immediate. Such leadership was described as transitory: "Indeed, in its pure form charismatic authority may be said to exist only in the process of originating. It cannot remain stable, but becomes either traditionalized or rationalized, or a combination of both" (Weber, 1947). Followers are drawn to him by the promise of future opportunity, either to throw off a past which they regard as distasteful, or to gain future rewards which they view as exceptional (Hoffer, 1958).

In spite of such identification of a unique form of leadership influence, the prevailing attitude by management writers has been that single individuals have relatively little influence on organization outcomes. One explanation for this attitude is that investigators have chosen to study organizations which have already gone through the process of institutionalization (Kimberly, 1976, p. 22). Thus, they tend to deny the importance of a single individual in creating and developing an enterprise.

There are numerous conspicuous cases in which individuals have had a major impact on the founding or redirection of organizations. Examples would include Gandhi's leading of India toward self-government (Collins & LaPierre, 1975), Durant's development of General Motors (Dale, 1960), Jesus' initiation of Christianity (Haley, 1969), and Hitler's responsibility for the Nazi war movement.[2] These and other promotors seem to engage in certain common activities which include the following:

First, promoters initiate organizations which have new or unique goals or functions. Virtually all descriptions of promoters describe them as creating new organizations which represent a sharp break with tradition or which exploit some kind of innovation (e.g., House, 1976; Kimberly, 1976; Collins, Moore & Unwalla, 1964; Weber, 1947; Filley, 1962). Newness is a pervasive element in all such organizations, whether new market, new method of distribution, new product, or new way to achieve personal peace, wealth, happiness, or salvation. The particular advantage of newness may be seen in a profit-making organization. If a product is desired by its customers and if the enterprise is the only source of the product, then profit margins are likely to be high and concerns for efficiency and good management are likely to be minimal.

Second, whether guru, revolutionary, or company president, the promoter promises outcomes which are highly valent to organization members. While the outcomes may be as disparate as personal ownership of land, or entering the Kingdom of Heaven, or achieving wealth, they are all attractive. As Berlew (quoted in House, 1976) has stated:

> The first requirement for . . . charismatic leadership is a common or shared vision for what the future *could be*. To provide meaning and generate excitement, such a common vision must reflect goals or a future state of affairs that is valued by the organization's members and thus important to them to bring about . . . All inspirational speeches or writings have the common element of some vision or dream of a better existance which will inspire or excite those who share the author's values. This basic wisdom too often has been ignored by managers (1974, p. 269).

Third, promoters enhance the likelihood that organization members will attain desired outcomes. The confidence and perceived competence of the promoter has been stressed in the literature (House, 1976). Of particular importance is Weber's (1947) observation that a leader must be attributed extraordinary powers by subordinates; in particular, he must not fail to succeed in their eyes. The functional importance of this attribution is explained by the fact that in a new venture with no experience to the contrary, it is not possible to objectively assess the odds of success. Consequently, the leader is able to generate high subjective probabilities of success among subordinates which cannot initially be disproven. In situations where the attainability of outcomes is unverifiable (e.g., going to heaven), such a situation may continue to prevail. However, where objective information concerning probabilities of outcomes is somehow available, it is likely that subjective probabilities will become increasingly veridical, often jeopardizing the leader and the organization.

Fourth, promoters enhance the self-esteem of their subordinates. Charismatic leaders seem to influence the sense of personal worth and personal competence among their followers. House (1976) has suggested that this influence occurs functionally through the role modeling of the leader and has reported studies consistent with this notion. Apparently by identifying with a leader perceived as superior and exceptional, followers gain similar values for themselves, even changing long-standing attitudes (Bandura, 1968).

Finally, promoters unify divergent interests. Such leaders demonstrate an ability to gather a heterogeneous group around them and to maintain cohesion among the membership. This functional process has been explained by Wilensky (1957). He reported two conditions under which the integration of organization membership was high. The first occurs when members interact frequently and have common ideologies. Such is certainly the case when the group has high mutual expectations. The second situation, according to Wilensky, is one in which group members do not interact frequently but exhibit homogeneity of values. Such values might be established by selection or by indoctrination (Selznick, 1957).

In sum, promoters perform functions which contribute to a high level of motivation: they do something new; they promise desirable outcomes; they enhance felt competence and confidence among organization members; and, they unify diverse interests. Their charismatic style thus creates an organization in which relationships are personal, nonbureaucratic, and idealistic, much as Wilensky described in his first motivational situation. In terms of expectancy theory, the members perceive themselves as quite able to perform what is required, the instrumentality of their efforts is high since there is no evidence to the contrary, and the personal outcomes to be gained are viewed as extraordinarily desirable.

The promotion type of organization is also vulnerable, containing within itself the seeds of its own destruction. The organization is inefficient, making it vulnerable to competitors who are efficient. The newness of its innovation is subject to imitation. The leader's charisma may well end, either through personal failure or through failure to achieve idealistic goals. Finally, the personal influence of the leader may cease when contact with organization members is no longer practical.

If such organizations are to continue their existence, it is likely that they will change to a different type. They become institutionalized and managed, most likely acquiring the characteristics which we have described as administrative. As Collins, Moore & Unwalla (1964) stated:

> The new phase of the firm will have set up new role demands and these the entrepreneur as leader may find increasing difficulty in playing. Either gladly or reluctantly he will pass on the power to the younger generation, a generation not of entrepreneurs but of administrators (p. 201).

Thus, the metamorphosis of the promotion stage is not that of a butterfly escaping its casing. Rather, it occurs through crisis when the promoter persists in an organization type which is no longer useful, or it occurs through planning when the need to change is apparent and acted upon.

### *Research on the Types*

It is of course essential to determine whether there are such patterns as those suggested and whether or not they are a simple function of such variables as size and technology. If in fact the clustering of variables into types is the spurious consequence of their common relationships to a single measure, a focus on that single measure would be most appropriate and parsimonious.

One relevant study by Haas, Hall & Johnson (1966) presented an empirically-based taxonomy of 75 organizations created by a cluster analysis of 99 variables. The ten clusters which emerged varied in terms of number of member organizations from 2 to 30 and of common attributes from 4 to 43. Significantly, inclusion of organizations in particular clusters appeared to be independent of size, goals, prime beneficiaries, and all other single characteristics considered.

In another study related to the types described earlier, Filley and Aldag (1976) presented a 63-item, Likert-type scale to 85 top executives of small- and medium-sized manufacturing firms. The items were based on the organization-type descriptions summarized earlier. A principal component analysis of responses with varimax rotation yielded a relatively clean three-factor solution that was consistent with the a priori conceptualization of types. Principal items in the subscales are shown in Table 1. Those 40 items were subsequently used to form a shortened version of the original scale, titled "Company Characteristics—Form M." That 40-item questionnaire was administered by Oliva and Peters (1976) to top- or upper-level executives from 120 different firms with wide differences in size. They found correlations of factor scores between their sample and that of Filley & Aldag of 0.96, 0.98, and 0.94 respectively.

A further study by Filley & Aldag (in press) administered the questionnaire to 211 chief executives of manufacturing firms and to 61 chief executives of retailing firms. That study also provided data with which to analyze predictions from this theory of types. Factor loadings were largely consistent with those expected. Average factor congruency between the two manufacturing samples was 0.48. All items loaded positively on the correct factor and only one item loaded more heavily on a different factor than expected. For 33 of the 40 cases, items loaded above 0.3 on only the correct factor. While not a perfect match with the first study, the analysis suggested that the factors are fairly stable. The retail sample also indicated congruence with the manufacturing sample factor structure. Relationships of type scores to size and technology indicated no clear evidence that types can be predicted simply by knowing organization size or technology. Thus, it does not appear that the complex constellations of characteristics associated with the types are simple functions of other readily identifiable single variables. As such, it seems that the typology represents more than an unnecessary elaboration of a more parsimonious explanation. In general, while not all of the characteristics included in the theoretical description of craft, promotion, and administrative types are included in the factor structures, the evidence does suggest the existence of three different patterns of organizational characteristics which are similar to those expected.

*Key Influences on Types*

The following discussion will indicate variables which appear to be associated with these types. It will be argued that such associations provide clues concerning causes and consequences of the types. In particular, it will be shown that the types seem to capture different constellations of variables which affect the growth and survival of organizations.

*Leader Influence and Role.* A review of the items in the three factor structures in Table 1 indicates that leaders have three quite different perceptions of the organization and of their own intentions. Leaders in the craft type emphasize a preference for "sameness" over time: same methods of operation, same customers, same goals. That is, they indicate a desire for stability. They believe that one learns the business by experience and such leaders are directly involved in performing the technology of the organization. Craft factor scores are positively related to age of the firm and adherence of the leader to the Protestant Work Ethic and negatively related to leader risk taking and to technological complexity and growth rates of the firm (Filley & Aldag, in press).

In contrast with the craft type, leaders in the promotion type stress the need for personal control, the rapid changes in the organization, rapid growth, an absence of formal organization, and the presence of an innovative advantage as the basis for success. Correlates of these factor scores also indicate that promotion firms are relatively small, are relatively short-lived, and show rapid growth rates (Filley & Aldag, in press).

Finally, factor items in the administrative type indicate the presence of formalizing techniques: forecasting, planning, budgetary control, written policies, and job descriptions. Administrative leaders apparently establish systems which minimize the presence of people deemed indispensable. Correlates of these factor scores indicate that such firms are larger in size, though their level of technological complexity and rate of growth may range from high to low (Filley & Aldag, in press).

The difference between these three types suggests quite different patterns of leadership influence and role. In the first two, the character of the business appears to be shaped directly by the leader. Both the craftsman and the promoter are directly involved in the organization: the former in the maintenance of stability, the latter in coping with change. The administrative type, on the other hand, seems to evidence the development of systems and the establishment of an institutional character. In other

*Table 1.* Factor Analysis of Company Characteristics

Factor I (Craft)

| *Item* | *Factor Loading* |
|---|---|
| We are doing things pretty much the way they have always been done in this type of business. | .365 |
| The fact that I make a comfortable living is enough success for me. | .305 |
| We are selling to the same customers this year that we did last year. | .309 |
| I practice the open door policy—I am always available to people in the company when they want to see me. | .387 |
| Next year we will continue to operate pretty much as we always have. | .647 |
| I prefer a safe income to a gamble. | .558 |
| I don't want to shoot for the moon in my business, just make a good living and see the business survive. | .442 |
| In evaluating employees I place great emphasis in their loyalty to me. | .444 |
| One has to learn this business by experience. | .413 |
| I'd have to say that business survival is my main goal at present. | .479 |
| I am more effective in handling technical or production problems than I am with management issues. | .327 |
| I am worried about what to do with some loyal supervisors who expect to be promoted but just don't have the ability. | .359 |
| I have daily contact with the workers in the plant. | .476 |
| My employees are good workers or craftsmen who appreciate the security of the job. | .547 |

Factor II (Promotion)

| *Item* | *Factor Loading* |
|---|---|
| I'm afraid to leave the office for a few days for fear of what I'll find when I get back. | .475 |
| We avoid the use of job titles in our company. | .326 |
| When I am away from the business I call in at least once a day to make sure that everything is all right. | .416 |
| Key personnel have left the company in the last two years because of conflicts with me or others. | .340 |
| We are successful because we have a unique product, source of supply, or method of distribution. | .599 |
| Right now our firm is growing so fast that it's almost impossible to plan and control the way I'd like to. | .607 |
| We are gamblers—it is impossible to plan for the future. | .425 |
| I am a promoter at heart. | .565 |
| My business policies change daily as conditions change. | .487 |

*Table 1. (Continued)*

| | |
|---|---|
| I hire people who will sacrifice pay and fancy titles now for the tremendous opportunity the job offers. | .542 |
| I have been self-supporting since high school age. | .408 |
| We can afford some internal confusion in this company because profit margins are high. | .499 |
| An organization chart would be outdated before we could get it on paper. | .386 |
| Things are changing so fast in my business that I have to take complete control to avoid confusion.* | .545* |

Factor III (Administration)

| *Item* | *Factor Loading* |
|---|---|
| I prefer an employee who will do his present job well and continue to develop for future advancement. | .409 |
| In our company, specialized employees handle planning and control. | .321 |
| Things are a lot more predictable in my firm than they used to be. | .345 |
| In our business there are no indispensable people, including me. | .413 |
| We estimate our sales for each year and prepare budgets for internal financial control. | .594 |
| We have sales forecasted for at least the next three years and are planning facilities and operations on the basis of that forecast. | .470 |
| I am able to take a two-week vacation with the assurance that business operations will continue effectively in my absence. | .434 |
| We have a company policy manual. | .536 |
| In my business, controls and operating methods are in writing. | .568 |
| I employ good workers who believe in the goals and values of the company. | .312 |
| We have a working board of directors in our company that meets at least four times a year. | .368 |
| When we hire new people, we compare their characteristics with written job descriptions. | .428 |

* Also loaded on Factor III at .329.

words, both the craftsman and the promoter seem to exercise personal control of the organization and its members while the administrator exhibits control through structure and systems.

The sharp differences in leadership role described may be explained, at least in part, by considering two questions: First, to what extent does the leader choose to deal with the organization's environment? And, secondly, in what form will the leader deal with the environment if that is the choice?

*Degree of Interaction with the Environment.* Many recent writers have stressed the influence of the environment upon the nature and effectiveness of organizational characteristics (e.g., Burns & Stalker, 1961; Lawrence & Lorsch, 1967). Our discussion up to this point suggests that the degree of environmental influence is in part a function of the decision of organization leaders about whether or not to adapt to and respond to the environment. That is, leaders decide, consciously or not, whether their organizations will be "open systems" or "closed systems."

If in fact the degree of the organization's response to the environment is a function of the leader's decisions, it might be expected that leader-related variables would moderate the relationships between environmental dimensions and such organization characteristics as size and structure. A study by Meyer (1975) supports this contention. Meyer examined 215 finance departments over a six-year period. In particular, he considered the moderating effect of the autonomy of the leader's role on the relationships between environmental change and organization structure and size. The leader's role was defined as autonomous when the leader was unchanged over the period and was relatively independent of control by superiors. Conversely, the role was less autonomous when leadership had changed during the period and was more dependent upon superior control.

Results indicated that when the leader was autonomous, the organization functioned as a closed system, relatively unresponsive to changes in the environment. Measures of structure, size, and environment were unrelated. In contrast, where leaders were relatively new and were more dependent on higher authority, they were more responsive to environmental change, and the organization functioned as an open system. In this case, measures of environmental change, structure, and size were highly related.

The relationship between leader role and tenure in the job suggests a useful speculation. It may well be that the characteristics which conventional wisdom and some studies tend to associate with age, such as risk aversion (e.g., Vroom & Pahl, 1971), reduced creative output (cf. Lehman, 1953), and decision latency (e.g., Taylor & Dunnette, 1974) are better predicted by time in the same role. The nonadaptive managers in the Meyer study had been present for at least six years while managers with shorter tenure showed greater responsiveness. Similarly, Pelz & Andrews (1966) found that creativity of scientists and engineers peaked at about six years in the same job.

In much the same way, the craftsman described earlier showed preferences for stability and evidenced little preference for or behavior associated with reacting to the environment. While our research does not show an expected relationship between craft factor scores and length of time in the job, it does show relatively longer existence for the craft firm itself (Filley & Aldag, in press). The insular behavior of the craft leader may well create a closed, nonadaptive system, sacrificing growth and profitability for security and survival. In contrast, the promotion leaders and their firms were short lived and the promoters were actively engaged in exploiting opportunities in the environment. If the leader may or may not choose to respond to the environment, the second question may be addressed as well: How do leaders react to their environment if they choose to do so?

*Nature of Interaction with the Environment.* The question of the way in which leaders react to the environment can be addressed by considering whether the organization relies primarily upon feedback or feedforward controls in its operation.

The use of feedback or feedforward control appears to be the same as what Boulding (1953) calls the "carpenter principle."

> In building any large structure out of small parts one of two things must be true if the structure is not to be hopelessly mis-shapen. Either the dimensions of the parts must be extremely accurate, or there must be something like a carpenter or a bricklayer following a 'blueprint' who can adjust the dimensions of the structure as it goes along (p. 337).

In other words, actions are controlled personally by adjusting for errors, i.e., *feedback control*, or else contingent actions must be planned so that error *does not* take place, i.e., *feedforward control* (Filley, House, & Kerr, 1976). It should be noted that both feedback control and feedforward control involve planning. However, as will be discussed later, they differ in terms of the nature and degree of that planning and, perhaps, in terms of who does the planning.

Feedback control has two primary weaknesses. First, it occurs only after an error takes place; that is, when there is a difference between what is intended and what actually occurs. Secondly, because of the time lag between assessment of a deviation and the consequent corrective action, actual performance tends to oscillate around the intended standard. Several conditions are suggested in systems of feedback control: (1) the controller must be able to exercise skilled judgment, (2) the capacity of a single controller is necessarily limited, (3) the lag time for correction must be minimized by constant review and rapid corrective action, and (4) finally, there is no "learning" in the system, such that compensatory action is planned before errors occur.

While some feedback systems may be made automatic (e.g., the thermostatic heat control system), most in organizations are not. When not automatic, the systems require a high degree of constant personal review and adjustment. In some cases, obviously, the newness or uniqueness of problems mitigate the use of systems other than feedback control. In other cases, however, feedforward control could be used even though it is not. As will be discussed later, technology plays an important part in determining the use of control systems. In addition, however—and quite aside from technology—the absence of control systems and of the people who create such systems means that the leaders in the craft and promotion types of organization must rely heavily upon feedback control in their organizations.

Feedforward controls, in contrast, anticipate deviations from intended action and take compensatory action before error actually takes place. Such systems suffer from two main limitations. First, feedforward control requires an accurate determination of the relationship between two or more processes. For example, Tustin (1955) discusses how feedforward control may be used to maintain the temperature of a room at a constant level. A thermometer measures outside temperature and, as the outside temperature changes, the thermometer activates a device to release a predetermined amount of heat into the room. Use of such feedforward control requires precise specification in advance of the functional relationship between inside and outside temperature. A second limitation of feedforward controls is that they cannot deal with unusual circumstances. That is, relevant causal links must be fully specified in advance. As such, only those variables for which the need to specify those links has been anticipated can be accommodated. Several conditions are implied by this discussion of feedforward systems: (1) such control systems require detailed planning, (2) the planning is often done by "staff" people different from those creating products or services directly, (3) the capacity to control is not limited by the personal capacity to monitor error, and (4) there is "learning" in the system as more accurate predictions can be made and those are built into more refined control mechanisms.

As may be seen, aside from careful planning and the necessity for dealing with exceptions, feedforward systems are more impersonal than most feedback systems. In a business, for example, feedforward control may be used to adjust levels of inventory, production volume, purchase schedules, and staffing as sales volume increases or decreases. It is the decision to use feedforward control to the extent possible which characterizes the administrative form of organization. In large batch and mass production firms the use of such control systems affects the success and growth potential of the organization (Khandwalla, 1974). Even in job shop or batch production, however, feedforward systems may be used to assure the proper allocation of raw materials, machines, and people (Reeves & Turner, 1972).

*Consequences of Control Systems.* If the previous logic concerning personal versus system control of the organization is correct, one would expect two phenomena to occur. First, the development of impersonal feedforward systems of control should overcome the natural limits to growth which would be imposed when a single craftsman or promoter is responsible for directing, integrating, and controlling the operations of the enterprise. Secondly, where an enterprise is willing to utilize impersonal control systems and to employ specialists to develop and operate such systems, the resulting efficiency should more than offset the cost of added employment.

Again, the first expected consequence of feedforward control is that natural constraints on organizational growth associated with reliance on a single individual for personal control will be overcome. This notion has rather consistent theoretical and empirical support. Coase (1952) argued that decreasing returns to a fixed entrepreneurial function limited continued efficiency and continued growth of a firm. This limit could be overcome through techniques of professional management. A similar argument has been advanced by Kaldor (1960). Present evidence is also consistent with the notion. In their study of success and failure in 95 small manufacturers, Hoad & Rosko (1964) noted reliance on a single individual for direction of the organization as an important cause of failure. Finally, Filley & Aldag (in press) found a positive relationship between the administrative type of organization and measures of size. While administrative attributes may to some extent be evoked by size, it is also likely that those attributes (systems) permitted the achievement of greater size.

The second expected phenomenon is that the additional staffing for planning and control purposes required by feedforward systems should create profit advantages which more than compensate for the additional cost. Here the evidence is sparse but consistent. Ware (1975) investigated performance of 74 firms in the food and beverage industry, half of which were owner controlled and half of which were run by professional managers. While it is not argued here that owner control is necessarily identified with nonprofessional management, the difference may be an adequate proxy. Ware's results indicated that manager-controlled firms had higher profitability (net income/net worth), even though they had lower sales per employee. The two differences were shown not to be a function of firm size. Such findings suggest that the greater efficiency of the firms led by professional managers offsets the added costs of employment by creating more profit.

If this analysis is correct, it does much to explain the difference between craft and promotional leadership on the one hand and administrative leadership on the other. The former rely heavily on feedback control; the latter emphasizes systems of feedforward control. Both craft and promotional firms might be expected to experience limits to growth, since they must rely upon personal control by their leaders. The administrative firms, in contrast, are not administrative because they are large; they are administrative because their leaders use more impersonal controls.

To this point we have argued that types of organizations may be identified which differ on a number of dimensions. We have further seen that organizations may move from one type to another, though not necessarily in any particular order, and that such transitions are in part a function of the nature of leadership. The types do not appear to be a simple function of technology or size. Instead, it has been argued that type characteristics, such as predominant control systems, influence the ability of the organization to overcome limits on growth. The following sections provide an expanded discussion of the nature and measurement of growth and further consider how the leader, through his definition of organization domain, initiates a causal sequence which defines types and influences growth potential.

## INFLUENCES ON GROWTH

### *Organization Growth*

When organizations change their size they are said to have grown or declined. Such changes are commonly measured with simple criteria such as sales, number of employees, assets, number of stores, and so on, compared with some measure of the past. Unlike organisms, organizations do not exhibit natural patterns of growth or decline against which actual experience can be measured.

In the following discussion we shall explore some elements of the growth process in organizations. The discussion must be viewed as highly speculative since both theory and research into the organizational growth process are sparse. The venture seems worthwhile for at least two reasons. First, without understanding how organizations do grow, it makes little sense to talk about how they should grow. The normative literature about organization development seems to be based on the standards of those who influence organizations rather than upon criteria which are inherent in organizations or are imposed by their leaders.

A second reason for attempting an explanation of growth and change is the likelihood that it may lead to a better understanding of how to accomplish dramatic changes in the behavior of organization members and the accomplishment of organization goals. We have already indicated that the kind of leader—craft, promoter, or administrator—can have a more profound impact on the performance of an organization than different degrees of, say, initiating structure. That is, the differences are those of kind rather than degree. Similarly, the consequences of selection, systems, structure, and organization domain may also provide for or explain far greater variations in organizational performance than the learning or level of motivation of organization members.

*Kinds of Growth.* Boulding (1953) identified three classes of growth: simple growth, populational growth, and structural growth. *Simple growth* is the "growth or decline of a single variable or quantity by accretion or depletion" (p. 326). While not an issue of concern for Boulding, it may be seen that if simple growth—more or less of something—is treated as an independent variable, then one has a typical analysis of correlates of size. Examples are studies of group size which show the changes in shared leadership, formality, or concern for member needs as groups get larger.

The second class of growth is *populational* in which the composition of a growing population is not considered to be homogeneous. The concern in this case is with the composition of the population. A *population* is defined as "an aggregation of disparate items, or 'individuals' each one of which conforms to a given definition, retains its identity with the passage of time, and exists only during a finite interval" (p. 328). Thus, on one level, the growth of an organization means that more people are entering employment than are leaving it. On a more complex level, these changes may be further considered in terms of the age, sex, or skill composition of the members as well.

While Boulding considers the growth (positive or negative) of the population in terms of entry and exit (births and deaths), it is easy to see that such an analysis can also be used to consider the composition of the population itself. For example, the growth of job satisfaction may be determined by considering the number of members in an organization who meet the definition of "satisfied" and the number who meet the definition of "dissatisfied," or the number who are high or low on different facets of satisfaction.

The third class of growth is *structural*. Structural growth involves an aggregate of parts. As the aggregate "grows" a complex structure of interrelated parts changes in size and shape and the relations and size of the parts change as well. Since most investigations of organizations occur after a growth process has taken place, relatively little is known about the actual development of organizations.

Boulding offers a few general principles about structural growth upon which the present authors have expanded.

(a) First, any structure has a minimum set of conditions which are necessary for "nucleation." In part, these conditions involve necessary resources: people, materials, money, etc. In addition, nucleation probably requires a particular mixture of these resources through composition and structure. Finally, nucleation seems to involve perceptual elements as well, which encourage or discourage the subsequent success of the organization. For example, the successful nucleation of the Meteorological Satellite Program at NASA appeared to depend upon a generous budget created by the Sputnik scare, a unique matrix organization and joint agency effort involving NASA and the weather bureau, and perceptions of weather observation by satellite rather than on the surface of the earth (Delbecq & Filley, 1974).

(b) Second, structural growth involves nonproportional change; the proportion of parts and significant variables cannot remain constant. This principle has two corollaries. *The first corollary is that the growth of a structure always involves compensatory change in the size of the parts.* Drawing on a biological analogy, Boulding points out that ". . . those functions and properties which depend on volume tend increasingly to dominate those dependent on area, and those dependent on area tend increasingly to dominate those depending on length" (p. 335). Without attributing life to the organization, the physical properties are parallel to the growth of any object, living or not; the mass expands by a cube function while the surface increases by a square function. If the surface supports the mass in any way, say for providing information, then compensatory changes in information processing will have to occur as an organization grows. Stated more generally, size, shape and function change their relationship with growth (Haire, 1959, p. 274). And, further, there should be an optimum relationship between the three which is calculable (Haire, 1959, p. 303). If this line of reasoning is correct, then it seems likely that as an organization grows in mass, increases in both its level of authority and its differentiation of function should occur even though they may increase at different rates in different forms of technology. This concomitant change in levels and differentiation has been noted in Reimann's study (1973) of 19 manufacturers. Nonproportional change in the growth of four firms has also been demonstrated by Haire (1959).

*The second corollary of nonproportional change is that the inability to compensate for changes in the parts constitutes a limit to growth.* We have already alluded to the limits created by having an organization dependent upon a single individual for decision making and coordination. On a more complex level and as noted by Boulding (p. 336), compensatory change may either take the form of solving problems created by growth or it may take the form of attempting to avoid such problems. It is this difference which appears as a recurring theme, largely implicit, in discussions of growth.

In the case of information processing, Galbraith (1977) has noted that organizations may deal with increasing complexity (as with growth) either through increasing the capacity for information processing (through sophisticated information systems) or through reducing the need for information processing (by creating smaller, autonomous units). Similarly, we have already shown that organizations may either plan and standardize operations to utilize feedforward controls to avoid errors or use feedback controls which require that people take action to reduce or correct errors as they occur.

It is precisely this difference which seems to account for the natural constraints on size created by technology. As will be discussed later in more detail, job shop technologies require compensatory efforts at both the top and the bottom of the organization which create some natural limits to growth. On the other hand, standardized, mass production technologies are less constrained by the limits of size, relying as they do on additional structure and systems to compensate for increases in size.

(c) At any moment, the form of an organization is a result of its growth up to that moment. The purchase of a machine, the employment of a staff specialist, or the addition of a system will shape the structure in a manner which portends the future. Thus, it seems likely that organizations might exhibit a constant struggle between inherited and required form. If individuals or groups have the power to maintain unnecessary but inherited organization components, they may make the organization maladaptive. This difficulty has been demonstrated in the administration of school systems, which maintain administrative positions regardless of changes in the size of the schools themselves (Freeman & Hannan, 1975).

Even if prior development of the structure was adaptive, it may create maladaptive consequences for the future. As Boulding says, "growth creates form, but form limits growth" (p. 337). Since innovation and creativity seem to occur where the system is not pinned down and highly organized, the result of a closed system, then, is opposition to further development. We saw this situation illustrated in our earlier discussion where the years of service by leaders and their lack of stimulation by imbalance created closed, nonadaptive systems. Thus, while newness and openness in a system can limit growth through the demands upon processing information, the opposite—closure—can also act as a deterrent to future growth.

(d) As a structure grows, either the dimensions of the parts must be extremely accurate or else the divergence from a plan must be adjusted by a skilled person. (This is Boulding's "carpenter principle," mentioned earlier in connection with control systems, p. 337.) Planning is involved in either case. The former involves precise planning before action so that error does not take place, e.g., in assembling products which are mass produced. Implied here is prediction, control, and a person developing such plans and controls. These planners are not likely to be the same people as those doing the assembling. In the other case, where someone is exercising judgment to compensate for errors, a plan is also present. In such a situation the person making the plan is also likely to be the person doing the work. It is the errors and omissions in this case which call for adjustive action. Obviously, in the growth of organizations both systems will be found to some extent. However, the survival needs and consequent technology of the organization will require a preponderance of one or the other: one case is required for efficiency; the other case is required for adaptability and creativity.

(e) Activities or parts of an organization will move toward areas which are of greater advantage and away from areas which are of lesser advantage. Boulding calls this the "principle of equal advantage." In a market economy, such movement is easily seen when businesses develop where the potential payoff is high and not in locations or forms of enterprise where the payoff is low (e.g., during prohibition illicit enterprises were started). Within organizations, this principle is much more complex. In part it explains why units will absorb key functions. In part, also, it suggests why key professionals might want to resist efforts to reduce their monopoly of power.

The principle of equal advantage is most readily seen in the development of an organization when its size increases. There are undoubtedly some activities which are required simply as a function of organization size. When size dictates the need for the activity, then its relative advantage will increase and the odds for the emergence of the activity will increase. In addition, relative advantage seems also to occur in a sequential fashion for some activities. That is, activity A must be developed to some point before activity B has a relative advantage. For activities which develop in sequence, firm size does not predict relative advantage; rather, the advantage is determined by the evolutionary history of the firm.

The first situation, in which emergence of activities is a simple function of size among organizations of similar technology, may be illustrated in the emergence of staff functions. For example, DeSpelder (1962) surveyed 155 automotive parts manufacturers, finding emergence of such activities as production scheduling, personnel management, and purchasing as size increased. This is not to say that such activities automatically emerge as organizations grow; only that the relative advantage of adding such functions seems to change with size.

The second situation, in which there is a sequential advantage of functions with organization growth, may also be demonstrated. As Filley and House (1969, p. 462) have noted, it appears that manufacturing firms start their development with an emphasis either on the production or on the sales function, then expand the second of these two, then develop their financial function. After finance is developed, they then separate their administrative activities from direct operations. Part of this sequence was observed by Wickesberg (1961) in his study of 106 small manufacturing companies. From his cross sectional research, he inferred that the firms first emphasized formalization of the production unit, then sales, followed by purchasing, quality control, research and development, and credits and collections. Finance was not included in the study. Similarly, McGuire (1963) noted a time lag between hiring more production workers and expansion in employment of other types of personnel. More recently, a study of the growth history of three manufacturing companies (Feldbaumer, 1973, p. 188) found that the proposed sequential patterns held quite closely.

In sum, the previous discussion of structural growth suggests that there are some describable limits upon the growth of organizations: (1) the absence of the resources necessary for nucleation; (2) the inability of an organization to compensate for changes in its parts, e.g., the limited capacity to process information; (3) a divergence between inherited form and required form; (4) perfect closure which makes a firm nonadaptive; (5) the absence of either precise planning or a skilled decision maker who can correct errors; and (6) the failure to add activities or functions when, either because of sequential development or because of size requirements, they are advantageous.

*Measurement of Growth.* If the previous analysis is correct it suggests that changes in size are both causes and effects; and that both on the basis of size and developmental criteria it should be possible to develop a probabilistic model of organization design as well as predictions about relative effectiveness. That is, it is quite conceivable that one should be able to predict a structural configuration which is likely to develop, and that one should be able to predict the relative effectiveness of different configurations. If the latter, it should also be possible to be prescriptive about *proper* organization designs.

To argue, as some have done, that structural configurations may exhibit equifinality (Reimann, 1973, p. 476) is, except at the highest level of abstraction, difficult to accept. Since any normative conclusion requires a statement of one or more criteria, the argument is specious without defining such criteria. Perhaps if one uses as the criterion, "the existence and demise of an organization" then all configurations would meet the condition. If one takes criteria such as profitability, degree of adaptability to the environment, or innovativeness, then it is likely that organizational configurations may be shown to meet the standard more or less well.

Why then is so little known about proper organization design as enterprises develop? One reason seems to be the heterogeneity of sampling, perhaps because those doing research assume similarity of organizations which are vastly different in structural imperatives. For example, Aldrich (1972) has demonstrated that the conclusions about technology drawn from the Aston studies are simply the result of comparing service and manufacturing organizations. In effect, what has happened in that case and elsewhere is like using peas and oranges as a proxy for "size of sphere," and color as a dependent variable, and concluding that as spheres get larger they turn orange.

A second reason for such a dearth of knowledge is simply the absence of normative data. The simplest kind of questions like: When should I hire an accountant? or, How many managers should I have? or, What kinds of departments should I have? are unaided by reports of the degree of concomitant variation, since such reports often do not provide averages or dispersion of raw data. In addition, comparisons of norms for effective and ineffective performers are even more rare.

A third reason for an absence of information about organization design is the availability and use of cross-sectional studies. The only way that one can justify the use of cross-sectional data to make inferences about organizational development is to have assurance that the sample is homogeneous and that patterns of development hold true across all organizations of that type (Aldrich, 1972, p. 27). Otherwise, to study the growth and development of organizations one must rely upon histories or longitudinal investigations in individual organizations.

Finally, it seems likely that problems have resulted from the knotty issue of size versus sequential development mentioned earlier. If sequential development is not a function of size and if cross-sectional studies are used to infer characteristics of growth, then the absence of relationships in such studies does not mean that variables do not change predictably with growth. Instead, they mean that growth related variables are dispersed along the size continuum, masking important information.

*Why Organizations Grow*

Given a rather selective perception of organizations which focuses upon a few visible enterprises, it is generally assumed that organizations must and do grow. Yet it is likely that the majority of the enterprises in the business population do not grow, except perhaps in the most trivial sense. The world is full of service stations, grocers, repair shops, and the like which show little evidence of growth except for sales price increases. To some extent growth is a passive response by the enterprise, as when more customers seek the company's products. In other cases, growth is the result of conscious decision making. For example, companies seek to increase their share of a present market, to add related products or to move to new territories. Further, companies may grow by acquisition and merger with similar or different enterprises.

As Pfeffer and Salancik (1978) point out, most of the research which seeks to explain why organizations grow has "addressed the consequences of growth, with the expectation that motivation can be inferred from consequences" (p. 132). As they indicate, such processes can be misleading since the failure to experience intended outcomes does not mean that the goals were not present in the first place. In their view, growth is simply one means for coping with interdependence with the environment, either to reduce uncertainty or external control.

Penrose (1959) argued that growth occurs through an attempt to utilize unused managerial resources. These resources may expand through managerial experience as well as through the addition of management talent. She also pointed out several constraints upon growth: (1) the fact that some kinds of business are unsuited to large size, such as local jobbing operations or some kinds of service establishments; (2) the protection of small operations by large firms through protective pricing; (3) situations in which easy entry and high mortality rates preclude the achievement of much size; and (4) the inattention of large firms to smaller competitors which could be driven out of business, and the consequent reward for smallness.

Other reasons for growth include simple economies of scale, a desire for increased profits, or the personal motives of executives for power or compensation. As Pfeffer & Salancik point out in their review, however, profit rates generally increase to some modest company size and then remain stable or decline, and large firm size does not necessarily generate significant economies of scale. Executive incomes are, however, related to size of enterprise.

A more persuasive case for growth motives can be made on the basis of organizational stability and viability. Profits of larger firms vary less than profits of smaller firms (Caves, 1970; Samuels & Smyth, 1968); share of market is more stable where shares are concentrated (Gort, 1963); and survival rates are better for larger than for smaller firms (Steindl, 1945).

While these arguments suggest the value to an organization of increased size, they may be specious. For one thing, the smaller and larger firms may be different in kind (technology, type). For another, they do not address the advantages of growth; only, the average experience of size. As will be discussed in detail later in this paper when we address technology, it is likely that growth does provide the advantage for some firms as just suggested; for others, however, it is likely that growth is a distinct disadvantage, reducing the likelihood of success and survival. Whatever the motives for growth, it is clear that the growth experience is related to implicit or explicit decision making by people within an organization. One such decision is that regarding the domain of the enterprise.

*Domain and Growth.* A fundamental influence on the consequences of organizational behavior is a sort of imagery held by key decision makers in an organization about the boundaries of the enterprise. Simply stated, the domain of the organization is the answer to three questions: What do we do? How do we do it? For whom do we do it? In other words, as Thompson (1967, p. 40) suggests, the domain identifies the products or services of the enterprise, the technology involved in their creation, and the market to which they are offered. Consider the difference, when identifying products or services, between saying "We make wooden study desks for children" and saying "We are wood fabricators." Consider the difference, when determining technology, between saying "We are a production shop" and saying "We are a job shop." Finally, consider the difference between saying "We serve the Madison area" and saying "We sell throughout the world."

An example may further clarify the importance of domain. One of the authors has been acquainted with the owner of a company that initially made a product which was used exclusively on ships in the Great Lakes. When asked what kind of business he engaged in, the owner said "We make this product for Great Lakes ships." The growth of the enterprise was stifled when it managed to service about 80 percent of the commercial vessels on the Great Lakes. Later, the owner redefined the business as one of removing sediment from liquids in commercial processes. The business expanded to industrial and sewage treatment applications. Later the business domain was defined as a water treatment company. It grew to provide a complete line of industrial water treatment processes. Finally, the owner defined the business as the application of industrial chemicals. Each time the domain was redefined the company provided different products and services to new markets with new technology. Had the domains not been redefined, company growth would have been constrained.

In much the same way, the domain redefinition of film companies from making movies to the entertainment business, or of rail companies from running railroads to transportation, has altered the operations and potential growth of those enterprises. The change must also be accompanied by a consensus with environmental agents (customers, suppliers, etc.) about the domain as well, if such external exchanges are to fit with the internal operations of the enterprise.

A clue to the consequences of differing definitions of domain has been provided by Meyer (1975), who investigated the finance function of 227 public bureaucracies. Three domains were identified: (1) the comptroller's office, (2) the finance department, and (3) the department of administration. The comptroller function is mainly regulatory and the domain is well defined. People employed in such offices regulate expenses and engage in traditional forms of bookkeeping. In contrast, the finance department is supposed to manage all fiscal activities and its domain is not well defined. Besides accounting and auditing, the finance office includes such activities as revenue collection, purchasing, and investments. In addition, unlike the comptroller, the finance director is involved in policy making. Finance departments developed because the change from line-item budgets to program or performance budgets tended to blur the distinction between financial and nonfinancial matters. The third type, the department of administration, claims a broad domain which encompasses not only finance activities but those like personnel or property management which do not fit neatly into other departments.

Meyer's findings indicate that where domains are narrow and services performed are essential, the organization remains stable over time. This was the case with comptrollers' offices. In contrast, where domains are broad but not well defined (i.e., the finance departments) their functions were shifted to other departments unless computer technology was present. The use of computers helped to fix the activities of the finance department and avoid contraction of their services. Finally, the broad domains claimed by the departments of administration and their desire to adopt new functions and new technologies has led to growth of their departments. In addition, the heads of such departments have achieved positions of power and prestige in governmental administration.

In general, the domain is defined by a person or persons in an organization and it identifies what is to be done, how it is to be done, and the clientele. Such definitions, if they are to be effective, must agree with the expectations of agents external to the organization who interact with it as contributors or claimants. The definition of domain appears to have an important impact on the growth potential of the organization and upon its vulnerability to external influence.

*Technology and Growth.* The domain which is established by the administrators of organizations determines the extent to which the products or services provided are *standard* or *nonstandard*. As mentioned, if the domain established is to be effective it must be consistent with the expectations of external contributors and claimants. The customers of McDonald's are likely to be satisfied if they do not expect a varied menu and gourmet food. Indeed, the survival of organizations seems to depend, on the one hand, upon creating efficiency of operations, or on the other hand, producing an outcome which is relatively made-to-order.

Organizations which survive on the basis of efficiency typically produce in high volume, with a predictable demand for their products, and pay great attention to the management of their capital. In contrast, organizations which survive on the basis of a made-to-order outcome will produce in units or small batches or complete single projects. Since they adapt to a wide variety of customer requirements, they must rely heavily upon meeting customer requirements and upon pricing their products or services so that an appropriate margin is maintained. Home builders, and manufacturers of special tooling, fashion items, or highly technical products are examples of such made-to-order enterprises.

The technology required to meet these objectives seems to vary more in kind than in degree. While many descriptions of technology have been offered, one of the more useful and universal is that of Thompson (1967). Thompson differentiated between standard and nonstandard technologies in his writing. The standard technologies which he identified were called "long-linked" and "mediating." As Thompson saw it, long-linked technologies involve a serial interdependence between activities, a fairly limited number of products, repetitive activities by those performing the direct labor, and a constant rate of performance by the organization. Long-linked technologies seek to perform at capacity, aided by a stable and predictable output and by protection of work activities from external interference. As Thompson pointed out, "[t]he constant rate of production means that, once adjusted, the proportions of resources involved can be standardized to the point where each contributes to its capacity; none need be underemployed" (p. 16).

Organizations with long-linked technology expand through vertical integration, e.g., through control of suppliers and sources of raw materials, and through control of market channels. Complexity of inputs or outputs may, however, serve as barriers to forward or backward integration.

While Thompson chooses to emphasize mass production as the primary focus of his discussion, it should be noted that his examples include steel producers, the educational domain of hospitals, and larger retailers. The chief attributes of these examples seem to be more the standardization and full utilization of resources than the mass production of goods.

The second standardized technology discussed by Thompson was called "mediating." Mediating organizations were described as linking clients or customers who wish to be interdependent. Banks, insurance companies, the post office, and employment agencies are offered as examples. Mediating technology requires that organizations operate in standardized ways, relying on rules and procedures. Organizations with mediating technologies also serve clients or customers which are widely distributed in time and space. When these organizations grow they attempt to increase the populations served through establishing multiple locations.

Like long-linked technology, the mediating technology is also characterized by a standard and often limited range of products or services, repetitive activities by direct labor, and attempts to minimize disruptions to efficient operations by fluctuations in demand. Long-linked technologies do this by isolating the technical core, by attempts to smooth out services in peaks and low demand, and by careful planning for forecasted changes in demand. Mediating technologies are in more direct contact with clients and are thus less able to buffer output activities. They can, however, engage in smoothing and forecasting.

A chief difference between Thompson's long-linked and mediating technologies is the direction of their expansion. Long-linked firms will expand vertically by gaining greater control of distribution or by controlling suppliers. Mediating technologies expand horizontally to new territories through the duplication of enterprises. In addition, it appears that the former lends itself to large scale enterprise because its technology can be isolated at both the input and the output side. In contrast, mediating technologies cannot remove themselves from direct contact with the client and must limit the size of their operations territorially.

The two technologies have much in common. First, they have detailed planning for contingencies and this planning is provided by clearly identifiable staff or administrative segments of the employee population. Second, they have a relatively narrow range of products or services provided with a relatively predictable volume of output. Third, their emphasis is upon efficiency through full utilization of resources. Finally, standardized technologies attempt to fit direct labor to a system, permitting the use of less expensive labor at lower levels of the organization. Relatively unskilled individuals are employed and trained to meet organization needs and may be advanced as their skills increase. The organization can tolerate turnover as long as a supply of new employees is available.

Thompson also identified a form of nonstandard technology. Calling it "intensive," he said that "intensive technology is a custom technology" (p. 18). This form operates through feedback from the object worked on and is applied either by skilled individuals working alone or in mixed-skill teams. This difference is not trivial, since the organization form varies between the two. Technologies which have skilled individuals working alone, as in a job shop, commonly group people doing the same kind of work in the same departments, but their efforts are distinctly nonrepetitive and require much individual judgment. In contrast, project groups are typically mixtures of skills, working on a project which is new or unique to the experience of the organization, with a definite end outcome which determines the success or failure of the project. Research teams, therapeutic teams, and construction projects often meet these characteristics (Delbecq & Filley, 1974).

Both individual efforts and team efforts in nonstandard technologies have much in common. First, the ability to plan for contingencies is limited. People doing the work are highly skilled and make adjustments in the process as deviations occur. Secondly, products or services have uniqueness about them which precludes mass production or standardization. Third, machines and people are justified on the basis of their contribution to the desired outcome rather than on the basis of full utilization. Finally, both require the acquisition of skilled personnel whose skills are difficult to develop and relatively transferable within the organization and to other employers as well.

It is difficult to believe that standard and nonstandard technologies differ merely by degree. Anyone who has observed the two must surely see this. In standardized technologies every effort is made to keep the operations in balance and functioning smoothly. Machines and people are single-purpose and change infrequently. Operators are employed, trained, and advanced as skills increase. The separation between planners and doers is marked. Power is centralized at the top of the organization, though decision making may be decentralized. Schedules are set, with necessary contingencies, and are followed with precision.

In contrast, nonstandard technologies appear to be confused and inefficient. In job shops, for example, machines are idle, sometimes for weeks. Skilled personnel use general purpose machinery and exercise personal judgment about how to do what. Expediters are busy determining or changing work progress and supervisors do their best to keep work moving. Employees will complain if necessary tools and equipment are not present or if they are not considered to be adequate by their own standards. Reeves & Turner (1972) described the complexity of information processing under such conditions. Some scheduling is done initially to insure adequacy of raw materials and more scheduling is done on the shop floor. Programs listing availability of raw materials are used to pull work through the system and problems are handled by daily meetings to discuss shortages and progress. Reeves & Turner added that in custom shops no one assumed that operations and their control were efficient, only that the best possible job was being done to keep machines and people busy.

It may be seen that for standard technologies, growth poses fewer problems than for nonstandard technologies. As was pointed out earlier, larger size firms have less variance in their operation. As a result, growth is advantageous for firms with standard technologies. For nonstandard technologies, however, growth poses serious problems. Coordination and information processing problems grow exponentially with size, suggesting either a need to duplicate facilities in small independent units in manageable proportions or to change the domain to standardized methods (Galbraith, 1973).

*Coordination and Growth.* The effects of technology on growth may be seen if one considers the forms of coordination which may be used. Drawing on the work of March & Simon (1958), Thompson (1967) and Galbraith (1977) described three forms of coordination: standardization, by plan, and by mutual adjustment.

Standardization involves establishing rules which are internally consistent and apply to situations which are relatively stable and repetitive. For example, such rules might specify times for people to begin or stop work. While rules govern the behavior of individuals, they also allow individuals to predict the behavior of other members of the organization. In this manner, a stoplight tells an individual when to go or stop and also allows him to anticipate the behavior of other drivers (Filley, 1976).

The second basis for coordination is by plan, which together with rules is feedforward control. Essentially, coordination by plan says, if or when this happens, do this. Thus, staffing, expenses, or inventories may be adjusted as sales change, without waiting for errors to occur. Such planning requires predictability and calibration for contingencies to be accounted for and, as we have earlier stated, suffers from an inability to deal with unusual situations.

The third basis for coordination, which Thompson calls mutual adjustment, is feedback control. As he states, "the more variable and unpredictable the situation, . . . the greater the reliance on coordination by mutual adjustment." Such actions are seen in the job shop through constant monitoring and adjustment by top management as well as by workers, the efforts of expeditors, the constant change in schedules, and the meetings of supervisors to discuss priorities. Thompson further asserts that these forms of coordination follow the complexity of an organization. In simple organizations standards may suffice; in more complex organizations, both standards and plans are needed; and in very complex organizations all three are required (pp. 55–56). If this is the case, then a growing organization will be overwhelmed with information-processing difficulties if it does not establish standing orders in the form of rules and contingency plans. Viewed in another way, nonstandard technology will have difficulty in dealing with growth: it will standardize what it can, forecast and plan anticipated contingencies, and then rely upon complex systems of mutual adjustment to handle the rest.

This discussion also suggests the different ways in which centralization and decentralization may be construed in standard and nonstandard technologies. Standard technology has power centralized at the top of the organization. Lower participants are trained to meet system requirements and are relatively easy to replace. On the other hand, particularly where the enterprise is large, effort will be made to decentralize decision making as feedforward controls permit. In contrast, nonstandard technologies have relatively more power at lower organization levels since such employees are skilled, mobile, and expensive to replace. Yet the decision-making responsibility of top management is great, since reliance on feedback controls necessitates constant redirection by top management.

The latter reasoning is quite consistent with Khandwalla's (1974) study of 79 manufacturing firms. He found that the more the technology of the firm was characterized by high volume and mass production and the larger the firm's size, the more it was vertically integrated. Further, the more vertically integrated, the greater the decentralization of decision making. Finally, the more decentralized, the more the firm used sophisticated controls for quality, costing, inventory, scheduling, finances, and the evaluation of executive performance. Equally important, more profitable firms tended to follow this pattern more than less profitable firms. Considering these findings from the standpoint of nonstandard technology, the information suggests that such firms do not integrate vertically and have more centralized control by top management. Sophisticated feedforward controls are simply not possible in such firms (Reeves & Turner, 1972; Reimann, 1973; Khandwalla, 1974).

If we consider the levels of coordination and the two classes of technology, some conclusions about barriers to growth are apparent. First, since standing orders in the form of rules reduce information processing, the failure to provide them will require reliance on personal direction and will inhibit growth. For example, unless there are rules for employee termination, each case will be handled on an ad hoc basis. When handled on an ad hoc basis, such cases must be considered at higher organization levels, using time better spent on nonrepetitive issues. Next, unless contingency plans are developed to integrate operations and avoid errors, further executive time must be used. Standard technology will make such contingency planning more possible than nonstandard operations, but the potential for growth will be limited in either case if such plans are not developed. This logic is consistent with the typology discussed earlier. The craft and promotion types exhibited personal control by the chief executive; the administrative type exhibited a variety of systems (budgets and plans) for control purposes.

*Organization Structure and Growth.* There appear to be two fundamental designs for organizing, which apply to total organizations or to units within total organizations. The first pools relatively large numbers of people doing the same kind of work in the same organization unit. The second basic strategy for organizing is one which organizes teams of people doing different kinds of work into joint effort. To date, these two structural forms have been largely undifferentiated from standard and nonstandard technology. As a result, discussions have generally dealt with two generic forms of organization design: one which enhances efficiency and one which enhances creativity (e.g., Walker & Lorsch, 1968; Galbraith, 1971; Swinth, 1974; Filley, House, & Kerr, 1976). As Swinth states, ". . . the organizational design for achieving efficiency is different than the design for achieving adaptability. . . . the demand for some combination of efficiency and adaptiveness potentially generates incompatible design needs" (pp. 79–80). The designs which are proposed for efficiency pool relatively large numbers of people doing the same kind of work in the same organizational unit. The consequent structure is generally called "functional." The designs which are proposed for creativity or adaptability have mixed skill groups working together as teams or project groups. This structure is generally called a "team" or "goal-oriented" organization.

While this dichotomy is approximately correct, it does not explain functional structures which do creative work (e.g., job shops) nor does it explain mixed skill groups which are standardized (e.g., fast food chains, fast copy services). As shown in Figure 2, these may be accounted for if the technical process alternatives (standard or nonstandard) are distinguished from job grouping alternatives (homogeneous groupings versus heterogeneous groupings of jobs in organization units). For the present, we shall concentrate on two combinations: standard technology and homogeneous organization units, and nonstandard technology and heterogeneous units. For convenience we shall retain the labels of "functional" and "team" structure to represent the two.

The consequences of each may be illustrated as follows: Seiler (1967) reports a case study of the Utility Power Company in which regional offices had teams of engineers and draftsmen which were responsible for design and cost estimates of equipment for substations. Later, the organization was redesigned pooling the draftsmen in a single department for more standard operation and organizing several groups of engineers. The second design was more efficient in making greater use of manpower but was less creative and adaptive. The new arrangement was also less satisfying to individuals doing the work. Eventually, the structure was redesigned in the old form. In short, the structure went from team to functional and back to team.

*Figure 2.* Effects of Job Grouping and Technical Process on Organizational Structure

| Job Grouping | Technical Process: Standard | Technical Process: Nonstandard |
|---|---|---|
| Homogeneous | Functional | Jobbing |
| Heterogeneous | Process | Team |

Similarly, a research and development division might be organized into departments of data processing, chemistry, and engineering (functional); or into project groups containing a mixture of skills (team). Or, the secretarial services of an enterprise might be provided through a secretarial pool (functional) or through individual private secretaries (team). Finally, the sale of paint and hardware might be provided through larger retailers, e.g., Sears (functional) or through small home decorating centers (team).

A major difference between the two alternative structures is that the functional structure is designed for efficiency in processing its product while the team structure is designed for adaptability or creativity to achieve necessary outcomes. The reason for the efficiency of the functional structure and the creativity or adaptability of the team structure may be explained as follows (Filley, House, & Kerr, 1976). In the functional organization resources are used fully. Personnel or equipment are typically added when they can be justified on the basis of full-time utilization. People within the department can be divided into subspecialties and can be advanced within their specialty (Galbraith, 1971). Relatively unskilled people may be employed and advanced as their skills increase. By working in the same department as others doing similar work, people can learn additional skills from each other.

The supervision of functional departments is by specialists in the function who can evaluate the professional or technical competence of people within the unit. The functional organization does not provide experience in general management, however, since only the chief executive is required to manage a variety of functional units. Functional organizations typically have a very formal organization and relatively inflexible operations. Employees are likely to take a short-term view of their work since they focus on day-to-day job requirements (Walker & Lorsch, 1968). Conflict between departments is common. Departments are mutually dependent on each other and scheduling and coordination between them is difficult. When size permits, functional organizations employ relatively large numbers of staff specialists to develop and utilize systems of feedforward control.

A team structure, on the other hand, has quite different characteristics. Unlike the functional organization, a team structure is adaptive and creative, not especially efficient. Equipment and personnel may be duplicated between different teams because of justified need, not full utilization. Team units require all members to be highly skilled in their profession or specialty since they provide the complete resource base required; skills are not subdivided and unskilled personnel are rarely trained. Such specialists cannot advance in their specialty nor can they gain knowledge from others in their specialty.

Each manager of a team unit is a generalist or professional manager who serves to integrate the efforts of the various specialties. Structure within teams is relatively informal. If successful, team members take a relatively long-run view of their efforts since they are concerned with outcomes—construction of a building, completion of a piece of equipment, release of an advertising campaign, completion of a feasibility study, etc. On clear-cut projects teams meet schedules effectively; in complex organizations with shifting priorities, however, much effort must be exerted to keep priorities under control. Thus, multiple team structures will minimize organization levels to permit central coordination (Swinth, 1974, p. 82). Performance of teams is evaluated on the basis of outcomes, rather than process.

The consequences of growth are quite different in functional and team structures. In general, the increase of size in a functional organization provides stability of operations, the use of more sophisticated plans and controls, greater control of the environment, and the like. In contrast, the growth of a team structure creates difficulty since standing orders and feedforward controls cannot cover contingencies. Even if a larger project can be subdivided into subprojects, the necessity of feedback control and complex information processing makes growth a hardship. Complex organizations called "matrix structures" have been developed to contain multiple project groups but they are expensive to operate and relatively rare.

We may now consider the other two cells in Figure 2: standard technology in heterogeneous structural units, and non-standard technology in homogeneous units.

Standard technology in heterogeneous job groupings will occur in cases where localized outputs or inputs to the firm's technology are necessary for the success of the organization. The operations are standardized but different jobs are grouped together. People in the organization follow specified procedures and such people are relatively easy to train or replace. We shall identify this form as a "process" design. One example of a process firm is a manufacturer of mobile home furniture which at one time had over twenty small plants. Plant locations were adjacent to mobile home manufacturers and output of furniture went directly into the plant of the customer at a rate dictated by the customer. The mobile home manufacturer thus minimized working capital needs and the furniture plant had guaranteed sales. A second example of a process design is a chain of hamburger stands. Local convenience and standardized operations provide profitable business that would not be possible with a large single location. Growth of the enterprise depends upon extending services geographically through additional locations.

As shown in Figure 2, the fourth design is one dictated by nonstandard technology and homogeneous job groupings. The job shop is a convenient example and the design might conveniently be labled as "jobbing." Jobbing operations typically group similar work into departments. Employees are skilled and work processes depend upon feedback from the object worked on. For example, a job shop design in one company known to the authors contained sales, grinding, and spray coating departments. Each item worked on had unique requirements and would be spray coated according to need and then ground to specification by skilled workers. As the company grew, the difficulties with coordination generated discussions of either opening another plant or of standardizing operations to a more limited range of technology and becoming a production shop.

In general, it may be seen that a choice of domain involves choosing a technology and choosing a structure. If the organization requires efficiency as the basis for survival then it will require either a functional or a process design. If both the inputs and outputs of the technology can be buffered, it will benefit by having a large enterprise. If the input side or the output side of operations involve direct contact with suppliers or customers, then multiple small enterprises will be dictated.

If, on the other hand, the enterprise survives on the basis of creativity and adaptiveness, it will require nonstandard operation and either a jobbing or a team form. If items can be worked on serially, a jobbing operation is appropriate. If teams of people must work together in some way, a team structure is more effective. Jobbing operations will be limited in size by the limits on information processing requirements. Teams also have size constraints though they may be housed in a matrix structure.

## SUMMARY

The preceding discussion has indicated several factors which will influence the form and growth of organizations. We described three types of organization—craft, promotion, and administrative—which affect the way in which an organization deals with its environment. The first two are controlled by individuals rather than systems. The craft firm performs its functions in a traditional, nonadaptive way. The promotion firm exploits a short-run advantage. In contrast, the administrative type relies upon systems for planning and control. Such systems help to overcome limitations placed upon growth by personal control. In other words, when an organization has no formalizing methods such as standards and contingency plans, its entire operation depends upon feedback control.

Both theoretical discussions and empirical findings support the existence of such types. Similarities among the typologies reported by a large number of writers suggest a homology among levels of human organizations. Studies by the authors and others have revealed factor structures consistent with this notion and have shown type scores to generally relate in expected ways to leader characteristics, technological complexity, and other variables. Further, available evidence indicates that organization types are not a simple function of size or technology. Instead, we have argued that the role of leadership form is pervasive. Figure 3 summarizes that argument.

Organization leadership, in conjunction with expectations of relevant external agents, is seen as defining the domain of the organization—the specification of what is to be done, how it is to be done, and for whom. Where such a specification calls for standard organizational outputs, demands for efficiency are great. Where specified products or services are nonstandard, flexibility is required.

These differing objectives of efficiency and flexibility require technologies varying in kind rather than just degree. Standard technologies emphasize efficiency through full utilization of resources. Nonstandard technologies deemphasize efficiency to meet demands for flexibility. Similarly, the degree of homogeneity of structural units may be seen as a response to relative demands for efficiency as opposed to adaptability.

Where technology is standardized, subsequent growth will necessitate the simplification of information processing by providing rules which apply to everyone, and by the use of contingency planning. Where technology is nonstandard, subsequent growth depends upon the use of

*Figure 3.* Determinants of Organization Types and Behavior

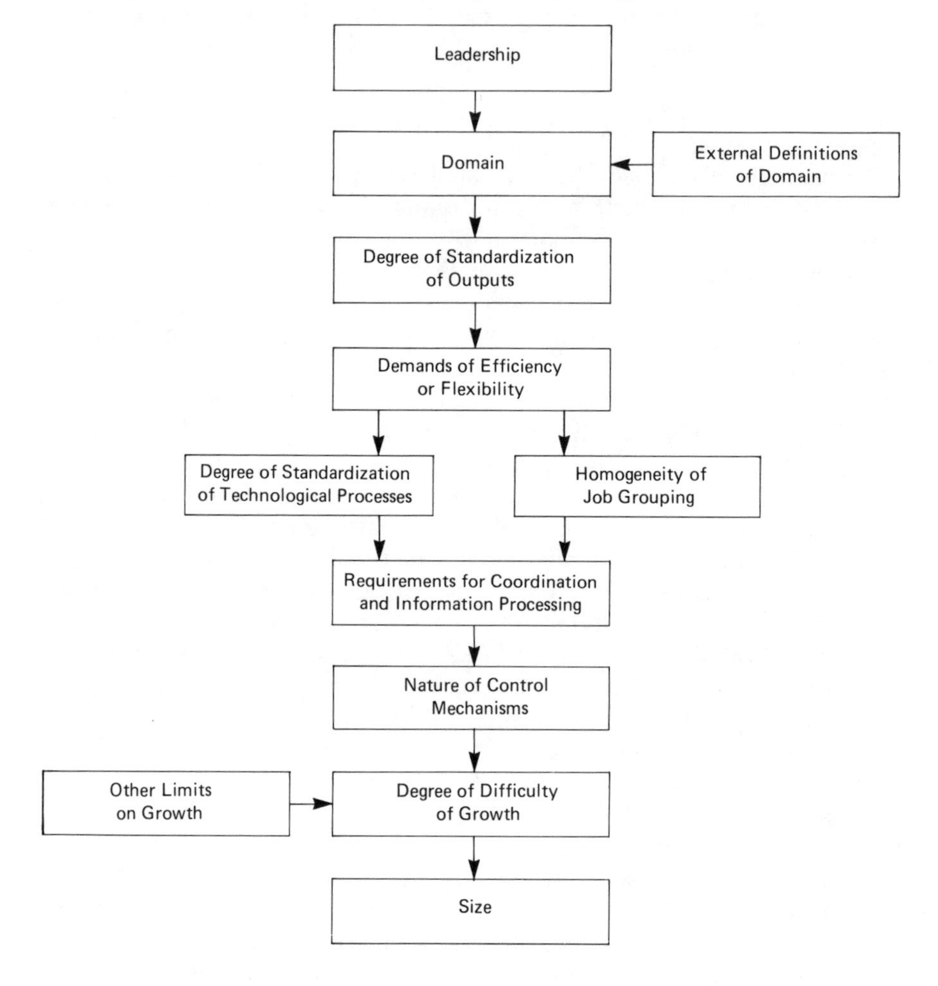


standards and contingency plans, plus the use of feedback controls by skilled personnel.

If inputs and outputs of the technology can be buffered, the structural form of standard technology will be functional, with growth in scale and vertical integration. If the need for close contact with suppliers or customers demands the use of small process units, growth will occur through duplication of standard units. In either case, growth offers continued advantages to such standardized operations, since they can operate with greater efficiency and economy.

The structural form of nonstandard technology is either through the homogeneous groupings of skilled employees in jobbing operations or through the use of teams. In either case, the purpose of the organization is to adapt to varying client needs with made-to-order technology. Team size will be limited to the ability of the group to function effectively together, though multiple teams may be established in matrix structures. If joint effort is not required, then the information processing requirements of jobbing operations will place increasing strains on performance as the organization grows. Thus, job grouping and technological process impact on requirements for coordination and information processing and on resultant standards and control mechanisms. The use of feedforward controls facilitates growth while the need for feedback controls constrains such increases in size.

The term "organization type" does not appear in Figure 3. Instead, as suggested above, the term refers to the configuration of characteristics defined by the sequence presented in that figure. The authors have argued elsewhere (Filley & Aldag, 1976) that recognition of a typology of organizations has important implications for the design and interpretation of research. The views presented in this paper illustrate that the notion of organization type lends useful insights into the issue of organization growth. That is, organization growth, and consequently size, are seen to be a consequence rather than a cause of organization type. Further, the arguments we have detailed suggest that organization leadership, in contrast with expressed attitudes of many management writers (cf. McCall & Lombardo, 1978), is neither passive nor impotent. Instead, through domain definition and the subsequent causal chain, leadership is the key determinant of the constellation of characteristics referred to here as type. A focus on leaders of small firms, rather than on managers of large organizations which have already completed the process of institutionalization, has permitted a more adequate specification of that determining role.

## FOOTNOTES

1. The literature search for this article included PSYCH (psychology and sociology) and ERIC computer banks plus a manual review of the psychological and sociological abstracts. Index words for computer searches were: industrial structure; management systems; organizations (groups); leadership qualities; organizational structure; management style; qualities; growth; development; industrial management; change; promotion; business organizations; organizational development; success; objectives; entrepreneur; administrative. These yielded 1889 references of which approximately 11 had to do with organizational growth and 3 with organization types. Other references on organization design were of course useful.

2. Such examples raise the question of how such individuals were able to influence very large organizations when personal control should constrain growth, and when both logic and evidence suggest that this is often the case. There seems to be agreement that, however it is managed, such leaders provide organizations which are distinctly nonbureaucratic (Weber, 1947; Tucker, 1968). Various explanations for overcoming size constraints have been of-

fered: (1) the promoter may change his leadership style; (2) the promoter may extend his capacity through the use of personal assistants, e.g., Jesus and the twelve disciples (Haley, 1969); and (3) the charisma of the leader may be institutionalized (Weber, 1947).

## REFERENCES

Aldrich, H. E. (1972) "Technology and organizational structure: A re-examination of the findings of the Aston Group," *Administrative Science Quarterly, 17,* 26–43.

Bandura, A. (1968) "Social learning theory of identificatory process," in D. A. Goslin (ed.) *Handbook of Socialization Theory and Research,* Chicago: Rand McNally.

Bennett, J. W. & R. W. McKnight (1956) "Approaches of a Japanese innovator to cultural and technical change," *Annals of the American Academy of Political and Social Science, 305,* 101–114.

Berlew, D. E. (1974) "Leadership and/or organizational excitement," in D. A. Kolb, I. M. Burbin, and J. M. McIntire (eds.) *Organizational Psychology,* Englewood Cliffs, New Jersey: Prentice Hall.

Blau, P. M., & W. R. Scott (1962) *Formal Organizations,* San Francisco: Chandler Press.

Boulding, K. E. (1953) "Toward a general theory of growth, *Canadian Journal of Economics and Political Science, XIX, (3),* 326–340.

Burns, T. & G. M. Stalker (1961) *The Management of Innovation,* London: Tavistock.

Carneiro, R. & S. Tobias (1963) "The application of scale analysis to the study of cultural evolution," *Transactions of the New York Academy of Sciences, Series II, 26,* 196–207.

Caves, R. E. (1970) "Uncertainty, market structure and performance: Galbraith's conventional wisdom," p. 283–302, in J. W. Markham and G. F. Papanek (eds.) *Industrial Organization and Economic Development,* Boston: Houghton-Mifflin.

Charan, R. (1976) "A model for change from entrepreneur to professional management," paper delivered at the 12th annual meeting of the Midwest Business Administration Association, April 1, 1976, St. Louis, Missouri.

Child, J. (1973) "Predicting and understanding organization structure," *Administrative Science Quarterly, 18,* 168–185.

Coase, R. M. (1952) "The nature of the firm," pp. 331–351, in G. J. Stigler and K. E. Boulding (eds.), *American Economic Association Readings and Price Theory,* Chicago: Irwin.

Collins, L., & D. LaPierre (1975) *Freedom at Midnight*, New York: Simon and Schuster.

Collins, O. F., D. G. Moore, & B. B. Unwalla (1964) *The Enterprising Man,* East Lansing: Michigan State University.

Dale, E. (1960) *The Great Organizers,* New York: McGraw-Hill.

Davis, R. C. (1951) *The Fundamentals of Top Management,* New York: Harper and Row.

Delbecq, A., & A. C. Filley (1974) *Program and Project Management in a Matrix Organization: A Case Study,* Monograph #9, Bureau of Business Research and Service, University of Wisconsin-Madison.

DeSpelder, B. E. (1962) *Ratios of Staff to Line Personnel,* Research Monograph 106, Bureau of Business Research, Ohio State University.

Etzioni, A. (1961) *A Comparative Analysis of Complex Organizations,* New York: Free Press.

Feldbaumer, W. C. (1973) "The management of an entrepreneurial crisis," Ph.D. Dissertation, College Park, University of Pennsylvania.

Filley, A. C. (1976) "The parable of the traffic light: On the uses and misuses of rules," *M.S.U. Business Topics, 24,* 57–59.

———. (1962) "A theory of small business and divisional growth," Ph.D. Dissertation, Columbus, Ohio State University.

Filley, A. C., & R. J. Aldag (*in press*) "Characteristics and measurement of an organizational typology," *Academy of Management Journal.*

———, & R. J. Aldag (1976) "Implications of an Organizational Typology," *Proceedings of the 19th Annual Conference of the Midwest Division of the Academy of Management,* St. Louis, pp. 1–12.

———, & R. J. House (1969) *Managerial Process and Organizational Behavior*, Glenview, Illinois: Scott, Foresman and Company.

———, R. J. House, & S. Kerr (1976) *Managerial Process and Organizational Behavior*, Second Edition, Glenview, Illinois: Scott, Foresman and Company.

Freeman, J. & M. T. Hannan (1975) "Growth and decline processes in organizations," *American Sociological Review, 40,* 215–228.

Galbraith, J. R. (1971) "Matrix organization designs," *Business Horizons, 14,* 29–40.

———. (1973) *Designing Complex Organizations*, Reading, Mass.: Addison-Wesley Publishing.

———. (1977) *Organization Design*, Reading, Mass.: Addison-Wesley Publishing.

Gort, M. (1963) "Analysis of stability and change in market shares," *Journal of Political Economy, 71,* 51–63.

Haas, J. E., R. H. Hall, & N. J. Johnson (1966) "Toward an empirically derived taxonomy of organizations," in R. V. Bowers (ed.) *Studies on Behavior in Organizations*, Athens, Georgia: University of Georgia Press, 157–180.

Haire, M. (1959) (ed.) *Modern Organization Theory,* New York: Wiley.

Haley, J. (1969) *The Power Tactics of Jesus Christ and Other Essays,* New York: Grossman.

Hall, R. H. & C. R. Tittle (1966) "A note on bureaucracy and its correlates," *American Journal of Sociology, 72,* 267–272.

Hoad, W. M. & T. Rosko (1964) *Management Factors Contributing to the Success or Failure of New Small Manufacturers,* Michigan Business Reports, No. 44, Ann Arbor, Michigan: Bureau of Business Research, Graduate School of Business Administration, University of Michigan.

Hoffer, E. (1958) *The True Believer,* New York: The New American Library of World Literature.

House, R. J. (1976) "A 1976 theory of charismatic leadership," Southern Illinois University, Fourth Bi-Annual Leadership Symposium, October 26–28, 1976, Carbondale, Illinois.

Kaldor, N. (1960) *Essays in Value and Distribution,* Chapter 2, London: Duckworth.

Khandwalla, P. N. (1974) "Mass output orientation of operations technology and organizational structure." *Administrative Science Quarterly, 19,* 74–97.

Kimberly, J. R. (1976) "Contingencies in the creation of organizations: an example from medical education," Copenhagen: Joint Eiasm/Dansk Management Center Seminar on entrepreneurs and the process of institution building.

Lawrence, P. R. & J. W. Lorsch (1967) *Organization and Environment: Managing Differentiation and Integration,* Cambridge, Massachusetts: Division of Research, Graduate School of Business Administration, Harvard University.

Lehman, H. C. (1953) *Age and Achievement,* Princeton, New Jersey: Princeton University Press.

Lester, R. A. (1958) *As Unions Mature,* Princeton: Princeton University Press.

March, J. G., & H. A. Simon (1958) *Organizations,* New York: John Wiley and Sons.

McCall, M. W., Jr., & M. M. Lombardo (1978) (eds.) *Leadership: Where Else Can We Go?* Durham, N. C.: Duke University Press.

McGuire, J. W. (1963) *Factors Affecting the Growth of Manufacturing Firms,* Bureau of Business Research, Seattle: University of Washington.

Meyer, M. W. (1975) "Leadership and organizational structure," *American Journal of Sociology, 81,* 514–542.

———. (1975) "Organizational domains," *American Sociological Review, 40*, 599–614.

Oliva, T. A., & M. H. Peters (1976) "Filley and Aldag's organization typology: A replication," *Proceedings of the Southeastern Meeting Society for General Systems Research.*

Pelz, D. C. & F. M. Andrews (1966) *Scientists in Organizations,* New York: Wiley.

Penrose, E. T. (1959) *The Theory of the Growth of the Firm,* New York: Wiley.

Pfeffer, J., & G. R. Salancik (1978) *The External Control of Organizations,* New York: Harper and Row.

Reeves, T. K., & B. A. Turner (1972) "Theory of organizations and behavior in batch production factories," *Administrative Science Quarterly, 17,* 81–98.

Reimann, B. C. (1973) "On the dimensions of bureaucratic structure: an empirical reappraisal," *Administrative Science Quarterly, 18,* 462–475.

Reisman, D., R. Glazer, & R. Denney (1950) *The Lonely Crowd,* New York: Doubleday.

Rostow, W. W. (1960) *The Stages of Economic Growth,* Cambridge: Cambridge University Press.

——— & M. Millikin (1957) *A Proposal,* New York: Harper and Brothers.

Samuels, J. M., & D. J. Smyth (1968) "Variability of profits, and firm size," *Econometrica, 35,* 127–139.

Schumpeter, J. A. (1935) "Analysis of economic change," *Review of Economics and Statistics, 17.*

Seiler, J. H. (1967) *Systems Analysis of Organizational Behavior,* Homewood, Illinois: Irwin.

Selznick, P. (1957) *Leadership in Administration,* New York: Harper and Row.

Starbuck, W. (1968) "Organizational metamorphosis," p. 113–122, in R. W. Millman and M. P. Hottenstein (eds.) *Promising Research Directions*, Academy of Management.

———. (1965). "Organizational growth and development," in J. G. March (ed.), *Handbook of Organizations*, Chicago: Rand-McNally, 451–533.

Steindl, J. (1945) *Small and Big Business,* Oxford: Blackwell.

Swinth, R. L. (1974) *Organizational Systems for Management: Designing, Planning, and Implementation.* Columbus, Ohio: Grid.

Taylor, R. N. & M. D. Dunnette (1974) "Relative contributions of decision-maker attributes to decision process," *Organizational Behavior and Human Performance, 12,* 286–298.

Thompson, J. D. (1967) *Organizations in Action,* New York: McGraw-Hill.

Tucker, R. C. (1968) "The theory of charismatic leadership," *Daedalus, 97,* 731–756.

Tustin, A. (1955) "Feedback" in *Automatic Control,* New York: Simon and Schuster.

Udy, S. H. (1958) "Bureaucratic elements in organization," *American Sociological Review, 23,* 415–418.

U. S. Small Business Administration (1977) *The Study of Small Business*, Washington, D.C.: Office of Advocacy, U. S. Small Business Administration.

Vroom, W. H. & B. Pahl (1971) "Relationship between age and risk taking among managers," *Journal of Applied Psychology, 55,* 399–405.

Walker, A. H. & J. W. Lorsch (1968) "Organizational choice: Product vs. function," *Harvard Business Review, 46,* 129–138.

Ware, R. F. (1975) "Performance of manager vs. owner-controlled firms in the food and beverage industry," *Quarterly Review of Economics and Business, 15,* 81–92.

Weber, M. (1947) *The Theory of Social and Economic Organization,* translated and edited by A. M. Henderson and T. Parsons, London: Oxford University Press.

Wickesberg, A. K. (1961) *Organizational Relationships in the Growing Small Manufacturing Firms,* Minneapolis: University of Minnesota.

Wilensky, H. (1957) "Human relations in the work place; an appraisal of some recent research," in *Research in Industrial Human Relations,* New York: Harper and Row.

Woodward, J. (1965) *Industrial Organization,* London: Oxford University Press.

# INTERORGANIZATIONAL PROCESSES AND ORGANIZATION BOUNDARY ACTIVITIES[1]

J. Stacy Adams

UNIVERSITY OF NORTH CAROLINA AT CHAPEL HILL

ABSTRACT

Organizations are components of macroscopic ecological systems. Their behavior and that of their subsystems cannot be explained without reference to the behavior of their environment and to their interactions with the environment. The organization-environment interactions occur at organization boundaries, which are preponderantly activities undertaken by persons.

As activities, organization boundaries are the functional linkages between an organization and elements of its environment. Consequently, boundary activities are discussed in the context of organization-environment interac-

Research in Organizational Behavior, Volume 2, pages 321–355

ISBN: 0-89232-099-0

tions. A general model of structural and dynamic characteristics of interorganizational relations is presented first, following which five essential classes of boundary activities are discussed. These are: 1. transacting the acquisition of inputs and disposal of outputs; 2. filtering inputs and outputs; 3. searching for and collecting information; 4. representing an organization to its environment; and, 5. protecting and buffering the organization from external threat and pressure. The organizational and role implications of the various boundary activities are examined.

## INTRODUCTION

*It is better to analyse in terms of doings or happenings than in terms of objects or static abstractions.*

*Percy W. Bridgman, 1959, p. 3.*

The study of organizations has been historically ruled by a number of paradigms. Most ignored Bridgman's sage enjoinder; many implied closed system characteristics. *Organization* is defined in this essay as a bounded, adaptive, open, social system that exists in an environment, interacts with elements of it, and engages in the transformation of inputs into outputs having effects on its environment and feedback effects on itself. So-defined human social organizations of all types are components of macroscopic ecological systems. Their behavior and that of their components cannot be explained without reference to the behavior of their environment and to their interactions with the environment. The organization-environment interactions take place at organization boundaries. More precisely, the loci of the interactions are wherever in space-time they occur because the boundaries of organizations are in significant proportion *activities* undertaken by persons (Adams, 1976; Katz & Kahn, 1978) as distinguished from physical structures such as walls, fences, and gates and from purely conceptual structures, exemplified by property and territorial boundaries.

As activities, organization boundaries—the plural form is significant, as will become apparent—are the functional linkages between any given organization and its environments (the plural is again intentional). It follows inescapably that an examination of boundary activities must be undertaken in the context of organization-environment interactions. It is useful to begin the task at the higher system level, the level of organization environments and interorganizational relations. A general model of structural and dynamic characteristics of interorganizational relations therefore will be presented first. Discussion of five classes of organization boundary activities will follow.

## STRUCTURE AND DYNAMICS OF INTERORGANIZATIONAL RELATIONS

The organizational literature is replete with theoretical and empirical, even prescriptive "armchair," articles and volumes focusing on *intra*organizational structure and processes, nearly to the exclusion of *inter*organizational considerations. The fact is astonishing in the light of organizational realities and the publication of Emery & Trist's (1965) and Terreberry's (1968) landmark papers and of the works of, for example, Burns & Stalker (1961), Forrester and his M.I.T. group (1961, 1971), Katz & Kahn (1978), Lawrence & Lorsch (1966), Miles, Snow, & Pfeffer (1974), Thompson (1967), and many economists.

A thorough treatment of interorganizational structure and dynamics is neither possible nor necessary in this essay; a general model of principal features of organization-environment structure and dynamic processes is, however, important as a context for the discussion of organization boundary functions. General aspects of structure are discussed first since, as will become evident, interorganizational dynamic processes are partially dependent on the structure of relationships among organizations.

### *Interorganizational Structure*

A given organization *F*, exists in a number of environments consisting of elements which in some degree and manner affect it. For present purposes, let us consider only elements that are other organizations. Little is lost by this simplification because, with only a moment's pause, one concludes that most nonorganization elements are in some measure under the control, or at least the influence, of organizations. Most individuals are members of one or more organizations that typically exercise some influence on some of their behavior, which is not to say that persons do not ever behave independently. Most land is under control of some organization or state; vast sea and air space is subject to the control of sovereign nations. Infrastructures are controlled by organizations or governments.

The environments of *F* (for "focal," in Figure 1) may be ordered in relation to interorganizational distance. One recalls in this context Bavelas' (1950) centrality measure of positions in communication networks. The first-order environment consists of organizations that interact directly with *F*; they correspond to Evan's input and output organization sets (1966). Second-, third-, and higher-order environments are composed of organizations the influence of which on *F* is mediated by one, two, or more other organizations. In Figure 1, *A* and *E* are in the first-order environment of *F, B, C,* and *D* in the second-order, and *G* in the third.

*Figure 1.* Interorganizational structure of organization *F* and organizations in its environments.

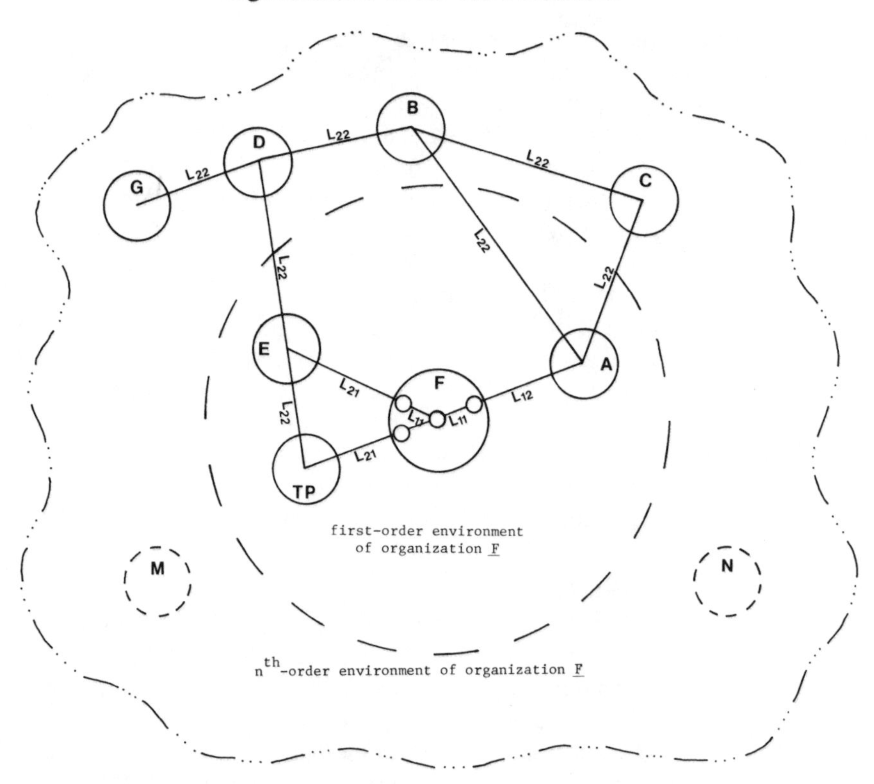


There is no implication in the ordering or environments for the degree of influence exercised by organizations on *F* (consider the immense effects of Saudi Arabian oil producers upon a small local fuel distributor, as well as upon numerous organizations intervening), but the ordering represents one measure of interorganizational distance from *F*, as well as a configurational aspect of interorganizational structure. The organization denoted as *TP* represents a "third party" interventionist organization, such as a mediation, arbitration, or regulatory agency. The organizations labeled *M* and *N* "exist" in objective reality, but, having no direct or indirect relationship to or influence on *F*, are not a part of *F*'s environment. The notation of Emery & Trist (1965) is useful in further describing the structure of interorganizational relationships and has been adopted in Figure 1. *L* designates a relationship between two organizations; the suffixes 1 and 2 refer to the focal organization, *F*, and to organizations in *F*'s environment, respectively. The serial order of the suffixes indicates the direction of influence: *From* a first-order environmental organization *upon F* ($L_{21}$) or *from F* upon an organization in *F*'s first-order environment ($L_{12}$). $L_{22}$ relationships identify influential connections between organizations in the $n^{th}$-order environments of *F*. It is the great contribution of Emery & Trist (1965), and later of Terreberry (1968), to have pointed out the singular importance of $L_{22}$ relationships: They are the interorganizational links that give the environments of organizations their potential "causal texture." More precisely, $L_{22}$ relationships make possible the existence of closed loop, mutual causal processes (Adams, 1975, 1976; Maruyama, 1963), about which more will be said later when dynamic interorganizational processes are discussed. $L_{11}$ relationships are intraorganizational. They link the components and subsystems of an organization, such as the managerial and boundary subsystems. A schematic representation of the four types of relationships are illustrated in Figure 1.

*Interorganizational Process Dynamics*

The purely structural characteristics of relations among a focal organization and organizations in its environments have implications for *potential* interorganizational process characteristics if the following conditions are met:

1. Each organization has at least one variable attribute that is causally related to at least one variable attribute in at least one other organization;
2. The direction of causation between pairs of variable attritutes (or clusters of variable attributes) is known;
3. The sign of the causal relation (as in a correlation coefficient) between pairs of variable attributes (or clusters) is known;
4. At least one pair of linked organizations includes the focal organization.

When these conditions are met, it is possible to determine the *general* nature of dynamic processes in the set of structurally related organizations that comprise a focal organization and its environments. The *specific* character of processes requires identification of the attributes and quantification of causal effects. The general conditions are met for illustrative purposes in Figure 2, the structural configuration of which is identical to Figure 1.

Figure 2, in contrast to Figure 1, has "causal texture," to borrow the language of Emery & Trist (1965) and Terreberry (1968). It is quite different, however, from the texture diagrammed by Terreberry for a system in a turbulent field environment; hers does not contain closed causal loops. What is the general causal texture illustrated? It may be determined by "tracing" the causal relations among the organizations, following the directions indicated. Beginning quite arbitrarily with the

*Figure 2.* Interorganizational causal processes among organization *F* and organizations in its environments.

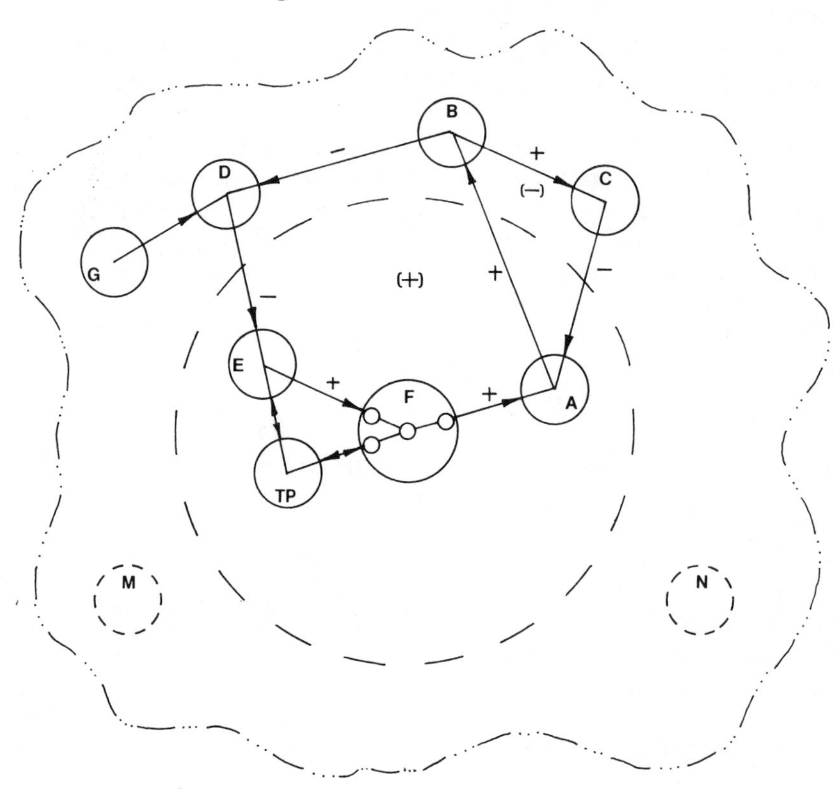


focal organization, *F*, and disregarding signs for the moment, we note that *F, A, B, D,* and *E* are causally and continuously connected. Since *E* and *F* are similarly connected, these five organizations form a closed loop of mutual causal processes, *FABDEF*. Consequently, the effects of a change in the relevant attribute of any organization in the loop may affect the four other organizations and, furthermore, may do so cycle after cycle. Organizations *A, B,* and *C* are causally linked in a similar manner, forming the loop *ABCA,* and it is evident that this closed loop of relationships will influence processes in the *FABDEF* loop. Finally, a causal influence of *G* on *D* is observed. However, since there is no causal influence on *G* by any other organization, no mutual causal processes involving it are possible, although *G* may have an effect on the mutual causal relations mentioned previously. It might, for example, trigger a change in *D*. Organization *G* is exogenous to the system of interrelated organizations *A, B, C, D, E,* and *F*. To that extent, loops *FABDEF* and *ABCA* are not truly "closed."

The nature of causal processes may be further defined if the signs (which may be loosely interpreted as correlation coefficient signs) that describe the causal relation between the variable attributes of two organizations are taken into account and the algebraic sign multiplication rule is used to assess the aggregate effects of all pair relationships in a particular loop. A positive result from that rule indicates that the processes in a closed loop are amplified—that a change in an attribute that is begun in an organization will be amplified in a cycle—a form of positive feedback. A negative result indicates attenuation or damping of such a change, as in negative feedback. In Figure 2, mutual causal processes are amplifying in the *FABDEF* loop and damping, or counteracting (Maruyama, 1963), in the *ABCA* loop. The fact that the causal relation between organizations *A* and *B* is common to both loops *FABDEF* and *ABCA* implies, in this illustration, that the amplifying characteristics of *FABDEF* are attenuated or nullified by the counteracting effect of *ABCA*. Of course, many more mutual causal process loops of both types and many more related organizations—some not an integral part of closed mutual causal process loops—might be observed in the study of a particular focal organization and its environments. Particular cases are not significant here; the vital, general case is: A given organization that is linked to other organizations in mutual causal process loops is *both subject to and a cause of the process effects*. To borrow the language of Emery & Trist (1965), organizations dynamically part of mutual causal processes are partially captives of the "turbulent field" forces generated by the *set* of interconnected organizations.

Special note may be taken of *TP*, the interventionist organization alluded to earlier. It is shown as having causal relations to both *E* and *F* in Figure 2; it might, in fact, have a direct influence on either or both. Whatever the particular case, the function of *TP*, as conceived here, is to modulate purposefully the functional relationship between *E* and *F*, for example the effects on *F* of price increases initiated by *E*. Government wage and price controls are intended to have this very type of effect precisely because it is known that organizations are linked in amplifying closed loops, as in economic inflation. (As a prime interventionist in such matters, the federal government, especially the Congress, conveniently ignores that it is a part of the inflationary loop and a major contributor to price inflation.)

Interactions between any organization and organizations in its first-order environment ($L_{12}$ and $L_{21}$ relations) entail boundary activities (Adams, 1976; Aldrich & Herker, 1977; Katz & Kahn, 1978; Keller & Holland, 1975; Miles, 1979). Interactions among organizations in a focal organization's $n^{th}$-order environments also require boundary activities, especially, as will be shown later, in intelligence collection. The frequency of these activities increases as the number of first-order interorganizational relations increases, either as the number of organizations dealt with directly increases, the number of transactions per organization increases, or both. Furthermore, if interacting organizations are connected in mutual causal process loops, the rate of their boundary activities will tend to increase from that fact alone since a change in a relevant causal attribute in any organization in the set will induce effects in first- and $n^{th}$-order environment members of the set, possibly through many cycles. Attempts by an organization to reduce the frequency of boundary transactions by such means as cooptation, long-term contracts, and control of supply (Katz & Kahn, 1978; March & Simon, 1958; Pfeffer & Salancik, 1978; Thompson, 1967) generate their own boundary activities and may, paradoxically, induce different mutual causal processes, which, since these processes feedback on the organization, may require adaptive boundary activities anew.

There is little disagreement that complex organizations in modern societies operate in what Emery & Trist (1965) classified as turbulent field environments. It is this writer's contention that the essence of turbulent environments is the presence of mutual causal process linkages, especially of the amplifying type, among organizations. If this be so, the significance of understanding boundary activities looms very large. These activities are discussed next.

## ORGANIZATIONAL BOUNDARY ACTIVITIES

Organizational boundary activities, suggestively described by some authors as boundary-spanning, have been given currency in the literature of the past few years (Adams, 1976; Aldrich & Herker, 1977; Katz & Kahn, 1978; Miles, 1979; Miller & Rice, 1967; Organ, 1971; Pfeffer & Salancik, 1978; Starbuck, 1976; Thompson, 1967; to cite only quite recent references). A concise comprehensive definition is lacking, although there is agreement that the principal role activities of negotiators, receptionists, salesmen, purchasing agents, police officers and guards, consumer complaint clerks, information gatekeepers, lobbyists, and the like are of an organizational boundary nature. There emerges from the literature the following general definition of *boundary activities*: activities of members or agents of an organization that serve to functionally relate the organization to its environments. Five classes of such activities are distinguishable: (1) transacting the acquisition of organizational inputs and the disposal of outputs; (2) filtering inputs and outputs; (3) searching for and collecting information; (4) representing the organization to its external environments; and (5) protecting the organization and buffering it from external threat and pressure.

Although conceptually distinct, the activity classes are often correlated as a matter of fact. Transacting exchanges frequently entails a simultaneous, continuous search for information, and intelligence gathering may require the purchase of information. Negotiating agreements inevitably shapes opinions and may partly be founded on the formation of impressions. Filtering activities are likely to be partly determined by the outcome of transactions. Whether in mixed or pure form, the activities constitute a significant part of an organization's boundaries. Since boundary activities often occur at some distance from an organization's "sovereign" territory—in another firm's place of business, on foreign soil, or on a regulatory agency's premises, for example—and since, as noted earlier, the activities define some of the boundaries, it follows that an organization's boundaries may be extended and detached. It follows, too, that, since boundary activities may, and typically do, vary in the location and time of their occurrence, the loci of organization boundaries vary spatially and temporally. A large proportion of the boundaries of organizations are not fixed as in biological systems, a fact noted by Katz & Kahn (1978); they are dynamic, varying in space and time.

One or more boundary activities constitute a boundary role performed by a person (Adams, 1976; Kahn, Wolfe, Quinn, Snoek, & Rosenthal, 1964; Katz & Kahn, 1978). More than one role may, of course, be performed by a specific individual. The structure of the interorganizational and intraorganizational relationships of boundary roles enacted by persons, which are in some respects organizationally unique, are depicted in Figure 3. Four unique organizational properties of boundary roles are noteworthy. First, and most obviously, are the five classes of activities of which the roles are composed. These are discussed in detail in the sections of the paper that follow. Second, boundary role persons (BRP) are typically more distant psychologically, organizationally, and, often, physically from other members of their organization who have different roles than the latter are from each other. The greater psychological and organizational distance is a function of the need for BRP to represent the points of view, values, and needs of nonmembers in their interorganizational interactions, as well as a function of transmitting within the organization information that is perceived by others as disturbing equilibrium because of its implications for change. Greater physical distance is a result of the fact that many activities occur at the extended boundaries of the organization, some of which are, indeed, viewed as isolated outposts. In turn, the physical distance of BRP may by itself induce considerable psychological distance. Third, there is an interpersonal power distinction between BRP and members whose roles include primarily intraorganizational activities. By reason of having to deal with outsiders, BRP usually possess no legitimate power or counterpower over nonmembers. Exceptions of lim-

*Figure 3.* The structure of inter- and intraorganizational relationships for boundary role persons, with component and relationship attribute classes of which boundary role behavior is a function.

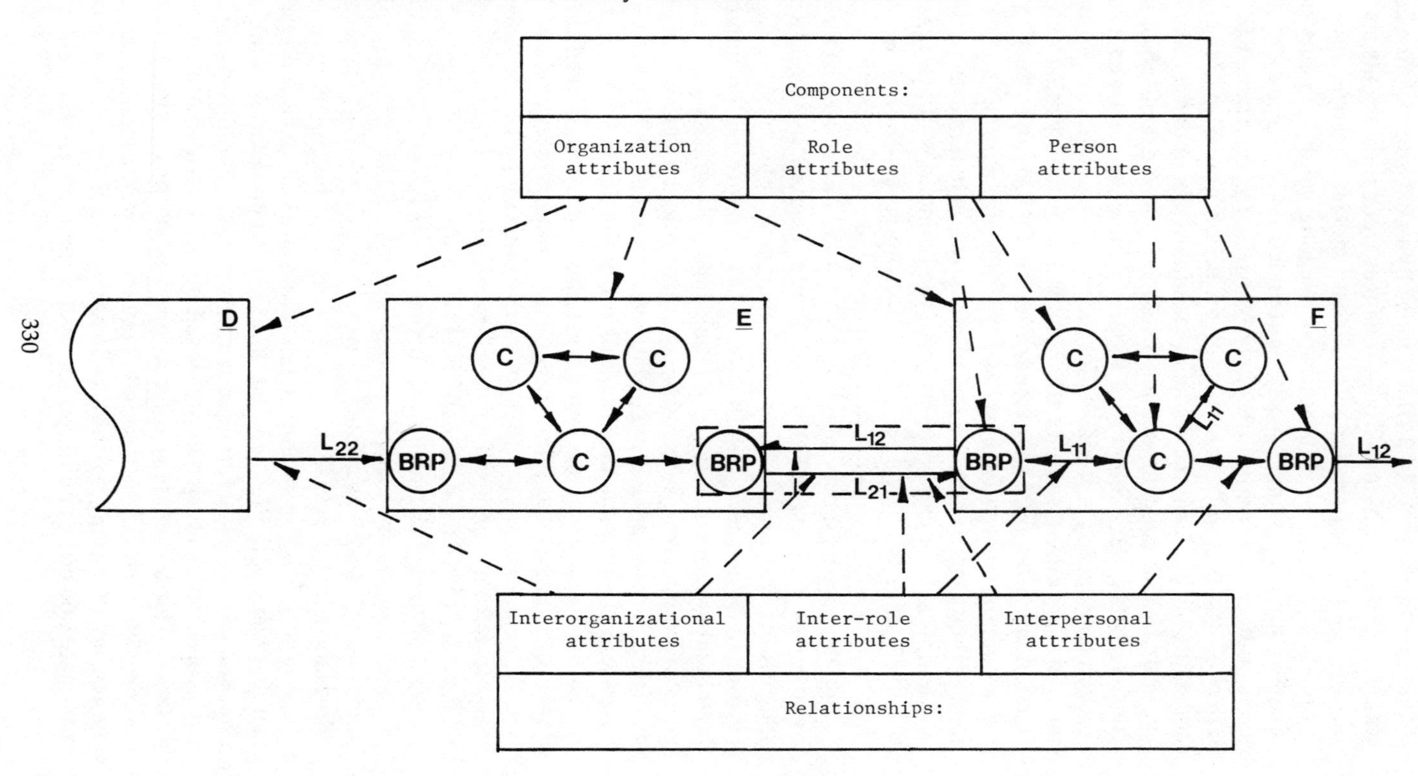

ited scope occur due to legal or contractual requirements. BRP must rely on referent and expert power but may have limited reward and coercive power in their transactions with outsiders (French & Raven, 1960).

The fourth unique organizational property of boundary roles arises from the structural position of BRP in interorganizational relationships: Occupants of boundary roles experience dynamic, dual conflicts with outsiders, on the one hand, and with insiders, on the other hand. In the performance of boundary activities, the BRP must influence—typically, alter—the behavior, attitudes, or preferences, perhaps the values, of members of other organizations, who, most frequently, are BRP themselves. Simultaneously, the BRP of a focal organization are the targets of influence or counterinfluence, even if in no form other than resistance, by the outsider whom the BRP is attempting to affect. Thus, a BRP experiences conflict by being both the influencer and the influenced. There is a second conflict experienced in boundary roles: BRP necessarily deal also with members of their own organization in the pursuit of their boundary activities. They are influenced by their constituents (C), as other organization members will be labeled, and the BRP must, in turn, influence their constituents. External influences must be effectively transmitted to constituents, much as influences from constituents must be transmitted to outsiders. This, then, presents the second conflict. BRP are at the crunode of the dynamic, dual conflicts in which elements of one conflict become inputs to a second conflict, of which some elements become new inputs to the first, perhaps for several cycles (Adams, 1976). Frey & Adams (1972) have described these coexisting, dual BRP in-group and out-group conflicts among negotiators.

The structural characteristics of BRP roles are diagrammed in the central portion of Figure 3, which is a "magnification" of a segment of Figure 2. As a member of the focal organization, *F*, the BRP interacts with his counterpart in organization *E*. The two persons constitute, at least transitorily, a boundary social system with its own norms (Adams, 1976; Walton & McKersie, 1965). The two persons also interact with one or more of their constituents (Holmes, 1971). The behavior of the BRP is a function of attributes of the components of the system depicted and of the attributes of the relationships between the components: organization, role, and personal attributes of components and interorganizational ($L_{12}$, $L_{21}$, $L_{22}$), interrole ($L_{11}$, $L_{12}$, $L_{21}$), and interpersonal ($L_{11}$) relationship attributes. The structural illustration makes evident the particular significance of BRP in organizations as open systems: The interrole relationships between the BRP of two organizations are also the vital linkages and binding forces between the organizations. They provide, too, the essential bonds among sets of organizations which permit mutual causal processes to arise, as stated earlier.

A significant deduction may be made from the above general model of the structural properties of boundary roles and their associated dynamic characteristics: Explanations of the behavior of organizations should be sought in the interactions of components and of component relationships, rather than in the main effects of either components or relationships. Support for such a deduction is found in the research of Adams (1976), Frey & Adams (1972), Holmes (1971), and Wall & Adams (1974), and others, who have repeatedly found significant proportions of BRP performance variance explained by interactions. The next five sections, in which classes of boundary activities are discussed, provide specific evidence.

*Transacting Input Acquisition and Output Disposal*

As open systems, organizations are dependent upon a wide variety of inputs that they must acquire externally for survival. Indeed, the proportion of externally acquired inputs has historically broadened. Rosenberg (1972) notes, for example, that, whereas farms in the 1870s produced approximately 90 percent of their input requirements, present-day farms purchase nearly 90 percent of their needs. With the extension of greater external dependence there is also associated an exponentially increasing variety of inputs available and rates of change in input characteristics, especially regarding products based on science and technology. The case of microelectronic devices, which influence change in other inputs, is dramatic (Noyce, 1977). These devices have had a profound impact on organizations; they have stimulated new products, new technologies, and, even, new industries. In turn, the new products, technologies, and industries generate their own effects.

The dependence of organizations on the ability to dispose of outputs has received less attention than input acquisition, perhaps because the need to dispose of primary products (goods sold, patients discharged, students graduated, for example) is obvious and because, until recently, waste disposal was scarcely worthy of attention. Perhaps, too, organizational scientists have not yet adopted a comprehensive open system perspective. In fact, the disposal function is as vital as input acquisition.

Acquisition and disposal transaction activities are central in such diverse boundary roles as negotiators and bargaining agents (Adams, 1976; Wall, 1975, 1976a, 1976b, 1977a, 1977b, 1977c), pruchasing agents (Strauss, 1962), salesmen (Pruden, 1969; Pruden & Reese, 1972), personnel and student recruiters, bank loan officers, stockbrokers, placement specialists, and student and patient admission staff. Closely related, but nevertheless somewhat different, are the activities of bank tellers, checkout counter cashiers and baggers, theater ticket sellers, receptionists, reservation clerks, parking lot operators, and similar occupational boundary roles. They do not so much acquire inputs or dispose of outputs as perform an activity essential to each; nor do they typically entail negotiation and bargaining. These roles are often more routinized and do not present the BRP with the inherent dual-conflict mentioned previously.

Transacting the acquisition of resources and the disposal of products implies a form of negotiation insofar as the focal organization and an outside organization have different preference orderings, which is most often the case. The focal organization's negotiating BRP, therefore, is subject to the dual-conflict arising from being influenced by both his constituents and by his counterpart negotiating on behalf of an external organization and, in turn, by having to influence constituents and counterpart (as diagrammed in Figure 3). As a member of the focal organization, the BRP is given role prescriptions and goals to achieve, is subject to formal authority, and has internalized his organization's values and norms to some degree. To perform his boundary role successfully, he must also be able to represent accurately and influentially to his constituents the needs (preference orderings) of the outside organization with which he is dealing, as well as its values and norms. If he does so successfully (which he must), however, he opens himself to the possible perception of having a suspect loyalty to the organization (Adams, 1976). Indeed, by coming to appreciate and understand an outside organization, and therefore being "partially included" psychologically (Allport, 1933; Katz & Kahn, 1978) in both that organization and his own, constituents have some basis for suspicion. As Kahn, et al. (1964) have noted, the rate of turnover or "defection" is higher among BRP than among other organizational roles.

Constituent distrust of the BRP has been shown to have powerful effects by Frey & Adams (1972), Frey (1971), Holmes (1971), and Wall (1972). In a complex laboratory experiment by Frey (1971) in which the BRP negotiated a labor-management contract, the BRP who were distrusted by their constituents, in contrast to those who were trusted, displayed very significantly more competitiveness and exploitativeness, even when their labor counterpart was initially cooperative. This increases the competitiveness of the opposing BRP, which, in turn, decreases the probability of contract settlement, as was demonstrated later by Holmes (1971). In an imaginative experiment, Wall (1972) examined BRP behavior that might induce distrust among constituents. He observed that salesmen (BRP) who achieved poor results and/or failed to obey constituent bargaining directives were distrusted and unlikely to be chosen again as negotiators. Wall (1976a and 1976b) has suggested that bargaining between BRP and constituent distrust are linked in a cyclical, mutual causal process, as illustrated in Figure 4. Wall describes the process:

*Figure 4.* Constituent—BRP—opponent cycle (Adapted from Wall, 1976b, with permission.)

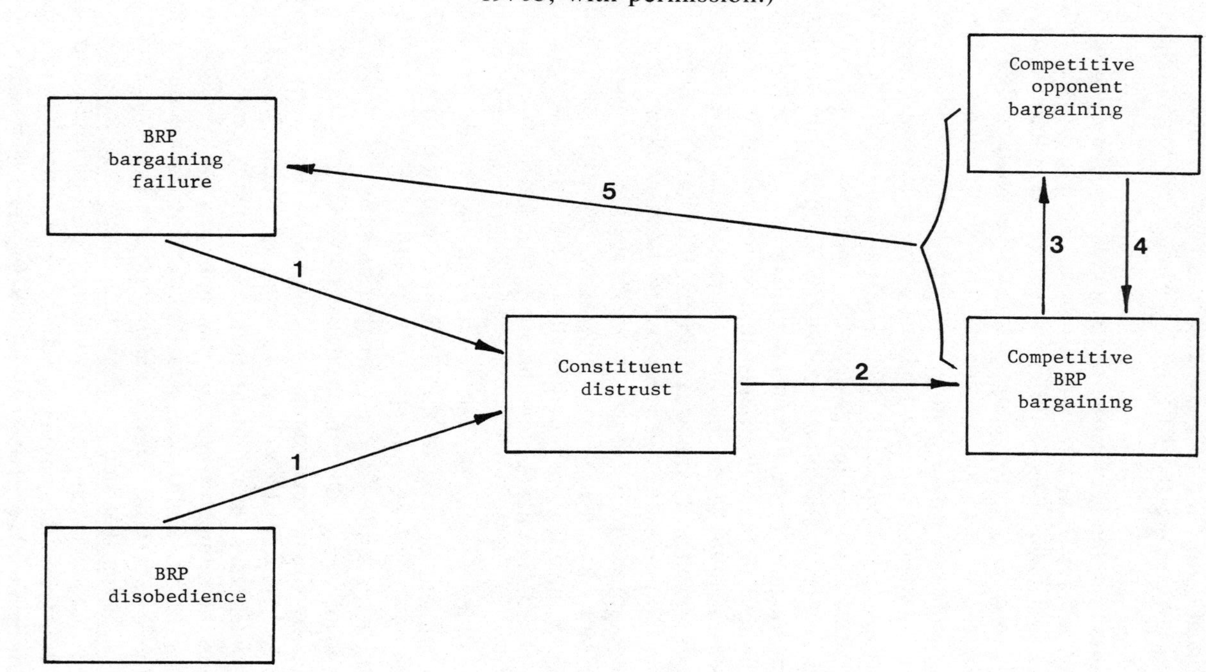

> BRP bargaining failure or disobedience leads to constituent distrust of the BRP (Arrows 1). In turn, distrust of the BRP leads to competitive bargaining on his part (Arrow 2). . . . As a result of the BRP competitiveness, the opponent bargains competitively (Arrow 3). . . . His retaliation reinforces the BRP's competitiveness, encouraging him to continue his competitiveness (Arrow 4). As a result of this mutually reinforcing competitive encounter with the opponent, the BRP is unsuccessful in his bargaining (Arrow 5), and, as a result, the constituent's distrust of the BRP rises further." (Wall, 1976, p. 30.)

The process discussed by Wall attests to a conclusion drawn earlier in this chapter that variations in BRP behavior may be best explained as a dynamic, interactive process. The conclusion is reinforced by an experiment by Holmes (1971), in which the structure of constituent groups in the focal organization and another organization (*F* and *E* in Figure 3) were systematically varied. Following a suggestion by Kelley & Stahelski (1970), Holmes developed a negotiation paradigm with which he investigated the performance of a focal organization's BRP as a function of (1) constituent instructions to bargain cooperatively or competitively, (2) cooperative or competitive behavior of the BRP in the external organization, (3) constituent consensus or dissensus in the focal organization, and (4) constituent consensus or dissensus in the external organization.

Using the number of negotiating agreements achieved as a measure of BRP performance, Holmes observed what he labeled a general "triangle effect": Agreements reached were directly related to constitutent instructions to cooperate or to compete, being either frequent or infrequent, respectively, when the external BRP bargained cooperatively; but agreements attained were uniformly few when the external BRP behaved competitively. More importantly, in the context of the present discussion, the triangle effect obtained only if the three constituents in the focal organization were in dissensus or if the constituents in the external organization experienced dissensus; the effect was not observable if either constituent group were in consensus (Holmes, 1971). The findings are partially illustrated in Figure 5, the lower half of which shows the "triangle effect" clearly. The results also lend validation to the general structural model offered in Figure 3.

A feature of the structure of BRP interorganizational relationships that has not yet been discussed is the transaction subsystem consisting of two (or more) BRP acting on behalf of their respective organizations, for example, organizations *F* and *E* in Figure 3. In the process of acquiring organizational inputs and disposing of outputs, negotiators form transaction subsystems that develop their own rules and behavioral norms, as well as values that are legitimated and sanctioned by professional and quasi-professional associations. Membership in a transaction subsystem requires adherence to its norms, if negotiation between BRP is to be

*Figure 5.* Bargaining agreements reached as a function of constituent role pressure, opponent behavior, and ingroup structure. [Adapted from Holmes (1971) with permission.]

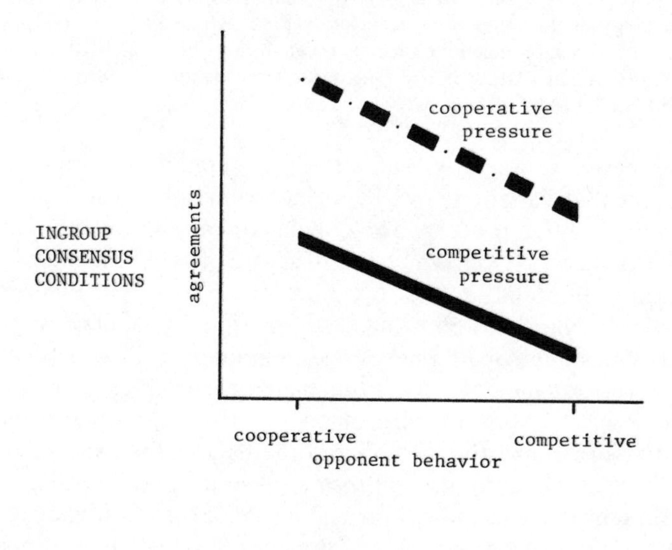


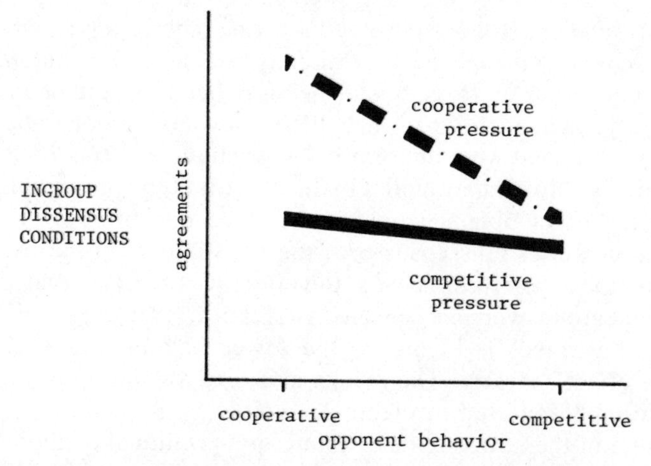


viable and if the members of the subsystem are to achieve a contract or settlement (Walton & McKersie, 1965). Frey (1971) has shown, for example, that, if a bargaining agent's verbal behavior is cooperative and conciliatory but his actual bargaining is competitive and tough, the negotiation concessions made by the other bargainer are vastly smaller than if there is consonance between verbal behavior and bargaining behavior. To account for variation in the behavior of BRP, it is insufficient to have knowledge of inter- and intraorganizational, role, and person attributes;

information about a boundary transaction subsystem is also required. Additionally, dynamic mutual causal processes of the type discussed in the first part of this essay could not be effectively traced since boundary subsystems provide the vital interorganizational linkages.

*Filtering Inputs and Outputs*

The acquisition of organizational inputs and the disposal of outputs is intimately associated with the filtering of inputs and outputs. Of course, not all inputs and outputs, as will be noted later, entail intentional transactions. Some information may not be "acquired", in the usual sense of the word; for example, unsolicited communications may be unexpectedly received and then screened with respect to content, relevance, and organizational implications (Rosen, 1976; Taylor, 1974). Generally, however, acquisition and disposal transactions are performed with knowledge of the filtering criteria or the boundary "codes" (Katz & Kahn, 1978). The specifications of materials purchased, for instance, are usually isomorphic with filtering criteria. Nevertheless, boundary filtering activities are distinct from boundary resource acquisition and product disposal activities. They are performed by different persons, who have roles suggestively named admissions officer, personnel interviewer, materials inspector, claims processor, customs inspector, final test inspector, certifying examiner, and the like; they are also typically performed in a different place and at a different time.

Boundary filtering role activities do not involve negotiation, as is the case in transacting input acquisition and output disposal. The essence of filtering activities is selective acceptance or rejection of inputs and outputs in accordance with criteria largely established by the organization in the instance of inputs and by external organizations in the case of outputs (Katz & Kahn, 1978; Levinson & Astrachan, 1974). There is variation in the source of filtering criteria, however. Student admission criteria are heavily influenced by statutes and governmental agencies; the quality criteria of certain products are set primarily by the manufacturer, rather than by retail outlets and consumers, as in the case of novelty and fad items that are not subject to regulation; the specification of certain outputs such as drugs are not so much set by consumers as by physicians and regulatory bodies like the Federal Drug Administration. If persons are the inputs and outputs, they may influence criteria, as is evident in the admission and discharge of patients in a hospital.

It is generally useful to distinguish the screening of persons from filtering materials, information, energy, and other nonhuman inputs and outputs. The distinction is important for two reasons. First, persons interact with the filtering process and the BRP performing the filtering task, and they react to both process and BRP. They may, therefore, affect the performance of the BRP in barring or allowing entry into or exit from the organization and may, consequently, alter screening criteria. Second, filtering errors have different implications in the case of persons.

Two types of errors are possible in relation to criteria: acceptance errors and rejection errors, referring, respectively, to admitting an input and emitting an output that should not have been—*false positive* (FP) error—and not admitting an input, and not emitting an output that should have been—*false negative* (FN) error (Dunnette, 1966).

Clearly, filtering errors of all inputs and outputs have organizational costs of two kinds: first, costs associated with direct and indirect losses resulting from the errors; second, opportunity costs. Organizations attempt to achieve an appropriate trade-off between FN and FP errors with regard to such costs, and BRP activities are specified accordingly. The trade-off points vary as a function of a number of variables, of which the potential aggregate cost of an error is most important (which is not necessarily closely related to initial expenditures). For example, to have rejected an employment application may involve a firm in endless, expensive litigation and, in addition, may result in the loss of reputation and consumers. Similarly, the award of a professional degree to an incompetent may have dire consequences for the institution, as might the discharge of an unrecovered patient.

Filtering errors relating to persons have unique consequences that are irrelevant in the case of nonhuman items. Persons often, if not usually, know when they are "errors". The woman whose employment led her to believe she was qualified for a job and could, therefore, succeed suffers anguish and loss of self-esteem when she is terminated because she failed to perform satisfactorily. Her relatives may grieve, too, and she may find that her "failure" follows her in subsequent attempts to seek employment. The high school recipient of one of 87 graduating awards is disillusioned when he learns that he is qualified only for menial or common labor jobs. The psychiatric patient whose disabling phobia disappears and is not discharged by the mental institution may become permanently ill. The effects of FN and FP filtering errors on persons are not typically, certainly not systematically, assessed by organizations and their BRP; they may be difficult to measure or even to observe.

Both types of filtering errors vary in their *visibility*, that is, whether or not their occurrence is or becomes known. This fact affects the performance of BRP, as Organ (1970) has demonstrated experimentally. FN input errors are usually fairly invisible. For example, a loan application falsely rejected by a bank loan officer is typically invisible; it is unknown that another lending institution later made a loan and the borrower faithfully repaid the principal and paid the interest due. The organization which turns down a woman applying for a vacant computer programming position does not learn that the applicant subsequently was a very successful programmer for another firm. FP input and output errors are comparatively quite visible: One "hears" rather quickly about their occurrence within the organization in the instance of inputs and from external sources in the case of outputs. A loan officer learns rapidly that a borrower has defaulted. A dealer whose customer reports that her refrigerator compressor failed six days after purchase will not delay his complaints to the manufacturer, whose product had passed final inspection. FN output errors—not emitting (shipping, discharging, graduating, etc.) outputs that should have been—have quite variable visibility. They are highly visible if they are manufactured goods or processed materials, provided there is a good management accounting system. In public service organizations, such output errors are frequently invisible. Mental health institution patients are often retained after discharge should have occurred, as Rosenhan (1973) has shown in a field experiemnt with pseudopatients who gained admission to psychiatric hospitals. Similar instances are believed to occur in the welfare and penal systems.

Visible filtering errors have repercussive effects that may alter the behavior of BRP and the distribution of FN and FP errors. Specifically, their error "preference" may change or, equivalently, their filtering criteria may become altered. Assuming that BRP will avoid making errors that have negative consequences for them, they will attempt to reduce such errors. So doing, however, increases the probability of making the other type of error. For example, assuming for the moment that FN and FP output errors are equally visible, a final test quality control inspector who is reprimanded for having approved too many faulty products for shipment (FP error) will attempt to reduce such faults by increasing the frequency of products rejected. This, however, will increase the probability of committing FN errors (not approving products that should have been passed). Given the general assumption that BRP will avoid punishment, they will also tend to bias their error when the two are unequally visible. Thus, loan officers could be hypothesized to exhibit a bias toward FN input errors.

Boundary filtering activities have been discussed up to this point as if there were only one screening activity with respect to a given input or output. That may be true in some instances, but frequently several successive filtering steps are required, quite analogously to a water filtration system. Filtering activities in such cases occur at different points in organizational space, or even beyond, in an organization's environments. As filtration progresses, the specificity of selection criteria increases and organizational preference for or tolerance of FN and FP errors at each filtering step may vary. For example, at the first, "wide mesh" screen FP may be preferred to FN errors, whereas at the last, "fine mesh" filter FP errors may be abhorred. Freedman & Mericle (1977) have examined the process in a detailed case study of the personnel department of a large public hospital. They identified five sequential boundary filtering activities in the employment of technical, professional, administrative, and other personnel in over 200 job classifications. Advertising at national, state, and local levels, depending upon the position, was the first screen. Advertisements, in effect, filtered out segments of the general population who did not read the notices, had no interest in the jobs, or did not perceive they met criteria set out in the advertisements. This left a pool of potential applicants. Some of the latter completed an application form which was evaluated by the personnel department; this represented the second filter. The third and fourth selection screens were interviews: a personal interview by a personnel officer and, if that was survived, an interview by the supervisor of the hospital unit in which there was a vacancy. For many applicants, this was the last filter, and a decision to hire or not was made. Other applicants underwent an additional screening for health and/or security reasons if, for example, the job required handling drugs. Freedman & Mericle (1977) note that there is a general tendency of the BRP performing filtering activities within the hospital to exhibit a preference for FN, rather than FP, errors "because the personnel subsystem is evaluated on hiring satisfactory employees." To this should be added the powerful effects of health organization value systems, which are intolerant of errors that endanger human life.

Despite the organizational and other costs of committing filtering errors, there appears to be a general bias toward FN errors, especially when they are less visible and/or when FP errors have dire implications for BRP and catastrophic consequences for the organization. There is also evidence that FP and FN *information* input filtering errors are especially susceptible to being committed deliberately with important long-term maladaptive consequences for organizations (Halberstam, 1969; Huneycutt, 1974; Rosen, 1976; Rosen & Adams, 1974; Taylor, 1974). Examples are discussed below and in a later section.

Modern organizations operate in increasingly complex, changing, "turbulent" environments (Emery & Trist, 1965; Terreberry, 1968) with dynamic characteristics described earlier. There is, therefore, a premium attached to filtering from masses of available information that which is vital to organizational adaptation (Leifer & Delbecq, 1976; Terreberry, 1968). Yet, BRP acting as information "gatekeepers" are guilty of omission, exaggeration, and selective bias in the performance of their roles under specifiable conditions (O'Reilly & Roberts, 1974; Rosen, 1976). A laboratory experiment by Taylor (1974) is illustrative; Rosen (1976) summarizes it as follows:

> The major focus of this investigation was the impact of certain combinations of environmental, organizational and personal variables on . . . filtering of information that conflicts with company policy.
>
> Eighty business students participating in a management simulation were provided with a brief history of a manufacturing company, an organizational chart, and information regarding a plant relocation decision. Each participant assumed the role of a company official responsible for keeping the president informed about important external affairs.
>
> Three independent variables were manipulated in the simulation through variations in written materials. External pressure for a change of organizational policy with respect to the relocation decision was manipulated by varying the source of protest letters coming to the company. Gatekeeper fate control was manipulated by informing participants that their performance would be evaluated by the personnel director (low fate control) . . . . [or by the president (high fate control)].[2] Perceived defensiveness of the recipient of upward communications was manipulated by varying the constituent's reactions to information critical of company policy. In the high defensive condition, the company president reacted with excessive rationalizations and sharp criticisms to opposition. No such reaction was given in the low defensive condition.
>
> Filtering and gatekeeping behavior was assessed by computing the frequency of negative or critical information that was transmitted to the company president . . . .
>
> Subjects in the high pressure condition transmitted significantly more negative information than did subjects in the low pressure condition. When message content was identical, gatekeepers were more likely to act upon information from nominally highly influential sources (high pressure) compared to less influential sources (low pressure).
>
> No significant main effects were found for constituent defensiveness or constituent fate control. However, the interaction of these two variables was significant. . . . The highest frequency of negative information transmission occurred in the high fate control, low constituent defensiveness condition. In order to meet role expectations of a constituent who controlled his fate, it appears that subjects in the high fate control condition experienced pressure to transmit negative information. This pressure, however, was moderated by anticipation of the constituent's likely reaction to negative or dissonant information. When the constituent had a history of defensive reactions to negative information, gatekeepers were less likely to transmit negative information. Negative information was transmitted at a high rate in the high defensive, low fate control condition. Subjects who were free from the apprehension that their persistence would result in lowered evaluations were more likely to transmit dissonant information to an unreceptive (defensive) constituent.
>
> Findings from the . . . study illustrate the powerful influence of information source on filtering behavior. Moreover, the findings demonstrate that gatekeepers are susceptible to bias based on the source of their rewards and personal characteristics of the constituent (Rosen, 1976, pp. 60–63).

In a subsequent investigation, Huneycutt's (1974) findings suggest strongly that independent sources of information available to organization decision makers can reduce the dysfunctional effects of filtering errors. Indeed, that is the very function performed by responsible mass media, as became evident in the "Watergate Affair", for example. Thus, the organizational difficulties and dilemmas of controlling the filtering errors to which information gatekeepers are prone (Rosen & Adams, 1974) may be moderated to some extent by independent external information—provided, of course, it is not processed by the same gatekeepers!

*Searching for and Collecting Information*

Nothing is as essential to organizational adaptiveness as information (Wilensky, 1967). The fact is well known to theorists and pragmatic decision makers alike. Although information about first-order environmental events is obviously vital, of even greater importance to a focal organization is knowledge about $n^{th}$-order environmental events and interorganizational relationships, as the seminal work by Emery & Trist (1965), later elaborated by Terreberry (1968), makes pellucid. As was pointed out earlier, organizations in turbulent field environments are linked in mutual causal process loops that tend to enhance and sustain change, which increases the need for and value of information and intelligence activities. It follows that a vital boundary activity is the search for and collection of information. *Interpretation* of such information is not considered a boundary activity but, rather, an adaptive subsystem activity (Katz & Kahn, 1978).

Organizations require two kinds of external information which differ principally in the purposes for which they are needed and in their search or monitoring implications. The first kind is that required for current decision making and policy formulation having short- and long-term effects. It might be called "operating information" and is discussed extensively in texts on management, operations management, operations research, and decision theory, as well as in works addressing discrete domains such as marketing and finance. The monitoring, search, and collection of this information is relatively focused. There is knowledge of what information is required and what the sources of it are. This does not imply, however, either that the information or its sources are perfectly reliable, that decision makers do not experience uncertainty, or that risk does not attend decisions.

The second kind of required external information is that concerning *unpredictable* events which *might* occur and *might* have relevance for the organization if they occurred. These are limited here to nonrandom events; they are simply unknown and, perhaps at a given stage of existing knowledge, unknowable, unpredictable as to place and time of occurrence, and uncertain in their implications for the organization if they should occur. Rachel Carson's *Silent Spring*, when it was published in

1962, approximated for the chemical and systemically-related industries and organizations the type of "unknowable" event being discussed. The *book* and its effects were the unpredicted event; the essential *content* was well known to biologists and ecologists, if not to industrial managers. Thus, there was some basis for predicting the book or its equivalent. Scientific and technological "breakthroughs" and many other events, some of them "natural", throughout history have had comparable consequences. The crucial point is that *if* there had been intelligence collection about the possibility of such events, the probability of any given organization's being vulnerable would have been reduced. It is in recognition of this that some organizations, the General Electric Company, for example, have incorporated the second type of information search in their strategic corporate planning (Dunckel, Reed, & Wilson, 1970; Wilson, 1974). GE has made use of experts in various fields and the Delphi technique to predict possible future events in science and technology, society, government, politics, and economics that would have an impact on it. Many other corporations are engaged in similar activities ("Piercing future fog," 1975), as are government departments, most obviously the Central Intelligence Agency.

If the search for operating information has quite specific foci and can be likened to viewing a distant ship with binoculars, the search for the unpredictable requires, analogously, radar and sonar scanning of the total surface, air, and undersea environments without prior evidence of the presence of vessels, aircraft, and other objects, and, of course, without knowledge of the implications for the searching ship if an object did appear. That is the extreme case; more limited scanning might be sufficient if there were some knowledge of the probabilities of various events, their likelihood of occurrence with respect to place and time, and/or their implications if they occur. The costs of search are obviously related to the unpredictability of events that have relevance to organizational adaptiveness and to the risks an organization is unwilling to incur by limiting search. For any given degree of unpredictability, search costs mount as risk aversiveness increases, the latter being reflected in intolerance for FN errors—not learning of events that would have significant organizational effects. Similarly, at a fixed level of risk aversiveness, search costs climb as unpredictability rises. These facts would be banal were it not for the nature of the search: The external events are at best ill defined, the probability of their occurrence is unknown, and their potential relevance to and implications for an organization are unknown. It is no wonder that organizations have been described as operating in a nearly impenetrable "future fog".

The problem implications for the BRP whose task is to search for and collect information that is poorly defined are, at best, enormous. First, their constituents, who require intelligence for decision making and policy formulation, exercise pressure on them for information that is at least proportional to their uncertainty and to the perceived importance of the information. In most instances, the constituents have authority over the BRP, but they are also very dependent on the BRP to obtain essential information, which gives the latter effective potential counterpower and is productive of dissonance. Indeed, Rosen & Adams (1974) have shown that managers are loath to discipline the BRP on whom they are dependent. Associated with this difficulty, at least under conditions of competition, is the need for secrecy in search activities. A consequence is further diminution of effective constituent authority and the arousal of doubt about BRP loyalty, which in special contexts has been labeled the "double-agent anxiety syndrome."

A second problem for the BRP, and by extension for constituents, relates to the possible overreliance of constituents upon the information provided them. When information is vitally required, there is a well-known tendency to treat it as more significant or more reliable than it may be, possibly with damaging consequences for the organization. BRP may impress the need for caution, but that may undermine their position in the organization. Alternatively, BRP may defer transmission of particular items of information until they have enough corroborating data to minimize the risks of overreliance by constituents; but that, too, may be risky for BRP because of the delays that are implied.

Decision-making constituents are not immune to problems, some of which were suggested above. One concerns information selectivity: if decision makers are averse to FN errors by BRP, they must bear the additional costs of search, processing, and of FP errors. They must also guard against BRP bias (Taylor, 1974) and what Tesser, Rosen, & Batchelor (1972) have referred to as the MUM effect, for which Taylor found supporting evidence—the propensity to not communicate information which is "bad news" to the recipient. Halberstam (1969) has documented the fact that President Lyndon Johnson's acerb reactions to information about the failure of U.S. involvement in the Vietnam War led his staff to convey to him intelligence that was selectively consistent with U. S. success. Halberstam relates, for example, that the President's national security advisor, Walt W. Rostow, culled the thousands of daily reports from Saigon to find the few positive ones for his boss. In turn, Rostow's aides searched for news items that would please *him*. Thus, selective bias of intelligence was self-reinforcing in an amplifying closed loop. Although the examples provided by Halberstam are of BRP information filtering activities rather than of raw intelligence gathering per se, the same biases are obviously possible in the latter activities.

*Representing the Organization*

Much as an organization requires information about its environments, it must selectively transmit information about itself to organizations, groups, and individuals, in part as a function of external demand and partly at its own initiative. It is to this that representation refers here, rather than to the usage of the term in the negotiation and bargaining literature (Rubin & Brown, 1975), in which "representative" is synonymous with "negotiator" and "bargainer." As used here, *representation* denotes presentation of information by an organization about itself to its external environments for the purpose of shaping the opinions and behavior of others. It is partially allied with Parson's (1960) and Katz & Kahn's (1978) institutional boundary function, serving to generate social support and legitimation for the organization. In addition to selective disclosure of facts, it includes impression management, "face-work" (Goffman, 1969), public relations (Montgomery, 1978), institutional and product advertising, and the like. Persons exemplifying such boundary activities are media representatives, lobbyists, public relations men and women, organization "spokesmen," and so on. The roles of other persons also include as a subset similar representational activities, for instance, the corporate president speaking to security analysts, the receptionist, the restaurant hostess, the airline steward, and company "reps." whose principal role activities are sales and consultation.

Representations may be veridical or not. If veridical, organizational relations with the external environment are potentially more stable, decision making may be simplified because external stability improves and the need for "gaming" is reduced, external trust may be engendered, and thus social support is more likely. Truthful representation may, however, also make the organization vulnerable to competitors and to the manipulation of other organizations. As was stated in the previous section, information is an essential, often strategic, organizational input. Consequently, information disseminated by an organization about itself and its operations and management may reduce a comparative advantage it might have otherwise enjoyed. The dominant tendency is for organizations to limit severely the outflow of information. This is evident even in voluntary disclosures in the speeches of company executives and government officials, which, on analysis, often have very little content, if they are not minor masterpieces of obfuscation. In effect, veridical representations are very selectively filtered.

False representations are commonplace, except when—or even when—limited by statute. This occurs as an essentially defensive response to pressure for information or as elicitation of erroneous response—as when information is "planted." It yields only short-run benefits, which nevertheless may be very important ones. On balance, untruthful representations, including denials of allegations, have self-defeating consequences in the long run. They lead to unstable interorganizational relations and increase environmental turbulence—for example, a false representation by *F* to *A* reduces *A*'s predictability of *F*'s behavior, may decrease *B*'s predictability of *F*'s and *A*'s behavior, and so on; because interorganizational relationships take on "gaming" characteristics, decision complexity and some operating costs are increased; distrust is generated when the fact of misrepresentation is known, which may also produce reactance (Brehm, 1966), and external social support and legitimacy are eroded; and, if it becomes known, any false representation imposes on other organizations the need to suspect all representations by the falsifying, and perhaps other, organizations, and it requires that the validity of every information release be determined. Notwithstanding their general maladaptiveness, misrepresentations may have prolonged beneficial consequences under special conditions. Consider the following case. Organizations *A* and *B* are adversaries, but both have an interest in controlling the level of conflict between them. A member of *A* commits a transgression against *B*, which, although minor, would cause *B* to retaliate in a manner resulting in possibly uncontrollable conflict between the two. The fact of the transgression by *A*'s member cannot be denied, but it cannot be proven that it was on behalf of *A*, although let us assume that it was. Through its representative spokesman (perhaps its president in such a case) *A* may usefully misrepresent the cause of the injury (it was an "unauthorized" act or the aberrant behavior of an individual, for example). Thus, escalating conflict between *A* and *B* may be avoided, especially if the transgression was publicly known. The prevarication permits both parties to "save face."

In the performance of their representative roles, BRP are dependent on internal constituents as sources of information, and their effectiveness is also partially dependent on the credibility attributed to them by the external targets of their representations. As in other boundary activities, three parties are engaged: the BRP representative, at least one constituent, and no less than one outsider (the target). Observations about each will be made in that order. If a representative gives information, either on his own initiative or upon instruction by a constituent, he is either knowingly or unwittingly truthful or deceptive. If knowingly deceptive and if he is carrying out constituent instructions, he is likely to experience a measure of person-role conflict, possibly intersender conflict, and some attendant tension (Kahn, et al., 1964). He will, in addition, experience anxiety about being found out and losing the credibility upon which effective representation depends, as was the case with President Gerald Ford's press secretary, Ter Horst, who resigned, refusing to mislead news reporters about White House policies (Rosen, 1976). If witting deception is founded on constituent instructions and if blaming a superior for false information is counternormative, the BRP must accept blame unfairly, which may generate emotional and interpersonal conflicts. Unwitting deceitful representations present similar problems.

Although targets of representation exercise power over representatives by virtue of their ability to grant or withhold credibility, BRP representatives are not without power over targets. To the extent that targets are dependent on the BRP for information and the latter is an exclusive, authoritative source, the BRP representative possesses potential expert power (French & Raven, 1960) and, by extension, so does his organization. A frequent organizational problem, however, is that information is "leaked" from within the organization, thus weakening the power of representatives over targets.

A constituent who is a representative's source of organizational information must act with circumspection in the interest of the organization as distinguished from self-interest. The constituent source must consider the consequences of giving the BRP false data, denying the existence of information when it exists, and withholding facts that would normally be transmitted to the representative; each of these entails a risk that the BRP would lose credibility among outsiders—thus diminishing his effectiveness—that the organization would lose external support, if not generate enmity, and that the representative would become alienated. Notwithstanding these risks, which would be normally avoided on rational grounds, it may nevertheless be necessary that a BRP representative be genuinely ignorant of certain information in order to limit the possibility that the information will become known to outsiders and in order to protect the representative's external credibility. Aside from the fact that the leakage is probably at least proportional to the number of persons possessing information, the practice can be justified on the grounds that representatives are the subjects of more systematic external pressure for information than most other organization members. It is reported that President John F. Kennedy withheld from Pierre Salinger, his press secretary, all information concerning Executive Office deliberations during the first stages of the Cuban missile crisis in 1962 (Kennedy, 1969; Sorensen, 1965). Although this allowed Salinger to report truthfully that he was ignorant of White House proceedings, his credibility was nevertheless somewhat endangered because newsmen were well aware of hurried military preparations, especially in the southeastern states. There can be little doubt, too, that he must have suffered severe role conflict.

Outside targets of representation may be relatively passive recipients of information, such as newspaper readers and TV viewers, or "active seekers," as in the case of mass media reporters, intelligence agents, and special interest investigators. The latter group's presumably greater felt need for accurate data results in the need to exercise considerable pressure on the BRP to provide valid information, to establish means of verifying it, and to develop effective power or counterpower over organizational representatives. It is a provocative paradox that targets who reward representatives by granting them trust and credibility for the purpose of building their own power will increase the probability that they will be manipulatively misinformed in the future! Of course, representatives are also entrapped in a paradox with respect to their credibility with and power over news seekers. Their power over the latter diminishes the more information becomes a "free good"; but withholding information risks a reduction in their credibility. The balance between representative and target is fragile and may depend at a given time on the relative need for credibility and information in the respective cases.

*Protecting and Buffering the Organization*

As open systems, organizations require that their integrity, internal equilibrium, structures, functions, personnel, and core technologies (Thompson, 1967) be protected and buffered from external disruption, trauma, and insult, much as biological organisms and communities have evolved comparable protective mechanisms in order to survive in the face of external threat. Boundary activities have a major—although not exclusive—role in the protection and buffering of organizations. Discussion is limited to these activities and will not meander into issues of maintaining internal equilibrium and efficiency, for example, which are protective, and have second-order relations to boundary functions.

The four previously discussed classes of boundary activities quite obviously serve to protect an organization. Without the ability to negotiate the acquisition of inputs and the disposal of outputs, including wastes, an organization could not survive. In the absence of selective filtering activities, a firm or public institution would experience gluts and famines and would suffer the costs of inputting and outputting errors. Maladaptive decisions are a consequence of inadequate information concerning external events. The inability to represent an organization advantageously to its environments leads to erosion of essential social support and legitimation. The class of boundary activities that are discussed in this section serve more directly to protect and buffer an organization.

The most apparent, yet sometimes subtly performed, kinds of protection are seen in the activities of security guards, receptionists, and secretaries (among whose other activities are control of ingress), ticket controllers, appointment secretaries, checkout clerks, and floor walkers. Some of these activities are performed at either the input or the output boundaries, as in airline passenger boarding and theater admission or as in grocery and other self-service store checkouts; or they are performed concurrently at both input and output boundaries, as in the case of security gate guards in prisons, military installations, and some research, manufacturing, government, and mental organizations. One may observe activities not normally prescribed by a role that are comparable in their function, especially in organizations that do not have BRP assigned to the illustrative roles above. Organization members seeing a stranger who appears not to belong on the premises may ask, "May I help you?", as a safety against theft or undesired intrusion.

Protective role activities vary in the amount of conflict and stress to which they expose the BRP. If the criteria for admission or exit are unambiguous, role conflict and ambiguity, and therefore role stress, are low (Kahn, et al., 1964), as among guards and ticket checkers. In contrast, considerable intersender conflict and stress may be experienced by personal secretaries and appointment secretaries, for whom admission criteria may be very subjective. Their bosses express a strong need to be protected from intrusions, while outsiders may be adamant about their desire to be admitted. Person-role conflict may additionally be experienced in boundary role activities such as parcel and briefcase checking, which may be perceived as invasion of privacy. This was evidently the case when airlines first began to search the purses, pocket contents, and carry-on baggage of boarding passengers.

Organizations often require, at least in the opinion of some members, that they be buffered from portions of their sociopolitical and ideological environments. Changes in norms, values, mores, political ideology, beliefs, and customs in the social and political environments of organizations result in attempts, frequently organized, to persuade organizations to adopt new, or to modify existing, practices and policies. Recent examples abound in regard to minority groups and women's rights, economic policy, taxation, the role of government, occupational safety, older citizens, social welfare transfer payments, defense policy, and abortion. The changes demanded are perceived as threatening equilibrium, "good order," performance, and goal achievement and as necessitating resistance or buffering.

Resistance can be accomplished in any number of ways, including representation, as discussed in the previous section, but organizations have created specialized buffer roles to deal with face-to-face representations and confrontations. The roles are commonly, but unimaginatively, entitled public relations manager or spokesman; Tom Wolfe (1970) has more descriptively and amusingly named the role occupants "flak-catchers." These buffer BRP *absorb* external pressures on behalf of the organization. Ideal role performance also induces external agents to believe they are being effective, when they are not.

Buffering may be organizationally adaptive or maladaptive. It is adaptive if it is a short-term tactic to prevent overload and/or to "buy time" to study the organizational implications of the changes that are pressed for. If it is merely a temporizing defensive tactic, it is likely to be counterproductive, as was the "stonewalling" of President Johnson in the face of the Vietnam War policies and of President Nixon when he was confronted by criticism and accusations about his role in the Watergate Affair, and as has been the stonewalling of numerous industrial corporations which hoped thereby to protect themselves from the discovery of various sins (Montgomery, 1978). Buffering is also maladaptive if it is the long-term and sole response to external pressure from significant portions of the environment because pressure for change constitutes information that may be vital to organizational survival—as vital, potentially, as technological, economic, legal and regulatory, and demographic information.

Role conflict and its concomitants, stress and tension, are experienced by buffers, as they are by BRP in general (Adams, 1976; Kahn, et al., 1964; Miles, 1976, 1977, 1978; Miles & Perreault, 1976). Intersender conflict arises, on one hand, from constituent expectations that the BRP will absorb external pressure, much like an effective automobile shock absorber, and that he will not transmit the force to the internal organization and, on the other hand, from the expectations of outsiders that the flak-catching BRP will *effectively* transmit demands. Person-role conflict may arise from at least two sources. The first source is inherent in the intersender conflict just mentioned, giving the impression to outsiders that their voice is being heard and given weight conflicts with the ethical principle that one should be truthful. Personal interviews conducted by Montgomery (1978) reveal a large number of such ethical conflicts among public relations men, among whom are buffer BRP. He states also that public relations positions ranked sixth among occupations having the highest rates of admissions to mental institutions in the U. S., but he does not provide direct evidence of a relationship between person-role conflict and hospital admissions. The data of Kahn and his associates (1964) reinforces the plausibility of a relationship, however.

The second source of person-role conflict among buffers is the frequent hostility of outsiders, especially members of "activist" groups, attempting to induce change in the organization. Their hostility is projected onto buffers, who are insulted, threatened, and personally humiliated. The buffer's organizationally prescribed role, however, requires that he absorb all this; he may not retaliate or succumb to the pressures for change that are exercised. He is a flak-catcher; he must subordinate personal feelings to the role demands.

# CONCLUDING REMARKS

This essay began with the exposition of a model of interorganizational structure and dynamics as a basis, indeed a rationale, for discoursing on organization boundary activities. It was argued that human *activities* constitute a significant proportion of organization boundaries and that these were the essential vital functional linkages between organizations. Five classes of boundary activities were examined with regard to their organizational functions and their implications for the persons who perform them.

Each of the five classes of boundary activities that were discussed and illustrated serves a particular, essential organizational function. Each also presents role incumbents with special problems, conflicts, and, on occasion, ethical dilemmas. As was suggested at the beginning of the discussion, and as became evident in the discussion proper, the conceptual distinctions among activity classes do not necessarily hold up cleanly in the ongoing boundary activities of a BRP. It was noted, for example, that representation and buffering often are concurrent. Indeed, the two may be the responsibility of the same person. Nevertheless, they have different purposes and are behaviorally distinct. Similarly, whereas transacting the acquisition of materials, intelligence gathering, and representation are conceptually, functionally, and behaviorally distinct, the three occur in the course of a purchasing agent's or salesman's negotiating a contract. Traditional role titles—ticket agent, salesman, public relations representative, for example—are misleading simplifications of the activities performed, the organizational functions served, and the psychological implications for the role incumbents, no matter how suggestive the titles may be. Research, therefore, should be undertaken on activities that have distinctive, differentiating properties, rather than on titular roles that typically encompass several different activities, including purely intraorganizational activities.

This essay might well end with this examination. This might, however, leave the misimpression that, having served an organization well or poorly, boundary activities had no further consequences, that once they were completed they ended a sequence of events. From the fact that boundary activities are the foci of interorganizational interactions, it follows that the external environments of organizations are directly or indirectly affected by them. Furthermore, the effect of the activities *upon* other organizations will, with some time lag, have return effects on the organization on whose behalf the activities were performed, if, as is common in highly-evolved, complex societies, organizations are linked in mutual causal process loops. This is true for each of the classes of boundary activities that have been explored. An organization, however, may not be able to predict with any accuracy what the feedback effects will be if many organizations are connected in closed loops, even less so if effects in the loop are subject to exogenous influences such as might be produced by organization *G* in Figure 2. It is seemingly paradoxical that boundary activities that have the intent of reducing feedback effects—the intent of stabilizing environmental effects and making the behavior of other organizations more predictable—may well have opposite effects. For example, negotiated wage and benefit agreements, intended to produce three-year labor and labor cost predictability, tend to set minimum industry patterns initially but result in increased demands toward the end of the three-year period. Since this affects consumer and other goods prices as well (which affects wage demands) the original organization is confronted by unexpected demands in the negotiation of a new contract. In such instances, generally, there is a noted tendency for new organizations to be created, usually governmental, for the purpose of regulating or mediating interorganizational relationships. These require the addition of new boundary staff in many organizations. One would predict, therefore, that interorganizational relations, and consequently boundary activities, would grow. Although acceptable data are not available for testing purposes, it can be reasonably hypothesized that over the past 25 to 30 years an increasing proportion of organization budgets should be observed to have been devoted to boundary activities.

## FOOTNOTES

1. The author acknowledges gratefully the sustenance of the Graduate School of Business Administration, University of North Carolina at Chapel Hill, and the collegial support of Deans Harvey Wagner and John P. Evans in writing this essay. The comments of James A. Wall, Jr., removed many deficiencies that would have offended the intellect of readers, but there would have been nothing to submit to his scrutiny had the author not had Daniel Katz's friendship, intellectual stimulation, and encouragement for many years.

2. Bracketed material added.

## REFERENCES

Adams, J. S. (1975) "The environmental context of negotiation between human systems." Paper presented at the Negotiation Conference, Center for Creative Leadership, Greensboro, N.C., July 1975.

———. (1976) "The structure and dynamics of behavior in organization boundary roles." In M. D. Dunnette (ed.), *Handbook of industrial and organizational psychology*. Chicago: Rand McNally.

Aldrich, H. & Herker, D. (1977) "Boundary spanning roles and organization structure." *Academy of Management Review, 2,* 217–230.

Allport, F. H. (1933) *Institutional behavior*. Chapel Hill: University of North Carolina Press.

Bavelas, A. (1950) "Communication patterns in task-oriented groups." *Journal of the Acoustical Society of America, 22,* 725–730.

Brehm, J. W. (1966) *A theory of psychological reactance*. New York: Academic Press.

Bridgman, P. W. (1959) *The way things are*. Cambridge, Mass.: Harvard University Press.

Burns, T., & Stalker, G. M. (1961) *The management of innovation*. London: Tavistock.

Carson, R. (1962) *Silent spring*. Boston: Houghton Mifflin.

Dunckel, E. B., Reed, W. K., & Wilson, I. H. (1970) *The business environment of the seventies*. New York: McGraw-Hill.

Dunnette, M. D. (1966) *Personnel selection and placement*. Belmont, Cal.: Wadsworth.

Emery, F. E. & Trist, E. L. (1965) "The causal texture of organizational environments." *Human Relations, 18,* 21–32.

Evan, W. M. (1966) "The organization-set: toward a theory of interorganizational relations." In J. D. Thompson (ed.), *Approaches to organizational design*. Pittsburgh: University of Pittsburgh Press.

Forrester, J. W. (1961) *Industrial dynamics*. Cambridge, Mass.: M.I.T. Press.

———. (1971) *World dynamics*. Cambridge, Mass.: Wright-Allen Press.

Freedman, S. M. & Mericle, M. F. (1972) "Boundary functions of coding and filtering: the hospital personnel department." Paper presented at the 37th annual Meeting of the Academy of Management, Kissimmee, Florida, August, 1977.

French, J. R. P., Jr. & Raven, B. H. (1960) "The bases of social power." In D. Cartwright and A. Zander (eds.) *Group dynamics: research and theory,* 2nd ed. New York: Row, Peterson.

Frey, R. L., Jr. (1971) "The interlocking effects of intergroup and intragroup conflict on the bargaining behavior of representatives." Unpublished doctoral dissertation, University of North Carolina at Chapel Hill.

———. & Adams, J. S. (1972) "The negotiator's dilemma: simultaneous in-group and out-group conflict. "*Journal of Experimental Social Psychology, 8*, 331–346.

Goffman, E. (1969) *Strategic interaction,* Philadelphia, Pa.: University of Pennsylvania Press.

Halberstam, D. (1969) *The best and the brightest*. New York: Random House.

Holmes, J. G. (1971) "The effects of the structure of intragroup and intergroup conflict on the behavior of representatives." Unpublished doctoral dissertation, University of North Carolina at Chapel Hill.

Huneycutt, M. J. (1974) "Effects of information distortion, personalized opposition, and defensiveness upon decision makers' responses to environmental pressure." Unpublished doctoral dissertation, University of North Carolina at Chapel Hill.

Kahn, R. L., Wolfe, D. M., Quinn, R. P., Snoek, J. D., & Rosenthal, R. A. (1964) *Organizational stress: studies in role conflict and ambiguity*. New York: Wiley.

Katz, D. & Kahn, R. L. (1978) *The social psychology of organizations* (2nd ed.). New York: Wiley.

Keller, R. T. & Holland, W. E. (1975) "Boundary spanning roles in a research and development organization: an empirical investigation." *Academy of Management Journal, 18,* 388–393.

Kelley, H. H. & Stahelski, A. J. (1970) "Social interaction basis of cooperators' and competitors' beliefs about others." *Journal of Personality and Social Psychology, 16,* 66–91.

Kennedy, R. F. (1969) *Thirteen days: a memoir of the Cuban missile crisis*. New York: Norton.

Lawrence, P. R. & Lorsch, J. W. (1967) *Organization and environment*. Cambridge, Mass.: Harvard University Press.

Leifer, R. & Delbecq, A. L. (1976) "Organizational/environmental interchange: a model of boundary spanning activity." Working papers 76-6. University of Wisconsin.

Levinson, D. J. & Astrachan, B. M. (1974) "Organizational boundaries: entry into the mental health center." *Administration in Mental Health,* 3–12.

March, J. G. & Simon, H. A. (1958) *Organizations*. New York: Wiley.
Maruyama, M. (1963) "The second cybernetics: deviation-amplifying causal processes." *American Scientist, 51,* 164–179.
Miles, R. E., Snow, C. C., & Pfeffer, J. (1974) "Organizations and environment: concepts and issues," *Industrial Relations, 13,* 244–264.
Miles, R. H. (1976) "Role requirements as sources of organizational stress." *Journal of Applied Psychology, 61,* 172–179.
———. (1977) "Role-set configuration as a predictor of role conflict and ambiguity in complex organizations." *Sociometry, 40,* 21–34.
———. (1979) *Macro organizational behavior*. Santa Monica, Calif.: Goodyear Publishing.
———. & Perreault, W. D., Jr. (1976) "Organizational role conflict: its antecedents and consequences." *Organizational Behavior and Human Performance, 17,* 19–44.
Miller, E. J. & Rice, A. K. (1967) *Systems of organization*. London: Tevistock.
Montgomery, J. (1978) "The image makers: in public relations, ethical conflicts pose continuing problems." *The Wall Street Journal,* August 1, 1978, *192,* 1; 14.
Noyce, R. N. (1977) Microelectronics. *Scientific American, 237,* 63–69.
O'Reilly, C. A., III & Roberts, K. H. (1974) "Information filtration in organizations: three experiments." *Organizational Behavior and Human Performance, 11,* 253–265.
Organ, D. W. (1970) "Some factors influencing the behavior of boundary role persons." Unpublished doctoral dissertation, University of North Carolina.
———. (1971) "Linking pins between organizations and environment." *Business Horizons, 14,* 73–80.
Parsons, T. (1960) *Structure and process in modern societies*. New York: Free Press.
Pfeffer, J. & Salancik, G. R. (1978) *The external control of organizations*. New York: Harper & Row.
"Piercing future fog." (1975) *Business Week,* April 28, 1975.
Pruden, H. O. (1969) "Interorganizational conflict, linkage and exchange: a study of industrial salesmen." *Academy of Management Journal, 12,* 339–350.
———. & Reece, R. M. (1972) "Interorganizational role-set relations and the performance and satisfaction of industrial salesmen." *Administrative Science Quarterly, 17,* 601–609.
Rosen, B. (1976) "Organization boundary roles: information gatekeeping and transmission." Paper presented at the 84th convention of the American Psychological Association, Washington, D.C. September, 1976.
———. & Adams, J. S. (1974) "Organizational coverups: factors influencing the discipline of information gatekeepers." *Journal of Applied Social Psychology, 4,* 375–384.
Rosenberg, N. (1972) *Technology and American economic growth*. New York: Harper & Row.
Rosenhan, D. L. (1973) "On being sane in insane places." *Science,* 1973, *179,* 250–258.
Rubin, J. Z. & Brown, B. R. (1975) *The social psychology of bargaining and negotiation*. New York: Academic Press.
Sorensen, T. C. (1965) *Kennedy*. New York: Harper & Row.
Starbuck, W. H. (1976) "Organizations and their environments." In M. D. Dunnette (ed.), *Handbook of industrial and organizational psychology*. Chicago: Rand McNally.
Strauss, G. (1962) "Tactics of lateral relationships: the purchasing agent." *Administrative Science Quarterly, 7,* 160–186.
Taylor, R. E. (1974) "The effects of certain environmental, organizational, and psychological variables upon the filtering of information inputs by boundary persons." Unpublished doctoral dissertation, University of North Carolina at Chapel Hill.
Terreberry, S. (1968) "The evolution of organizational environments." *Administrative Science Quarterly, 12,* 590–613.

Tesser, A., Rosen, S., & Batchelor, T. O. (1972) "On reluctance to communicate undesirable information (the MUM effect): a role play extension." *Journal of Personality, 40,* 88–103.

Thompson, J. D. (1967) *Organizations in action.* New York: McGraw-Hill.

Wall, J. A., Jr. (1972) "The effects of the constitutent's informational environment upon the constituent-boundary role person relationship." Unpublished doctoral dissertation, University of North Carolina at Chapel Hill.

———. (1975) "The effects of constituent trust and representative bargaining visibility on intergroup bargaining. *Organizational Behavior and Human Performance, 14,* 244–256.

———. (1976a) "Effects of sex and opposing bargaining orientation on intergroup bargaining." *Journal of Personality and Social Psychology, 33,* 55–61.

———. (1976b) "Organization boundary roles: the core experiments." Paper presented at 84th Convention of the American Psychological Association, Washington, D.C. September, 1976.

———. (1977a) "Intergroup bargaining." *Journal of Conflict Resolution, 21*, 459–474.

———. (1977b) "Operantly conditioning a negotiator's concession making." *Journal of Experimental Social Psychology, 13,* 431–440.

———. (1977c) "The intergroup bargaining of mixed-sex groups." *Journal of Applied Psychology, 62,* 208–213.

———. & Adams, J. S. (1974) "Some variables affecting a constituent's evaluations of an behavior toward a boundary role occupant." *Organizational Behavior and Human Performance, 11,* 390–408.

Walton, R. E. & McKersie, R. B. (1965) *A behavioral theory of labor negotiations.* McGraw-Hill.

Wilensky, H. L. (1967) *Organizational intelligence.* New York: Basic Books.

Wilson, I. H. (1974) "Socio-political forecasting: a new dimension to strategic planning." *Michigan Business Review, 26,* 15–25.

Wolfe, T. (1970) *Radical chic & Mau-mauing the flak-catchers.* New York: Farrar, Straus and Giroux.

## Research In Organizational Behavior

**An Annual Series of Analytical Essays and Critical Reviews**

Series Editor: **Barry M. Staw, Graduate School of Management, Northwestern University.**

**Volume 1. Published 1979 Cloth Institutions: $ 32.50**
**ISBN 0-89232-045-1 478 pages Individuals: $ 16.50**



Series Editors: **Larry L. Cummings, Graduate School of Business, University of Wisconsin, and Barry M. Staw, Graduate School of Management, Northwestern University**

**Volume 3. January 1981 Cloth Institutions: $ 32.50**
**ISBN 0-89232-151-2 425 pages Individuals: $ 16.50**

# Research in Organizational Behavior

**An Annual Series of Analytical Essays and Critical Reviews**

**Volume 4. January 1982 Cloth Institutions: $ 32.50**
**ISBN 0-89232-147-4 Ca. 425 pages Individuals: $ 16.50**

**CONTENTS: Contintuidies in Population Ecology: Research on Organizations,** *Howard Aldrich, Cornell University.* **Organizational Life Cycles,** *John Freeman, University of California, Berkeley.* **Functions of Organizational Theory in Creating New Forms of Activity,** *Kenneth Gergen, Swarthmore College and Michael Basseches, Cornell University.* **Path Analysis for Theory Testing in Organizational Behavior,** *John Hunter and David Gerbing, Michigan State University.* **Evaluation Processes in Organizational Evolution: Coping with Varying Environments,** *John Kimberly, Yale University.* **Organizational Roles of Leaders: Forces, Followers and Foils,** *Gerald Salancik, University of Illinois.* **The Effect of Complex Reinforcement Contingencies: A Functional Analysis of Problem Solving and Other "Cognitive Processes",** *William Scott, Indiana University.* **Worker Participation in Management: An International View,** *George Strauss, University of California, Berkeley.*